Death as a Process

edited by

John Pearce and Jake Weekes

OXBOW | books

Oxford & Philadelphia

Published in the United Kingdom in 2017 by
OXBOW BOOKS
The Old Music Hall, 106–108 Cowley Road, Oxford OX4 1JE

and in the United States by
OXBOW BOOKS
1950 Lawrence Road, Havertown, PA 19083

© Oxbow Books and the individual contributors 2017

Paperback Edition: ISBN 978-1-78570-323-2
Digital Edition: ISBN 978-1-78570-324-9 (epub)

A CIP record for this book is available from the British Library

Library of Congress Cataloging-in-Publication Data

Names: Pearce, John, 1969- editor, author. | Weekes, Jake, editor, author.
Title: Death as a process / edited by John Pearce and Jake Weekes.
Description: Oxford ; Philadelphia : Oxbow Books, 2017. | Series: Studies in
    funerary archaeology ; 12 | Includes bibliographical references.
Identifiers: LCCN 2017000884 (print) | LCCN 2017016253 (ebook) | ISBN
    9781785703249 (epub) | ISBN 9781785703256 (mobi) | ISBN 9781785703263 (pdf) |
    ISBN 9781785703232 (pbk.)
Subjects: LCSH: Funeral rites and ceremonies--Rome. | Human remains
    (Archaeology) | Funeral rites and ceremonies, Ancient. | Excavations
    (Archaeology) | Burial--History--To 1500. | Social archaeology. |
    Burial--Rome.
Classification: LCC DG103 (ebook) | LCC DG103 .D43 2017 (print) | DDC
    393/.930937--dc23
LC record available at https://lccn.loc.gov/2017000884

Printed in Malta by Gutenberg Press Ltd

For a complete list of Oxbow titles, please contact:

UNITED KINGDOM
Oxbow Books
Telephone (01865) 241249, Fax (01865) 794449
Email: oxbow@oxbowbooks.com
www.oxbowbooks.com

UNITED STATES OF AMERICA
Oxbow Books
Telephone (800) 791-9354, Fax (610) 853-9146
Email: queries@casemateacademic.com
www.casemateacademic.com/oxbow

Oxbow Books is part of the Casemate Group

*Front cover: Excavation (Canterbury Archaeological Trust); the tomb of Publius Vesonius Phileros in the niche
under the podium (A. Gailliot, MFPN/FPN).*
*Back cover: Tomb L14001, "Actiparc" site near Arras (France), 1st century AD (Service Archéologique d'Arras –
INRAP).*

Contents

Preface v
List of contributors.. viii

1. Introduction: Death as a process in Roman funerary archaeology...........................1
   John Pearce

2. Space, object, and process in the Koutsongila Cemetery
   at Roman Kenchreai, Greece..27
   Joseph L. Rife and Melissa Moore Morison

3. Archaeology and funerary cult: The stratigraphy of soils in the
   cemeteries of Emilia Romagna (northern Italy)..60
   Jacopo Ortalli

4. Funerary archaeology at St Dunstan's Terrace, Canterbury 83
   Jake Weekes

5. Buried Batavians: Mortuary rituals of a rural frontier community....................123
   Joris Aarts and Stijn Heeren

6. They fought and died – but were covered with earth only years later:
   'Mass graves' on the ancient battlefield of Kalkriese.............................155
   Achim Rost and Susanne Wilbers-Rost

7. Some recent work on Romano-British cemeteries 174
   Paul Booth

8. Funerary complexes from Imperial Rome: A new approach to
   anthropological study using excavation and laboratory data 208
   Paola Catalano, Carla Caldarini, Flavio De Angelis and Walter Pantano

9. Animals in funerary practices: Sacrifices, offerings and meals at Rome
   and in the provinces 226
   Sébastien Lepetz

10. "How did it go?" Putting the process back into cremation 257
    Jacqueline I. McKinley

11. Afterword – Process and polysemy: An appreciation of a cremation burial 287
    Jake Weekes

Preface

'Romans killed dozens of unwanted babies at English "brothel"'[1]

The Roman dead continue to speak to an audience beyond that of Romanists or burial archaeologists. In recent years' discoveries related to the excavation or analysis of burials have received widespread media attention. The extensive coverage in 2010 of the interpretation of infant burials from the villa at Yewden in Buckinghamshire as the unwanted offspring of prostitutes serving clients from Roman London is but one example. Through the medium of their tombs, real or highly contested, readers and viewers have also encountered famous names from antiquity, including St Paul, James, the claimed brother of Jesus, and the 'real life Gladiator', Marcus Nonius Macrinus, the discovery of whose monument on the via Flaminia not far north of Rome was an archaeological sensation in 2008. Colourful characters from Roman Britain met during the last 15 years include Southwark's 'gladiator girl', the Spitalfields' 'princess', and Catterick's 'eunuch'. Evidence for violent death or epidemic thrust other discoveries to prominence, for example gladiators from Ephesus and London, 'headless Romans' from York, and the occupants of the mass grave excavated in the catacomb of Sts Peter and Marcellinus, for which the Antonine plague is the soberest of the interpretations offered. Speculation over the contents of sealed coffins, like the 'burrito' from late Antique Gabii on the outskirts of Rome, recalls the antiquarian thrills of tomb opening in earlier eras. This sensationalising attention is a well-established tradition in (Graeco-)Roman funerary archaeology: Ian Morris commented two decades ago on the disproportionate attention paid by archaeologists to 'quirks' and 'oddities' in the study of Roman burial, and Eleanor Scott noted the propensity for 'ripping yarns' to emerge in interpretation of the dead, often with a strongly sexist dimension, using the early 20th century excavation of the Yewden villa and its later reception. This notwithstanding, it is essential to acknowledge key developments in the study of the Roman dead.

Fieldwork in the last two decades, especially through development-led archaeology, has enormously increased both the sample of excavated burials available for study and the diversity of contexts from which these derive. Much greater emphasis is given to the reporting and analysis of human remains. The much fuller integration of their study, including the new insights from biomolecular analysis, especially of stable isotopes, with the archaeological evidence for burial rituals, has given key new insights into diet, mobility and identity. Our volume benefits from both developments, but it takes further the development in Roman *funerary archaeology* as its primary focus, the exploration not just of burial practice as a single act of deposition, but rather as a complex sequence of acts on and around corpses and monuments in the rituals preceding and following death.

The Roman funeral has always attracted the attention of historians and anthropologists, not only for its spectacular character, the processions and pyres of emperors reminiscent of later royal rituals, but also because it lends itself to analysis as a rite of passage. It could be suggested that archaeological cemetery evidence, derived usually from the interment alone, is unlikely to be fully representative, or could even be misleading as a basis for extrapolating to the wider ceremony. We contend that this is too limited a view; stages of ritual before, during and after burial are also archaeologically accessible and a richer understanding of death as a process can be established from archaeological fieldwork in various contexts, including development-led and research excavations, as well as from texts. This, we argue, allows for a much fuller description and richer interpretation of Roman funerary rituals as practiced in their diversity across the empire.

Our shared interest in Roman funerary ritual and in the possibilities for documenting and interpreting ritual sequences prompted the starting point for this book, a session organised by us at the Roman Archaeology Conference in London in 2007. With subsequent expansion by additional contributions, we seek in it to assess what kinds of insights are being derived from this new work on cemetery excavations and the re-evaluation of older projects. New fieldwork results and the insights from osteological analysis feature prominently among the contributions, but we wish to bring to particular attention the reconstruction and interpretation of ritual sequence and to allow for comparison of the different perspectives taken by archaeologists working in different contexts. Since insights into methods for obtaining a richer documentation of a ritual sequence have emerged from projects in both a research- and development-led contexts, we have invited contributors who work in both spheres, with a view to considering how study might be further advanced, especially in the practice of development-led projects, from which the vast majority of today's data derive. The contributors also include specialists in particular methods and materials. In funerary archaeology in particular it seems useful also to compare the different approaches across Europe, where the study is marked by distinctive national traditions. This also provides a useful opportunity to bring to the wider attention of an English-reading audience important new work from continental Europe, including the results of projects at Rome, Pompeii, Kenchreai near Corinth and the site of the Varian disaster with its particular resonance in Roman memory. While the conference session took place at a time of a boom in information from development-related archaeology, work on its publication has coincided in the UK and elsewhere with a recession that has led to a greatly reduced volume of archaeological fieldwork. As and when the pace of new fieldwork linked to development quickens again, we hope that the volume will raise questions for consideration in the planning and execution of new projects as well as offering new insights into ancient behaviour and society.

We are delighted that Joe Rife and Melisa Morison, Achim Rost and Susanne Wilbers-Rost, Stijn Heeren and Joris Aarts, and Sebastien Lepetz all responded positively to the invitation to contribute to the volume alongside Jackie McKinley, Paul Booth, Jacopo Ortalli and Paola Catalano and colleagues who contributed to the original conference. As well as contributing a case study Jake Weekes considers approaches interpreting the meaning of ritual sequences in a final paper. The chapters were prepared in their present form between 2009 and 2012–2013, and then revised in late 2014. We would also like to thank our other speakers in the original conference session, Rebecca Gowland and Colin Wallace, and Peter Garnsey as respondent in the session.[2] We express our gratitude also to the RAC organising committee for accepting and facilitating the session, especially Ian Haynes, and the British Academy for supporting costs of non-UK based speakers, and to Federico Ugolini for much-appreciated practical help with preparing the volume. Above all we thank our contributors and Oxbow Books for their patience in bearing with the vicissitudes of its preparation where outside circumstances impinged on the editors' work and have meant a longer gestation than we had intended.

Notes

1. http://www.dailymail.co.uk/sciencetech/article-1289603/Romans-killed-100-unwanted-babies-English-brothel.html. S. Greenhill, 26.6.2010 [Accessed 28.06.2010].
2. Rebecca Gowland's paper is published elsewhere: Gowland, R. and Garnsey, P. (2010) Skeletal evidence for health, nutritional status and malaria in Rome and the empire. In H. Eckardt (ed.) *Roman diasporas; archaeological approaches to mobility and diversity in the Roman Empire*, 131–156. JRA Supplement 78, Portsmouth RI: Journal of Roman Archaeology.

List of contributors

Joris Aarts
Archaeological Centre VU University,
Faculteit der Letteren
De Boelelaan 1105
1081 HV Amsterdam
The Netherlands
j.g.aarts@vu.nl

Paul Booth
Oxford Archaeology
Janus House
Osney Mead
Oxford (UK)
OX2 0ES
paul.booth@oxfordarch.co.uk

Carla Caldarini
Soprintendenza Archeologica di Roma
Via di S.Apollinare 8, 00186, Rome, Italy
carla.caldarini@inwind.it

Paola Catalano
Soprintendenza Archeologica di Roma
Via di S.Apollinare 8, 00186, Rome, Italy
paola.catalano@beniculturali.it

Flavio De Angelis
Lab. of Anthropology, Centre of Molecular
Anthropology for Ancient DNA Studies
Department of Biology; University of Rome
Tor Vergata
Via della Ricerca Scientifica 1,
00133 Rome, Italy
flavio.de.angelis@uniroma2.it

Stijn Heeren
Archaeological Centre VU University,
Faculteit der Letteren
De Boelelaan 1105
1081 HV Amsterdam
The Netherlands
s.heeren@vu.nl

Sébastien Lepetz
Muséum national d'Histoire naturelle/UMR 7209
du CNRS
Archéozoologie, Archéobotanique:
sociétés, pratiques et environnements
Case postale 56
55 rue Buffon
75005 Paris (France)
lepetz@mnhn.fr

Jacqueline McKinley
Wessex Archaeology
Portway House
Old Sarum Park
Salisbury UK
SP4 6EB
j.mckinley@wessexarch.co.uk

Melissa Moore Morison
Department of Classics
Grand Valley State University
260 Lake Huron Hall
One Campus Drive
Allendale, MI 49401
morisonm@gvsu.edu

Jacopo Ortalli
Dipartimento di Studi Umanistici,
Sezione di Storia e scienze dell'antichità
Università di Ferrara
via Paradiso 12, 44100 Ferrara Italy
otj@unife.it

Walter Pantano
Soprintendenza Archeologica di Roma
Via di S.Apollinare 8, 00186 Rome, Italy
walterbpantano@libero.it

John Pearce
Department of Classics, King's College London
Strand
London UK

WC2R 2LS
john.pearce@kcl.ac.uk

JOSEPH L. RIFE
Program in Classical and Mediterranean Studies,
Vanderbilt University
PMB 0092
230 Appleton Place
Nashville, TN 37203-5721
joseph.rife@vanderbilt.edu

ACHIM ROST
Universität Osnabrück, Alte Geschichte:
Archäologie der Römischen Provinzen
Schlossstr. 8
49069 Osnabrück,
Germany
arost@uni-osnabrueck.de

JAKE WEEKES
Canterbury Archaeological Trust Ltd
92A Broad Street
Canterbury
Kent UK
CT1 2LU
jake.weekes@canterburytrust.co.uk

SUSANNE WILBERS-ROST
Museum und Park Kalkriese,
Department of Archaeology
Venner Str. 69
49565 Bramsche-Kalkriese
Germany
wilbers-rost@kalkriese-varusschlacht.de

Chapter 1

Introduction: Death as a process in Roman funerary archaeology

John Pearce

Recent decades have witnessed a boom in the study of Roman cemeteries. This is manifested in a dataset massively expanded by development-related excavation, the consolidation and proliferation of analytical techniques, and the exploitation of funerary data to explore the representation of individual identities and the dynamics of Roman society. This has also been a breakthrough period for the exploitation of human skeletal remains to explore ancient demography, seeing the first large-scale syntheses of evidence for health status from communities, regions and beyond, as well as much fuller integration with the study of funerary rituals (cf. Reece 1982). Funerary material now features more prominently, for example, in the characterisations of provinces offered in handbooks and similar volumes (e.g. Todd 2004; Millett *et al.* 2016; Ouzoulias and Tranoy 2010; Riggs 2012).

For most cemetery excavations, the evidence which is prioritized in documentation, analysis and dissemination comprises the burial itself, this being mainly the skeletonized human remains and the (inorganic elements of) their containers, and objects placed with the dead. More occasionally, depending on context, this also encompasses the monuments within which they were housed or by which they were marked. It is an archaeological piety that the rituals which precede or follow burial are likely to be illuminated mainly by texts and a limited corpus of images, both pertaining much more to Rome than to its empire. The cemetery itself is however a more productive entity than this might suggest. One of the major new directions in current scholarship is to expand the range of archaeological data which allows a ritual sequence to be reconstructed in much richer detail, including the phases preceding and succeeding the interment of the remains of the dead, as well as enabling the interment itself to be more fully reconstructed.

The focus of this volume therefore lies on exploring the new evidence from excavation of Roman cemeteries with a particular emphasis on ritual process. In this introductory chapter this focus is set in the context of current research interests into the Roman dead and their mourners. After briefly outlining the broader context of Roman cemetery studies in the most recent decades, the chapter explores the key interlinked developments which have contributed to this shift in perspective; some have been extensively applied, others are on a more preliminary footing. The enduring scholarly emphasis on the textual evidence for an extended sequence of rituals continues to produce insights into Roman funerary culture and offer changing perspectives on material remains. The ever-expanding range of techniques brought to bear on the study of burials, especially their organic component, is also illuminating stages of ritual before and after burial, while understanding of the spatial setting of burial is potentially enhanced through the application of archaeological prospection methods. Additionally, many excavations also now reveal a greater complexity of burial spaces as depositional environments, both the graves themselves and the diverse other features that are the product of funerary (and non-funerary) activities. Discussion of these factors provides the context for the individual papers which are introduced in the final part of this chapter.

Bodies of data – burials and monuments

The numbers of Roman burials documented in archaeological fieldwork have, in the last two decades, accumulated at a rate greater than ever before, accelerating a process already well established in the post-war period. Amongst the most striking advances in knowledge of cemeteries is the recent excavation and study of very substantial samples of burials from the city of Rome. Through large-scale rail and road infrastructural projects and development in the city's suburbs several thousand burials have been excavated, the majority from non-monumentalized cemeteries of early and mid-imperial date, including burial areas serving the urban poor and the farms and other work spaces of the wider hinterland (Catalano *et al.* this volume, with further references; *Dossiers d'Archéologie* 330, 2008). A similar phenomenon characterizes provincial settings. Of more than 1,100 burial and cemetery excavations in Britain, for example, reported since 1921 in the annual accounts of fieldwork in the *Journal of Roman Studies* and *Britannia*, almost 600 have taken place in the last 25 years, in the main the product of changes to planning policy in 1990 in England and Wales (PPG 16, revised as PPS 5 in 2010 and superseded by the National Planning Policy Framework in 2011), which requires proposed construction projects to take account of archaeological deposits (Booth and Boyle 2008; Pearce 2015b). The case of Britain also illustrates how the character of ancient burial practice, preservation conditions and the focus of contemporary fieldwork on towns and development corridors combine to produce a dataset skewed to urban cemeteries. In France, too, development has sparked an explosion in burial numbers, again with a focus on zones of urbanisation, industrialisation and *grands projets* for enhancing the communications network. While similar urban biases apply, as in Britain burials have also been excavated in ever larger numbers as small groups and singletons from the margins of farmyards, the ditches of garden plots and field boundaries (e.g.

Ancel 2012; Blaizot 2009; Raynaud 2006). By contrast in countries of the southern and eastern Mediterranean the growth in available data has been slower (e.g. Devreker *et al.* 2003; Mackinnon 2007; Pearce 2013b, 2; Rife 2012).

Research-led fieldwork directed at funerary contexts continues to be important. For example, collaborative work undertaken by the Ecole Française at Rome in collaboration with archaeological agencies and universities in Italy and Tunisia at Cumae, Pompeii, Ravenna and Pupput (Tunisia) has been especially important for methodological innovation (Scheid 2008; van Andringa *et al.* 2013; *Dossiers d'Archéologie* 330, 2008; Ortalli this volume; Lepetz this volume). Other major projects underway in a research context include the investigation of the cemeteries at Corinth and its port at Kenchreai (Rife this volume). Largescale geophysical prospection has also often yielded significant new data for the scale, layout and setting of burial areas (see below).

Through a combination of formal publication and 'grey literature', i.e. comprising 'accounts of developer-funded archaeological investigations which have not been published in a recognised journal or book' (Fulford and Holbrook 2014, 39), the data available from this fieldwork for the study of burial practice have also been substantially, if unevenly augmented. For example, cities from England including London, Colchester, Winchester, Cirencester and others to a lesser extent, are now served by a substantial sample of burials from multiple cemeteries and spanning the entire Roman period. Before 1990, this could scarcely be said of a single Roman town (Pearce 2015b). Syntheses of rural burial data reveal a similar growth in available data (e.g. Smith 2013). A similar expansion of the evidence base characterises Roman Gaul (e.g. Ancel 2012; Blaizot 2009). The continuing publication of older projects has also yielded substantial new information, in some cases completing major series of funerary publications, for example at Wederath and Krefeld-Gellep in western Germany (Kaiser 2006; Pirling and Siepen 2006), in others making complex (and often lacunose) archives from older fieldwork available. Examples of the latter include the cremation burials associated with the garrison at Brougham, Cumbria (Cool *et al.* 2004), small towns at Rheydt-Mülfort, Nordrhein-Westfalen, and Dillingen-Pachten, Saarland (Erkelenz 2012; Glansdorp 2005) or more than 1,300 inhumation burials from late Roman Nijmegen (Steures 2013). Archival resources on fieldwork in Greece, Lebanon and Egypt have been exploited to similar effect (De Jong 2010; O'Connell 2014; Rife 2012).

Likewise, funerary monuments continue to be more sporadically discovered, occasionally in situ, more commonly as spolia, from stelae to grand mausolea (e.g. Rossi 2012; Simmonds *et al.* 2008). These join the ever-proliferating body of corpora and thematic studies drawing on the near-inexhaustible body of funerary buildings, inscriptions and art documented over the last few centuries (Borg 2013, 1).

Interpreting Roman burials

The cumulative impact of this recent work is to enable the more detailed and precise mapping of the kaleidoscopic character of Roman funerary rituals, producing an ever more nuanced mosaic of practices within some common traditions. A striking example is the characterisation of the occurrence of inhumation. Across the western provinces, it can now be shown to have been practised on a widespread basis in the first to third centuries AD, sometimes as a minority ritual but in other cases, including Rome itself, accounting for the burials of a larger proportion of the population (Booth this volume; Catalano *et al.* this volume; Faber *et al.* 2004).

The transformation in analytical approaches is however as, if not more, important than the growth in data. Of this the most prominent characteristic is the greater emphasis on the analysis of human skeletal evidence; in this respect study of the Roman period follows a broader shift (Tarlow and Nilsson Stutz 2013). As well as many more site-specific studies of human skeletal remains, syntheses are now available at the regional level and beyond, with a particular focus on population health status, diet and mobility. This is the product of methodological change, a more sophisticated appreciation of the analytical potential of palaeopathological traits, an expanding suite of biomolecular techniques for analysing skeletal material, especially diet and geographical origin, and an enhanced emphasis on the insights into demographic and socio-cultural history that these can bring (Gowland and Garnsey 2010; Kilgrove 2014).

While long-standing questions in Roman funerary archaeology remain a common focus of study, for example concerning the religious or ethnic affiliations of individuals or populations, these have been extended or supplemented by consideration of identity as more broadly conceptualised. The availability of skeletal data is a significant contributory factor to the transformation of the study of burial practice for insights into identity. In particular, the application of a 'life-course' approach has given impetus to a much stronger focus on variability of funerary treatment according to age and gender. This has revealed a frequent close linkage between the two; the burial of 'gender-specific' artefacts, for example, especially of jewellery, shows significant variability according to age; in particular, girls and younger women are often distinguished by the richest and most diverse object assemblages (e.g. Cool 2010, 307–308; Gowland 2001; Martin-Kilcher 2000). The study of representation of social status in death has also been informed by the opportunity to compare funerary treatment with the diet and health of the deceased, as established from osteological and isotopic data (e.g. Müldner 2013; Pitts and Griffin 2012; Redfern *et al.* 2015). Closer attention to palaeopathological documentation has also permitted better characterisation of the atypical manipulations of corpses and or skeletal remains, commonly labelled 'deviant' burials and now documented widely beyond Britain (Belcastro and Ortalli 2010; Taylor 2008; Tucker 2013). Isotopic data and, to a lesser extent, the analysis of cranial morphology, have demonstrated the complex relationships between place of origin, ancestry and the group affiliations that burial ritual suggests (Eckardt *et al.* 2014).

The variety of perspectives applied to Roman burial data, conditioned in part by varying national scholarly traditions, make it difficult to map general theoretical shifts in the archaeological study of the dead and of the living who mourn them. The methodologies have also varied, with only patchy use, for example of statistical testing of associations between skeletal attributes or with grave good types (e.g. Catalano *et al.* this volume; Cool and Baxter 2005; Quensel-von-Kalben 2000). Some common trends can, however, be noted. Analysts of excavations from burials in France and Italy, for example, have often sought to interpret the rich evidence for cremation burials in relation to John Scheid's influential model for reading Roman funerary rituals as a sacrificial process, established by applying an anthropological perspective to the fragmentary textual sources (Scheid 2005, 161–188; see below).

Frameworks which characterise social identities as being continuously constructed through practice in domestic, public and ritual arenas are also especially relevant for analysing burial evidence (cf. Gardner 2013; Revell 2009). The scale of funerary ceremony, whether expressed in the quantities or materials of resources consumed on the pyre or as dispensed in hospitality, may, for instance, be manipulated in order to perpetuate political authority and social hierarchy. The lavishly furnished third and fourth century AD cremations and burials at Gamzigrad and Sarkamen, Serbia provide examples from an imperial context of the consumption through burial or burning of large quantities of gold and silver objects (Vasić and Tomović 2005, 269–294); the many burnt and broken ceramics (more than c. 86 kg) in a second century *bustum* from the Puy-de-Dôme are the residue of the likely obsequies of a local magistrate, presumably from the hospitality offered to the many participants (Trément and Humbert 2004). The wealth of contextual data can allow consumption of this kind to be calibrated through quantitative comparison between funerary and non-funerary contexts, for example of ceramics or personal ornaments (Cool 2011; Willis 2011). The same contextual evidence also informs understanding of the connotative range of symbols associated with the dead. I have proposed elsewhere, for example, that the objects found in the largest grave good assemblages documented in first to early third century northern Europe, especially dining vessels and artefacts related to bathing, literacy and sometimes hunting, cumulatively alluded to *humanitas*, the *savoir faire* necessary to membership of a Roman elite. Through its recurring presentation in the context of the dead, the status-defining character of this cultural expertise was reinforced (Pearce 2015a). Theuws and Alkemade (2000, 450–461) argue that the spears and axes of weapon burials of northern Gaul in the fourth and fifth centuries AD, are emblems of a 'settler' ideology, evoking the clearance of land which had been abandoned and become 'wild' during the disruptions of the third century and the symbolic control exercised over it by hunting.

Related approaches acknowledge the multiple 'readings' that participants might take from these occasions. In the early imperial cities of southern Baetica, for example, mortuary practices echoed those of the period before Roman control in objects, rituals or monument form. This reinvented traditional practice could be invested with multiple meanings, from evoking a local resistant tradition to sharing the commonly observed Roman respect for local *mores maiorum* (Jiménez 2008). As well as their generic allusive properties, funerary symbols possessed a mnemonic capacity to evoke a local and specific response from participants, drawing on their earlier encounters with those objects placed with the dead or used in death rituals (Williams 2004). Gosden and Garrow (2012, 196–255) suggest, for example, that decorated metal objects such as mirrors, weapons, and hearth furniture, placed with the dead in burials of the first centuries BC and AD in Britain, accumulated this capacity through their rich biographies. The making, use, adaptation and so on of these objects were all likely to have been associated with key moments in group history, repetitive events such as rites of passage or the festival cycle, or related to specific events. By their display at the funeral they evoked that earlier history, mediated through individual recall.

The reconstruction of ritual sequences

A major interpretive strand for Roman funerary practice which we wish to highlight in the volume is the increasing emphasis on the funeral as a process through time. After first noting the types of insights which may be derived into the ritual sequence from textual (and visual) sources, the following sections explore the methods and forms of evidence which enable this much fuller reconstruction of ritual process.

Textual sources

The textual sources which describe the Roman funeral have long been well known. The vivid accounts of Polybius or Herodian, or the images of the processions that accompanied the catafalques and corpses of Republican aristocrats and emperors to the pyre or tomb, are the best known of these, and they survive in sufficient number and variety to show that the act of interment was commonly one episode within an extended ritual sequence. After their laying out, the dead were accompanied by the mourners from the home to the pyre for cremations and to the tomb for interment; at that tomb groups would gather on subsequent occasions for commemorative ritual. For wealthier individuals, especially members of the urban elite, whose ceremonies would be conducted by undertakers, there might be a strong performative element associated with the funeral (and later commemoration). This drew large assemblies of people to speeches, games, distributions of food and so on (Lindsay 2000; Toynbee 1971, 43–61). Evidence of this type serves as a classic 'spoiler', prompting caution in the interpretation of archaeological evidence by revealing rituals which would normally escape archaeological attention: how would we know other than from a mid-imperial inscription, for example, that a certain Claudia Corneliana ensured the perpetuation of her memory through a property endowment supporting the performance of commemorative rites in a *vicus* in the territory of Brescia at the *Parentalia* (13th–21st Feb.), *Rosalia* (May–July), and *Vindemialia* (Sept.–Oct.) (Gasperini 1996)?

This evidence is also rich enough to apply the rites of passage framework widely used in history and anthropology to a Roman setting, and continues to expand through key discoveries, for example of the inscription recording the undertakers' contract awarded at Augustan Pozzuoli (Castagnetti 2012). Through a structured system of sacrifices, men, gods and the dead participated in the consumption of foodstuffs during the funeral, but through an increasing differentiation and separation in the materials, spaces or modes of consumption the dead and the living negotiated their liminal states, the former being assimilated with the ancestors and the *Manes*, the latter being returned to the society of the living. The treatment of the corpse as well as the bodies of the mourners, often transgressing behavioural norms, was the visible medium by which participants witnessed these transformations (Scheid 1984; 2005, 161–188). The richness of sources has allowed some exploration of how aspects of social identity, especially class and gender, and varying cultural context, conditioned the form taken by the funeral within a household and wider civic community, including some variation by date and geographical and/or cultural setting from the Republican period to late Antiquity (e.g. Hope 2007; 2009; Rife 2012; Scheid and Rüpke 2010; Schrumpf 2006; Šterbenc Erker 2011).

The literary evidence can also be exploited to explore the experience of Roman death rituals. E.-J. Graham (2011a; 2015) uses the textual attestations of interactions with the corpse through touch, smell and sight, to explain how memories of the event were established for participants and its mnemonic potential was activated. This sensory intensity of the occasion elicited associations with other events, previous funerals, other rites of passage and so on which had been shared with other participants, and developed a cumulative sense of group identity through embodied experience. Aided by digital visualisation of urban topography Favro and Johanson (2010; see also Graham 2011b; Johanson 2011) have explored how encountering the spaces of the Republican city while moving from house to Forum with the funerary procession re-animated an understanding of the history of the urban fabric and its community. It contributed new layers of association with the spaces through which the cortège passed and where the key events, for example the *elogium* or games, took place.

However, the limitations of the literary or historical descriptions of burial and of the legal and religious texts that bear on the status of the tomb and the corpse are well rehearsed. The description of the Roman funeral is a composite obtained from sources widely distributed in time and focused on an elite metropolitan milieu, especially the obsequies of emperors (Davies 2000; Morris 1992, 10–15). More is often revealed of aberrant practice, mutilated imperial corpses, for example, or disgraceful luxury; literary context conditions the references to death rituals (Hope and Huskinson 2011; Scheid and Rüpke 2010). Many aspects of ritual are described elliptically, for example the *os resectum* (Graham 2011b) and key behaviours go unobserved, for example preparation of the bodies of the dead for interment, or the placing of objects on the pyre or in the grave.

Archaeological methodologies

It would be wrong to exaggerate the differences in scholarship focused on textual and archaeological evidence. The approach developed by John Scheid has, for example, inspired a fieldwork agenda as well as an interpretive framework for many Roman period funerary projects in continental Europe, especially in France and Italy which have recovered significant evidence for the pre- and post-burial processes (e.g. Ancel 2012; Blaizot *et al.* 2009; van Andringa *et al.* 2013; Rebillard 2009; Ortalli this volume; projects published in issue 69.1 of the journal *Gallia* in 2012). The literary sources remain nonetheless limited in key respects. For the archaeological evidence, we wish first to signal the expanding repertoire of analytical techniques which bear directly or indirectly on the understanding of funerary process and/or environment. The following observations make no claim to be a systematic overview but illustrate different facets of this expanding potential.

The developing analytical potential of human skeletal remains represents a key development. The much closer attention to the study of cremated human bone has afforded not only better comprehension of the identity of the deceased individual (age, sex, pathology etc.) but also of the processes to which the body has been subjected (Duday 2009, 146–153; McKinley 2013; this volume). This has significant potential to inform understanding of the different stages of the cremation process, including the condition of the human remains at the time of burning, the arrangement of the body on the pyre and the temperatures reached and maintained during burning, even if some key factors related to change in bone during burning remain poorly understood (Thompson 2015). Allowing for taphonomic processes, there is also significant scope to examine the process of collecting and depositing human remains. The demonstration of this potential is now demonstrated in many instances in the publication of cremated remains (e.g. van Andringa *et al.* 2013, 861-908; Thompson 2015a; McKinley, this volume). The study of taphonomic processes undergone by inhumation burials also has significant potential for informing understanding of the burial process, though this has been more systematically applied outside the UK. The 'anthropologie de terrain' approach, pioneered by Henri Duday and collaborators and applied in excavations in France and elsewhere in continental Europe, is a significant innovation (Duday 2009). Close observation of the relationship of skeletal elements where limbs articulate allows the observer to extrapolate the decomposition process and by extension the form of the original burial, especially the organic elements of which no trace may now survive. Eight inhumation burials of fourth–fifth century date from a roadside settlement Uckange, Moselle, illustrate the insights gained by the technique. Here the disposition of the skeleton, combined with some staining from the timbers revealed the presence of a hollowed-out trunk used as a burial container, five instances of jointed wooden coffins made without nails and one complex arrangement where the body was placed in such a coffin within a larger pit protected by a wooden cover. Beneath the latter grave goods had been placed on a wooden platform adjacent to the coffin. In two instances, at least, it was possible to demonstrate that the heads of the dead had originally rested on pillows or cushions (Lefebvre *et al.* 2013).

The closer attention to the carbonized remnant of cremation ceremonies is also revealing much more systematic evidence of the display of the deceased on the pyre.

Examples include the scraps of objects, especially ceramic sherds and glass fragments, occasional dress items and hobnails of most pyre residues from the Colchester Garrison sites or the fragments of gold thread from two burials from Nijmegen (Koster 2013, 190–195; Pooley *et al.* 2011). The study of the burnt (and occasionally mineralised) remains of plants and animals in particular has enriched the understanding of funerary process, especially of the fuels and foodstuffs consumed in ritual. Along with other evidence, for example pollen, molluscs, soils and so on, this evidence has also enabled a more nuanced characterization of cemetery environments (e.g. Bui and Girard 2003; Bouby and Marinval 2004; Cooremans 2008; Kreuz 2000; Lepetz this volume; Wiethold 2000). Biochemical analyses have illuminated materials employed in funerary rites to achieve effects of colour and odour on or around the corpse, for example the scattering of Tyrian purple dye across a third century inhumation at Naintré (Vienne) (Devièse *et al.* 2011) or the application of plant resins and gums in cremation and inhumation burials from Britain and beyond (Brettell *et al.* 2015).

Archaeological prospection is also extending understanding of the configuration of spaces for the dead and their environs in many settings, especially where excavation is unlikely to be applied on a large scale. In particular, geophysical prospection undertaken for research and heritage management purposes now contributes significant new information. Prospection in some urban settings has revealed probable enclosures and masonry-built monuments, for example from the margins of cities in central Italy such as Faleri Novi and Forum Novum, (Gaffney *et al.* 2004, 209–211; Hay *et al.* 2010, 34–35). In Britain prospection around the towns of Silchester and Cirencester, in the latter case complemented by aerial photography, has also identified extensive (probable) burial areas (Creighton in prep.; Winton 2009). Surveys of *vici* on Britain's northern frontier have identified likely funerary enclosures and burials, confirmed by excavation in the case of Birdoswald (Cumbria) (Biggins and Taylor 2004a; 2004b; Wilmott 2010). The relatively small number of studies so far undertaken is in part related to the general limitation of geophysical surveys to intra-mural areas; there has also so far been limited testing of survey results against excavation, especially of the hypothesized traces of burials and related deposits.

Beyond the grave: archaeological features from cemeteries

In many cases, what survives of a complex ritual will be limited to a truncated remnant of the burial proper, the cremated or skeletonised remains of the dead, their container and associated grave goods. Nonetheless, even tombs with 'ordinary' preservation can yield significant information on burial process. The potential of graves, non-burial features and monuments to illuminate ritual process and wider engagement with the dead is explored in the following paragraphs.

Cremation and inhumation burials

Cremation burials offer a variety of evidence for the cremation act. Occasionally grave and pyre are the same, the so-called *bustum* burial, often rich in combustion debris

(Struck 1993). More commonly, pyre material may be buried in some quantity in the grave, whether mixed with or separate from human bone. Even sorted cremated bone often includes some scraps of charcoal, burnt objects or faunal remains. Combined with observations from pyre sites and from excavation in spits of cremation containers these data have given significant insights into the pre-burial rituals for many cemeteries (e.g. Blaizot *et al.* 2009; Cool *et al.* 2004; Cool 2011; see further below). In the continental European provinces, the depositional relationship between debris and cremated bone and the types of container has spawned multiple typologies and associated ethnic or status attributes (Bechert 1980; Blaizot *et al.* 2009; Bridger 1996, 220–226). However, the character of many of the deposits is uncertain. Closer attention to cremated bone in the latter part of the 20th century revealed that the quantity of bone found in many cremation burials is rarely more than a fraction of what can be expected from the burning of an adult (McKinley 2013; this volume). In the absence of a burial container or grave goods, distinguishing between one-use pyre sites, *busta*, graves where cremated bone is introduced partly or wholly unsorted from the matrix of charcoal, ash and burnt artefacts ('*Brandgrubengräber*') and the burnt remains of commemorative meals is near impossible without gridded excavation and detailed analysis of the quantity and distribution of cremated bone (e.g. Bel *et al.* 2008).

As Weekes emphasises in his study of St Dunstan's (this volume; see also Weekes 2014, 5–6), it is possible to infer some pre-burial treatments of inhumed corpses from the disposition of objects in the coffin if the body is placed within it at a stage prior to interment, for example laying out; this must allow of course for possible movement during the transport or deposition of the corpse and for taphonomic changes preceding or following deposition. Cases of exceptional preservation of organic material in association with inhumation burials, whether through aridity, waterlogging, or anaerobic conditions created by tomb sealing or treatment of the body (mummification, mineral packing), furnish an especially rich picture of the treatment of the body before burial. Evidence from temperate Europe pertaining to the presentation of the corpse during the funeral process is occasionally comparable to what might be potentially available from Roman Egypt (Gessler-Löhr 2012). Although not yet fully published, the later third century burials from Naintré, Vienne, illustrate the potential richness of evidence which may be preserved. An adult and a child had been interred in separate lead-lined stone coffins within masonry vaults which had remained largely airtight since burial. Alongside a rich object assemblage were a range of organic materials of which the presence could normally be little more than guessed at. Amongst the many textiles attested were damask silk and gold thread, as well as peppercorns and dates among the foodstuffs. Traces of Tyrian purple dye were detected across the face and body of one corpse. Preservation of organic materials was complemented by further chance circumstance, for example an imprint from a plant fibre sandal was preserved in plaster that fell from the vault while traces of pollen in the lead lid's corrosion product allowed a funerary bouquet to be identified (Bui and Girard 2003; Devièse *et al.* 2011; Farago-Szekeres and Duday 2008; Gleba 2008).

Since the principal source of evidence for most funerary analyses is the grave, its analytical potential needs no lengthy restatement. However, some underused evidence deserves emphasis. The potential to illuminate lost organic elements of the tomb through study of corpse taphonomy, whether it relates to the structures of the grave (timber linings, shelves, pillows etc.) or the original disposition of the corpse, has been noted above. The arrangement of objects horizontally and vertically is also understudied for its insights into the 'snapshot' moments formed during burial, at what may have been the most emotionally charged moment of death rites. At Lamadelaine by the Titelberg (Luxembourg), for example, in burials made during the first century BC, human remains were deposited first in the grave pits, followed by joints of meat laid out in approximate anatomical order, and then ceramics and other items, some being broken. The placing of the ensemble in particular exaggerated the apparent generosity of the meat offerings (Metzler-Zens and Metzler 1999, 367). The careful depositional sequence as reconstructed for the well-furnished cremation burials from mid-first century AD date from Britain, including the famous placing of pieces as if the game were underway at Stanway (Colchester), suggests that the process of deposition itself was a key part of the spectacle as objects accumulated in the grave (Allen *et al.* 2012, 322–354; Booth this volume; Crummy *et al.* 2007, 201–253).

The fill of the grave may incorporate further residues of graveside activity, especially the consumption of food and drink, although it is likely to be complicated by the incorporation of material from disturbed graves or from non-funerary assemblages. The substantial sherds from broken vessels, and concentrations of pig and chicken documented in grave fills from Lant Street, Southwark, for example illustrate the remnants of materials which are plausibly associated with the burial process (Ridgeway *et al.* 2013, 35–36, 54–55; further examples in Pearce 2013b, Booth this volume). In grave fills disarticulated human skeletal material is also often encountered. Its very variable documentation makes it difficult to assess its significance, but treatment extends from careful reburial of selected elements, typically crania and long bones (the so-called burials '*à reduction*'), to accidental incorporation of fragmentary material (e.g. Duday 2009, 72–88; Pearce 2013b, 461–463; Rife 2012, 169–170; Spanu 2000, 170–172). The analytical potential of disarticulated material for interpreting attitudes to the corpse and the dead is so far little used.

Grave fills also accommodated structures to enable access for the living to the remains of the dead, especially conduits to allow the admission of fluids. The arrangement in an inhumation burial at Castellaccio (via Laurentina outside Rome), where a tube formed by two *imbrices* was placed over an upward-reaching right arm, is a striking recent example of this type (Buccellato *et al.* 2008, 18). Exceptional preservation reveals the frequency of such arrangements at the Porta Nocera cemetery, Pompeii where most cremation burials were connected to the surface through a libation tube (van Andringa *et al.* 2013, 925–926). In the provinces, these arrangements are much rarer, though the necks of burial containers or accessory vessels standing proud of the ground surface may have served the same function

(Blaizot 2009, 236–237). Within the graves at Lamadelaine, Titelberg, some animal bones showed localized exposure to frost, from which the excavators inferred the existence of narrow conduits in organic materials through which liquids were introduced to the buried remains, otherwise too deeply interred to be affected by low winter temperatures (Metzler and Metzler-Zens 1999, 434–435).

Non-burial features

The grave is but one of the diverse contexts potentially to be encountered within burial areas which may relate to the funerary process. These other features have been very much the poor relation, taking second place in excavation priority and, as noted above, sometimes being confused with burials, especially cremation burials. Nonetheless it is possible at least to sketch the variety of features which might be anticipated. As well as redeposited material from disturbed burials, these include (i) features related to the cremation of the corpse, including pyre sites as well as deposits and spreads of pyre material; (ii) other deposits, usually of material related to the consumption of food and drink associated either with the funerary or commemorative ceremonies; and (iii) other deposits which are not related to funerary practice but inform the wider understanding of the spatial and chronological relationship of burial to other activity (see further Weekes this volume). It is important not to underestimate the significance of such deposits for the later Roman period too, which sometimes contain a rich material culture in contrast to minimal or absent furnishing in the grave (Blaizot *et al.* 2009; McKinley this volume; Pearce 2013a, 29–35; Pearce 2013b, 463–465; Weekes 2008).

Where a complex of features appears to relate to a single burial, the various deposits can be more easily differentiated and a sequence of rituals more straightforwardly established. Among the best-known examples from north-west Europe are the 'princely' graves of the first centuries BC and AD at sites in south-east England and north-east Gaul. Alongside lavishly furnished burials, sometimes in timber chambers, a recurring range of features has been documented. These include extensive ditched enclosures to demarcate the funeral space, timber chambers or scaffolds for exposure of the corpse represented by pits and post-hole settings, separate pyre areas marked by patches of scorched earth and assemblages of artefact-rich pyre material, deposited within pits or incorporated (in the case of Goeblingen-Nospelt Luxembourg), in the upcast of the burial mound, as well as deposits and/or structures relating to later commemorative rituals (Crummy *et al.* 2007; Niblett 2004; Metzler 2009; 470–508). Later periods also offer examples of a comparable range of structures and deposits related to single ceremonies where the scale of activity represented by sometimes many hundreds of vessels and remnants of multiple animal carcasses is amongst the largest for single episodes of consumption attested from the northern Roman world (e.g. Ancel 2012, 67–70; Koster 2013, 228; Schucany and Delage 2006, 113–130; Silvino *et al.* 2011, 266–267).

At the cemetery by the Porta Nocera at Pompeii the combination of short-lived funerary use and lack of later disturbance, allied with high-resolution documentation of pyre and similar material in situ, allowed a detailed reconstruction of ritual sequences associated with specific individuals for more modest interments. One of many possible illustrations is the burial of Bebryx, a six-year-old slave (tomb 201, enclosure 21), whose obsequies were marked by the consumption of small quantities of meat, oil and wine and the destruction of objects used to serve or pour the latter. The pyre treatment, deposition of the cremated remains and subsequent commemorative activity could be reconstructed from the burial, its fill, the marker and the finds in its immediate proximity, as well as from Bebryx's pyre. The latter lay outside the burial enclosure but could be linked to the grave by two joining bone fragments and by the complementary anatomical distribution and identical maturation stage of skeletal elements (van Andringa *et al.* 2013, 657–659, 687–689).

While it is uncommon to be able to reconstruct ceremonies with such precision, in other cases the general character of rituals can be reconstructed. Structural evidence for pyres, especially in permanent materials, may be rare; deposits of material associated with their function are more common. At Wederath-*Belginum* (Rheinland Pfalz) more than 120 *Aschenflächen* were documented, i.e. spreads sometimes extending over several square metres of charcoal, ash and cremated human bone and of burnt and fragmentary artefactual material including ceramics, nails and glass and metal objects (Kaiser 2006). At Soissons, ditches demarcating individual pyre sites, horseshoe-shaped, square and round, have been documented, their fills including pyre debris and earth baked by the flames (Sissinger *et al.* 2012, 26–34). Other examples from Germany, Luxembourg and Britain are recorded in varying degrees of detail (e.g. Barber and Bowsher 2000; Polfer 1996, 16–18). The comparison of pyre material and graves assemblages reveals major differences in the form and fabric of ceramics used at different stages of the funerary process (Blaizot 2009; Blaizot and Bonnet 2010).

These features are not easily separated, practically or conceptually from similar spreads and deposits of burnt material which are differentiated from pyre residues by the absence (or minimal presence) of cremated human bone. These features too are widely documented and important for revealing the scale of funerary consumption, whether substantial, as at the Roman imperial period Wederath-Belginum, or more modest, for example in the ten burial-related features excavated near Contrexéville, Vosges (later first to early second century AD), generally containing pockets of charcoal, sherds and scraps of animal bone (Abegg-Wigg 2008; Ancel 2012, 112–120). These in turn are not easily distinguished from mixed deposits of burnt and unburnt material of similar composition which may or may not relate to cemetery activity. For example, recent excavations at Colchester have revealed, alongside graves and monuments, multiple deposits in pits and roadside ditch sections of animal bones, burnt and unburnt, as well as oyster shells, all the likely residue of burial ritual (Brooks 2006, 4–5; Orr *et al.* 2010, 45; Pooley *et al.* 2011, 101–102, 227). The complete carcasses or limbs of animals are perhaps the most frequently occurring deposits of single or multiple objects without any immediate association with cremated human bone (Lepetz this volume, Maltby 2010, 302). The process and purpose of these not easily identified; some may represent rubbish deposition, others may be funerary equivalents of the placed deposits increasingly recognized in non-funerary settings and comprise the likely residue of a sporadically occurring sacrificial process in association with the dead (cf. Chadwick 2012). As the Porta Nocera project (see above) and Ortalli's (2008; this volume) work on Italian sites show, the exiguous traces of common commemorative ritual will only survive in those rare cases where the strata accumulated on cemetery surfaces have been spared truncation. Machine excavation of cemetery areas risks destroying surviving evidence of this type (Weekes 2007; this volume).

On intensively used urban margins in particular, the use of a given space for burial is typically interleaved with profane purposes both synchronically and diachronically. Textual and epigraphic evidence for the ambivalent character of these spaces is now complemented by many excavations (Goodman 2007). The diversity of peri-urban land use documented by excavation and palaeobotanical analysis on the northern and eastern margins of Londinium epitomizes space of this type. Alongside human burial there continued to take place crop cultivation and processing, quarrying, and dumping of rubbish, including animal carcasses, in extraction pits and the roadside ditches which parcelled up the land (Barber and Bowsher 2000, 50–59; Hiller and Wilkinson 2005, 47–49; Swift 2003). Instead of the park-like backdrops of some reconstruction images, reminiscent of the prospectuses of 19th century cemetery reformers, these zones are better imagined as variegated spaces, in which monuments and burial plots are as, if not more, often derelict and despoiled as they are well-maintained (cf. Barber and Bowsher 2000, frontispiece; Koster 2013, 227, fig. 130; Moretti and Tardy 2006, frontispiece). The study of funerary process is, as noted above, complicated by this other activity. However, the multiplicity of uses, often succeeding one another in the same spaces, also enriches the understanding of contemporary engagement with the dead during or following the use of a space for burial. It explains too the susceptibility of monuments to be spoliated once the scrutiny of *familia*, *collegium* or patron fell into abeyance.

Monuments – resta viator et lege

Some final comments in this section relate directly to the topic of monuments. Stone-built monuments, whether stelae or more substantial markers, and especially the texts they bore, have been central to writing Roman history from the foundations of the discipline. Texts, portraits and architectural form have served as an inexhaustible quarry for analyses which cannot be summarised here; their principal importance for this discussion lies in the modes of interaction between the living and the dead which they illuminate.

Inscriptions and surviving features, especially in central Italy, illustrate the facilities associated with the funerary process at monuments, for example ovens, masonry benches, and sources of water, as well as gardens to accommodate funerary activity and furnish an income to maintain monuments (e.g. Fiocchi Nicolai *et al.* 1999,

45; Koster 2013, 224–225). Occasional epigraphic evidence, such as 'the testament of the Lingon', illuminates the similar facilities and associated activity which can be conjectured for similar monuments and funerary enclosures elsewhere, for example gardens, assembly spaces, baths and so on (e.g. Haüssler 2010, 211–218; Koster 2013, 32–36, 228–229; Le Bohec and Buisson 1991). Local contextual evidence sometimes suggests other purposes not otherwise identifiable; Mattingly (2003) for example proposes tombs at Ghirza (Libya) as the *loci* for incubation in order to seek oracular guidance.

As well as the practicalities of use, the monuments provide key evidence for funerary 'visuality'. It is a topos of Roman culture that tombs' location on road frontages allowed a direct appeal to passer-by, sometimes explicitly acknowledged in speaking inscriptions. However even in the *Gräberstrassen* of central Italy, this simple formulation obscures significant variability; here the tombs of the late Republic and beginning of the imperial period, differentiated from one another in scale, form and materials gave way to homogenizing facades and a shift of embellishment to interiors (von Hesberg and Zanker 1987; Borg (2013) and Scott (2013, 77–109) report more recent work). Elsewhere these visual relationships are less intensively studied (Krier and Henrich 2011; Pearce 2011). The analysis of the enormous Bartlow Hills tumuli, Cambridgeshire, revealing surprisingly limited visibility, is one of the few applications of Geographical Information Systems (GIS) for this purpose in a Roman setting (Eckardt *et al.* 2009). The visual experience of tomb interiors has had similarly limited study; the contexts for sarcophagi, sometimes the principal and unencumbered visual focus of a tomb, more often occluded or invisible, illustrates the range of possibilities (Borg 2013, 236–240; Meinecke 2014, 62–76; Rife and Morison this volume).

The duration of engagement with the individual dead is also (potentially) accessible through examination of monument biographies. The re-use of many monuments as *spolia* within the Roman period itself clearly signals the limited span of commemoration, but chronological precision is rarely possible. From examples at Rome Bodel (2014) proposes a typical 'lifespan' for cemeteries of c. 150 years, although the use in these cases was ended by interventions related to imperial building projects. The Porta Nocera tombs at Pompeii (see above) allow an extraordinarily detailed reconstruction of the history of burial markers in this respect. Leaving aside the measures taken by Publius Vesonius Phileros to curse his erstwhile *amicus*, Marcus Ofellius, excavations show that most of the lava *columellae* grave markers were invisible within a decade or two of burial because of rising ground surfaces or that they had been removed (van Andringa *et al.* 2013, 620). The brevity of a marker's lifespan echoes Sidonius Apollinaris' report c. 400 years later of the violation of his grandfather's grave (Reece 1977). These cases must stand as possible proxies for the more ephemeral markers, mounds, posts, cairns, small slabs, and so on that must have commonly marked graves throughout the empire and have rarely survived (for some exceptions see: Ancel 2012, 168 (rural Gaul); Pooley *et al.* 2011, 26–34 (colonial Colchester); Rebillard 2009, 46–48 (urban Musarna); Ortalli, Weekes, Aarts and Heeren this volume). Some depositions of the rural dead in Britain, sometimes 'marked' through proximity to pre-existing features, entrances to fields, hedge or ditch lines, prehistoric monuments and so on, and sometimes occurring intermittently in the same place over several hundred years, hint at longer memories connected to the place of the dead (Pearce 2013b, 107–108).

Current and future directions

The papers collected here engage with several key strands identified above. They draw on the data furnished both by large-scale development-led projects and research excavations from smaller samples. The projects described by Ortalli at Ravenna and at Pompeii by Lepetz were explicitly developed to explore and extend the model for the structure of ritual elicited by Scheid from textual evidence. Some take a holistic view of the range of evidence from individual cemeteries, others range across a number of sites in order to discuss the insights which may be derived from specific analytical processes or materials. The first chapters in the volume deal with the funerary rituals of four very different burial communities across the empire, wealthy merchants at Corinth, townsfolk of more modest means in Canterbury and in Umbria and Emilia-Romagna, Batavian farmers and the members of Germanicus' expedition of AD 15 who performed the funeral rites for the dead of the *Varusschlacht* at Kalkriese. Those which follow consider larger samples drawn from recent development-led excavations from southern Britain and Rome, the former more focused on the archaeological evidence, the latter on the osteological. The final papers consider the methodological and interpretive potential of particular kinds of data, such as pyre sites, faunal deposits, and the character of the cemetery as a space, drawing on evidence from France, Britain and Italy.

Rife and Morison's analysis of burial rites at Kenchreai, the port of Corinth, demonstrates the rich reconstruction of funerary sequence enabled where stratified deposits survive in chamber tombs (and are closely documented). Despite damage and depredation by tomb robbers, abundant evidence survives for ritual sequence from the first to third century AD Koutsongila tombs from the promontory north of the city. As well as the burials themselves, cremation burials and inhumations, the chambers contain abundant evidence for commemorative feasting, especially ceramics and faunal remains. Good preservation of the funerary structures and their decoration, including frescoes on chamber walls, helps enable reconstruction of the sensory responses of participants to ritual, generating a connectedness of mourners to the long-term 'narrative' of both individual household and the mercantile community who used this burial area.

Ortalli's chapter too discusses evidence from well-preserved urban funerary environments investigated through research projects, from Sarsina, in the Umbrian Apennines, and Ravenna, as case studies for the investigation of cemetery space. He notes the methodological and theoretical deficiencies of focusing only on the monument or tomb, and of ignoring the 'connective tissue' between them where the living circulated. In both case studies, better preservation of cemetery surfaces through lacustrine inundation (Sarsina) and dune accumulation (Ravenna) allows detailed examination of the traces of post-funerary commemoration which survived among the layers that accumulated around these mostly modest tombs. This reveals, in these cases at least, its transient character, commonly occurring but modest in scale and brief in duration.

Weekes and Aarts and Heeren consider the insights into funerary process that can be derived from datasets more typical of provincial Roman cemeteries, where what survives and has been documented primarily relates to burial proper, i.e. the interment of the (usually cremated) remains of the deceased.

Drawing on excavations of first to fourth century AD burials from Canterbury, Weekes discusses how phases of the entire ritual sequence, from selection of the dead for (archaeologically) visible burial to commemorative activity, may be analysed exploiting human remains, materials associated with them and the features and deposits associated with burial. Using the St Dunstan's site, he illustrates a holistic approach, prioritising the funeral as the entity to which analysis should be directed, while explicitly evaluating the ambiguous relationship of different categories of evidence to the ritual sequence. He notes that deposition of cremated human bone after collection from the pyre and accompanying objects shows much greater heterogeneity of ritual than in earlier funerary phases. This manifests a reformulated personhood for the dead, as different agents involved in the burial process express their relationship to the deceased in material form. Weekes also considers fundamental interpretive issues further in an afterword at the end of the volume (see below).

Aarts and Heeren interpret the rituals of Batavian farmers of the first to third centuries AD at Tiel-Passewaaij more explicitly through a rites of passage model, demonstrating how its application can be enriched through close attention to non-burial elements of ritual. This allows for observations concerning pyre areas, pyre deposits and monuments to be given due weight alongside burial in evaluating ritual sequence. They also compare cemetery assemblages to the material culture of the living from the adjacent settlement in order to understand how grave materials were selected. This shows that dimensions of identity related to status, class or gender were clearly expressed by objects such as brooches or militaria among the living but much less so in burial, reflecting the assimilation of the dead to a more anonymous and collective body of ancestors.

Examination of human skeletal remains excavated at Kalkriese, near Osnabrück, has revealed complex post mortem histories for corpses of Roman soldiers killed during the annihilation of three legions in the Varian disaster of AD 9. Rost and Wilbers-Rost discuss the results from Oberesch where some of the major discoveries have been made. Some bodies of men and animals were found on the ancient ground surface, protected by the collapse of the rampart used to hem in the Roman soldiers. The disarticulated and fragmentary condition of human remains found within pits, mixed with animal bones, suggests that they represent what was reburied during

Germanicus' campaign to avenge the defeat in AD 14. Some articulated bone groups, in particular comprising hands, lead the authors to identify traces of bandaged limbs; the findspot at Oberesch makes for an unlikely field station but was perhaps where wounded soldiers died or their corpses were dragged for spoliation.

Booth's chapter discusses key recent excavations of Roman cemeteries in southern Britain, undertaken in advance of infrastructure and housing projects as well as gravel extraction. In these cases, intensive agriculture has usually truncated stratified sequences: again, the burial itself, the residue of the act of interment, is typically all that remains. The much greater number of rural burials in particular is transforming understanding of rituals on a population level, for example concerning the practice of archaeologically visible burial ritual after the Roman conquest, or of the relative importance of cremation and inhumation. Assisted by more frequent application of radio-carbon dating, inhumation is documented more widely in the early Roman period than previously considered. The enormous scale of many development-related excavations helps assess the spatial relations between living and dead over many generations.

Recent excavations around Rome have also furnished very extensive new data, putting the study of Roman funerary rituals and population health on a new footing. The cemeteries discussed by Paola Catalano and colleagues are those of lower-status groups, deriving predominantly from cemeteries where monuments are scarce. A consistent approach to the recording and analysis of the excavated skeletons across these projects allows comparison of skeletal samples on a large scale. Their paper introduces five key sites and applies correspondence analysis to skeletal data for more than 1,000 individuals. This clearly demonstrates that this is a heterogeneous sample in terms of the composition of the samples (age and gender) and their living conditions. Two sites especially are distinguished by high frequency of fractures and other trauma; in both cases the contextual evidence from cemetery setting suggests arduous working environments, one a fullery (Casal Bertone), the other salt pans (Castel Malnome).

The two papers which follow this discuss methodological and interpretive questions related to particular types of evidence. Sebastien Lepetz exploits research excavations at Pompeii and development-related fieldwork in France to consider the use of animals in Roman funerary ritual. Close examination of the character and distribution of faunal elements in a burial enclosure near the Porta Nocera exemplifies the wealth of information which can be harvested when circumstances allow. The principal species, pig and chicken, are those animals which are reared in the household for domestic consumption and smaller scale ritual. Remnants of invertebrates and amphibians, as well as faunal material showing signs of having passed through canine guts, shed light on the tomb as micro-environment. Likewise, recent work at Evreux, Chartres and elsewhere reveals equine remains, carcasses, limbs and other bones as regular finds among human burials. Lepetz suggests that this comprises the residues of the knacker's yard, dumped in little-regulated burial spaces.

McKinley's chapter focuses on cremation and its archaeological correlates, the pyre site and its residues, including the buried cremated bone. Exploiting both older observations and the results of recent development-related excavation projects from Britain, she discusses the criteria by which different types of archaeological feature with a direct or indirect relationship to the pyre and the cremation act can be defined, pyre sites and deposits of pyre material separate from the place of cremation, as surface spreads or in other cut features, as well as burials, where pyre material may or may not have been separated from the cremated bone. As other contributors also emphasise, non-burial features have been the Cinderellas of Roman burial archaeology and it is difficult definitively to classify them, even from recent excavations. Recommendations for improving both fieldwork methods and understanding of cremation rituals are made, where the opportunity occurs during excavation of burial spaces.

Conclusions

Readers must of course judge for themselves the value of the approaches outlined above and of the individual contributions. As has been argued the study of the Roman dead has become a much more dynamic and expanded field of enquiry in recent decades, and only some strands of recent scholarship are considered here. The focus of this chapter and of the contributions lies on the material remains of Roman death rituals, both the dead themselves and the traces of the ways in which the living engaged with the remains of the dead and the memorials to them at the funeral and beyond. We have also acknowledged that textual evidence continues to be central to the study of death rituals, both for generating new perspectives and, in the case of new epigraphic discoveries, for improving knowledge of them. Nonetheless, in the main better understanding of the dead and death rituals will come from archaeological evidence. It is already clear that analyses can be more than spoilers, achieving much more than mere demonstration of the inadequacy of an approach focused on the interment alone. Other phases of ritual and the broader interaction of living and dead can be accessed, enabling a more nuanced descriptive account of Roman burial rituals that is more sensitive to variation in time and space. This will enrich the understanding of the relationship between ritual and aspects of identity which has been the traditional concern of students of the Roman dead (Morris 1992).

It is important to acknowledge the challenges that the perspective outlined above raises. The majority of new funerary data will be generated in the context of development-led excavation, but some new approaches presented in this volume are often applied in the context of research-led fieldwork, with fewer demands on time at least, if not on other resources. Research frameworks for developer-led fieldwork, in the UK at least, tend to describe the inadequacies of current data in terms of the low numbers of burial from particular settings (rural, small town etc.). There is a tension between this and a view of burial spaces not only as repositories of assemblages of skeletal remains or artefacts, but also as dynamic and complex entities in which action continuously generates potential material traces, these then acting as a spatial and experiential framework for what follows (Weekes 2007). The widespread adoption of an approach focused on the evidence for the full sequence of rituals in development-led archaeology on urban and rural sites in France reveals however that the approach advocated here is practicable, at least where survival of archaeological features merits its application (e.g. Blaizot 2009). Dissemination of new fieldwork data is a major impediment to improving understanding of death rituals, not only because of its uneven occurrence across the Roman world. Even where numbers of burials are small, rich grave furnishing, or good preservation of evidence for ritual process and the proliferation of analytical techniques, especially for human remains, creates very large publications (e.g. Niblett 1999; Metzler 2009). It is especially difficult to disseminate in print skeletal data compiled during post-excavation analysis which are nonetheless necessary for comparative analysis between cemetery populations. There has so far been limited digital publication of such data (Pearce 2015b, 140–142).

As Weekes observes in his afterword, the rites of passage model for understanding funerary evidence provides only a loose framework and leaves considerable scope for interpretation, integrating the multivocality of symbols, extended liminality and semiosis. The understanding of the structure of the Roman sacrifice established by John Scheid and applied to a funerary setting has been highly influential, but faces the key challenge as to how far it may be applied to contexts at some distance in time and space. Other schemes, with alternative sources of inspiration can also be applied, as the analysis of the phases of ritual reconstructed from the Lamadelaine graves from the first century BC Titelberg illustrates, being linked by the excavators to a cosmological scheme involving the action of different elements in the destruction of the body (Metzler-Zens and Zens 1999). Consideration of the setting of tombs also shows that experiencing the dead transcended a purely ritualised context; the diversity of encounters with the dead, from the direct and deliberate through the mundane and the accidental, remains to be more fully explored through their material traces.

Acknowledgements

I am grateful to Ellen Adams and Jake Weekes for their reading of this paper in draft. Their comments have considerably improved it, but responsibility for any deficiencies remain my own. I also thank John Creighton for the kind opportunity to read the chapter on geophysical prospection on Silchester's extra-mural area in advance of its publication.

Bibliography

Abegg-Wigg, A. (2008) Die Aschengruben im Kontext der provinzialrömischen Bestattungszeremonien: Problematik und Analysemöglichkeiten. In J. Scheid (ed.) *Pour une archéologie du rite. Nouvelles perspectives de l'archéologie funéraire pour l'époque romaine impériale*, 249–257. Rome: École française de Rome.

Allen, T., Donnelly, M., Hardy, A., Hayden, C. and Powell, K. (2012) *A Road Through the Past: Archaeological Discoveries on the A2 Pepperhill to Cobham road-scheme in Kent.* Oxford: Oxford Archaeology.

Ancel, M-J. (2012) *Pratiques et espaces funéraires: la crémation dans les campagnes romaines de la Gaule Belgique.* Editions Mergoil, Montagnac.

Barber, B. and Bowsher, D. (2000) *The Eastern Cemetery of Roman London.* London: MoLAS.

Bechert, T. (1980) Zur Terminologie provinzialrömischer Brandgräber. *Archäologisches Korrespondenzblatt* 10, 253–258.

Bedat, I., Desrosiers, S., Moulherat, C. and Relier, C. (2005) Two Gallo-Roman graves recently found in Naintré (Vienne, France). In F. Pritchard and J.-P. Wild (eds) *Northern Archaeological Textiles, NESAT VII*, 5–11. Oxford: Oxbow.

Bel, V., Blaizot, F. and Duday, H. (2008) Bûchers en fosses et tombes bûcher: problématiques et méthodes de fouille. In J. Scheid (ed.) *Pour une archéologie du rite. Nouvelles perspectives de l'archéologie funéraire pour l'époque romaine impériale*, 233–247. Rome: École française de Rome.

Belcastro, M. G., and Ortalli, J. eds. (2010) *Sepolture anomale: indagini archeologiche e antropologiche dall'epoca classica al Medioevo in Emilia Romagna.* Florence: All'Insegna del Giglio.

Biggins, J. and Taylor, D. (2004a.) Geophysical survey of the *vicus* at Birdoswald Roman fort, Cumbria. *Britannia* 35, 159–178.

Biggins, J. and Taylor, D. (2004b.) The Roman fort and *vicus* at Maryport: geophysical survey, 2000. In R. Wilson and I. Caruana (eds) *Romans on the Solway*, 102–133. Maryport: Senhouse Museum.

Blaizot, F. ed. (2009) *Pratiques et espaces funéraires de la Gaule durant l'Antiquité. Gallia* 66.1.

Blaizot F. and Bonnet, C. (2010) L'identité des pratiques funéraires romaines: regard sur le centre et le sud-est de la Gaule. In P. Ouzoulias and L. Tranoy (eds) *Comment les Gaules devinrent romaines*, 267–282. Paris: Louvre/La Découverte.

Bodel, J. (1999) *Death on display: looking at Roman funerals.* In B. A. Bergmann and C. Kondoleon (eds) *The Art of Ancient Spectacle*, 259–281. London/New Haven: Yale University Press.

Bodel, J. (2014) The life and death of ancient Roman cemeteries. Living with the dead in imperial Rome. In C. Häuber, F. X. Schütz and G. M. Winder (eds) *Reconstruction and the Historic City: Rome and Abroad - an interdisciplinary approach*, 177–195. Munich: LMU.

Booth, P. and Boyle, A. (2008) The archaeology of Roman burials in England – frameworks and methods: a perspective from Oxford Archaeology. In J. Scheid (ed.) *Pour une archéologie du rite. Nouvelles perspectives sur l'archéologie funéraire*, 127–136. Rome: École française de Rome.

Borg, B. E. (2013) *Crisis and Ambition: Tombs and Burial Customs in Third-Century CE Rome.* Oxford: Oxford University Press.

Bouby, L. and Marinval, P. (2004) Fruits and seeds from Roman cremations in Limagne (Massif Central) and the spatial variability of plant offerings in France. *Journal of Archaeological Science* 31, 77–86.

Brettell, R. C., Schotsmans, E. M. J., Walton Rogers, P., Reifarth, N., Redfern, R. C., Stern, B. and Heron, C. P. (2015) 'Choicest unguents': molecular evidence for the use of resinous plant exudates in late Roman mortuary rites in Britain. *Journal of Archaeological Science* 53, 639–648.

Bridger, C. (1996) *Das römerzeitliche Gräberfeld 'An hinkes Weißhof', Tönisvorst-Vorst, Kreis Viersen.* Köln: Rheinland Verlag.

Brooks, H. (2006) *A Roman temple-tomb at Colchester Royal Grammar School, 6 Lexden Road, Colchester, Essex.* Colchester: CAT http://cat.essex.ac.uk/reports/CAT-report-0345.pdf (Accessed 10.14).

Buccellato, A., Catalano, P. and Pantano, W. (2008) Le site et la nécropole de Castellaccio. *Dossiers d'Archéologie* 330, 14–19.

Bui. T.M. and Girard, M. (2003) Pollens, ultimes indices de pratiques funéraires évanouies. *Revue archéologique de Picardie. Numéro spécial* 21, 127–137.

Castagnetti, S. (2012) *Le leges libitinariae flegree: edizione e commento.* Naples: Satura Editrice.

Cool, H. (2004) *The Roman Cemetery at Brougham, Cumbria: Excavations 1966-67*. London: Society for the Promotion of Roman Studies.

Cool, H.E.M. (2010) Objects of glass, shale, bone and metal, In P. Booth *et al. The Late Roman Cemetery at Lankhills, Winchester. Excavations 2000-2005*, 267–309. Oxford: Oxford Archaeology.

Cool, H.E.M. (2011) Funerary contexts. In L. Allason-Jones (ed.) *Artefacts from Roman Britain. Their Purpose and Use*, 293–313. Cambridge: Cambridge University Press.

Cool, H.E.M. and Baxter, M. J. (2005) Cemeteries and significance tests. *Journal of Roman Archaeology* 18, 397–404.

Cooremans, B. (2008) The Roman cemeteries of Tienen and Tongeren: results from the archaeobotanical analysis of the cremation graves. *Vegetation History and Archaeobotany* 17, 3–13.

Crummy, P. Benfield, S., Crummy, N. Rigby, V. and Shimmin, D. (2007) *Stanway: An Elite Burial Site at Camulodunum*. London: Society for the Promotion of Roman Studies.

Davies, P. (2000). *Death and the emperor. Roman Imperial Funerary Monuments from Augustus to Marcus Aurelius*. Cambridge: Cambridge University Press.

De Jong, L. (2010) Performing death in Tyre. The life and afterlife of a Roman cemetery in the province of Syria. *American Journal of Archaeology* 114, 597–630.

Devièse, T., Ribechini, E., Baraldi, P., Farago-Szekeres, B., Duday, H., Regert, M. and Colombini, M. P. (2011) First chemical evidence of royal purple as a material used for funeral treatment discovered in a Gallo–Roman burial (Naintré, France, 3rd century AD). *Analytical and Bioanalytical Chemistry* 401, 1739–1748.

Devreker, J., Thoen, T. and Vermeulen, F. (2003) *Excavations in Pessinus: the so-called Acropolis*. Gent: Academia Press.

Duday, H. (2009) *Archaeology of the Dead*. Oxford: Oxbow.

Eckardt, H., Brewer, P., Hay, S. and Poppy, S. (2009) Roman barrows and their landscape context: a GIS case study at Bartlow, Cambridgeshire. *Britannia* 40, 65–98.

Eckardt, H. ed. (2010) *Roman Diasporas: Archaeological Approaches to Mobility and Diversity in the Roman Empire*. Portsmouth RI: Journal of Roman Archaeology.

Eckardt, H., Lewis, M. and Müldner, G. (2014) People on the move in Roman Britain. *World Archaeology* 46, 534–550.

Erkelenz, C. (2012) *Die römischen Nekropolen des vicus Mönchengladbach Rheydt-Mülfort*. Rahden: Leidorf.

Faber, A., Fasold, P., Struck, M. and Witteyer, M (eds) (2004) *Körpergräber des 1.-3. Jh. in der römischen Welt*. Archäologisches Museum Frankfurt: Frankfurt am Main.

Farago-Szekeres B. and Duday H. (2008) Les tombes fastueuses de Naintré (Vienne). *Les Dossiers d'Archéologie* 330, 120–127.

Favro, D. and Johanson, C. (2010) Death in motion: funerary processions in the Roman Forum. *Journal of the Society of Architectural Historians* 69.1, 12–37.

Fiocchi Nicolai, V., Bisconti, F. and Mazzoleni, D. (1999) *The Christian Catacombs of Rome. History, Decoration, Inscriptions*. Regensburg: Schnell and Steiner.

Fulford. M. G., and Holbrook, N. (2014). Developer archaeology and the Romano-British countryside: a revolution in understanding. In D. J. Breeze (ed.) *The Impact of Rome on the British Countryside*, 38–44. London: Royal Archaeological Institute.

Gaffney, V., Patterson, H., Piro, S., Goodman, D. and Nishimura, Y. (2004) Multimethodological approach to study and characterize *Forum Novum* (Vescovio, central Italy). *Archaeological Prospection* 11, 201–212.

Gardner, A. (2013) Thinking about Roman imperialism: postcolonialism, globalization and beyond? *Britannia* 44, 1–25.

Gasperini, L. (1996) Ancora sul cippo di Arzaga (*I. It. Brixia* 817). In C. Stella and A. Valvo (eds) *Studi in Onore di Albino Garzetti*, 183–199. Brescia: Ateneo di Brescia.

Gessler-Löhr, B. (2012) Mummies and mummification. In C. Riggs (ed.), *The Oxford Handbook of Roman Egypt*, Oxford: Oxford University Press DOI: 10.1093/oxfordhb/9780199571451.013.0041.

Glansdorp, E. (2005) *Das Gräberfeld "Margarethenstrasse" in Dillingen-Pachten: Studien zu gallo-römischen Bestattungssitten.* Bonn: Habelt.

Gleba, M. (2008) *Auratae Vestes.* Gold textiles in the ancient Mediterranean. In C. Alfaro and L. Karali (eds.) Purpureae Vestes II *Vestidos, Textiles y Tintes: estudios sobre la producción de bienes de consumo en la Antigüedad*, 63–80. Valencia: University of Valencia.

Goodman, P. (2007) *The Roman City and its Periphery: from Rome to Gaul.* London: Routledge.

Gosden, C. and Garrow, D. (2012) *Technologies of Enchantment? Exploring Celtic Art 400 BC to AD 100.* Oxford: Oxford University Press.

Gowland, R. (2001) Playing dead: implications of mortuary evidence for the social construction of childhood in Roman Britain. In G. Davies, A. Gardner and K. Lockyear (eds) *Proceedings of the 10th Theoretical Roman Archaeology Conference*, 152–168. Oxford: Oxbow.

Gowland, R. and Garnsey, P. (2010) Skeletal evidence for health, nutritional status and malaria in Rome and the empire. In H. Eckardt (ed.) *Roman diasporas; archaeological approaches to mobility and diversity in the Roman Empire*, 131–156. Portsmouth, RI: Journal of Roman Archaeology.

Graham, E-J. (2011a) Memory and materiality: re-embodying the Roman funeral. In V. M. Hope and J. Huskinson (eds.) *Memory and Mourning: Studies on Roman Death*, 21–39. Oxford, Oxbow.

Graham, E-J. (2011b) From fragments to ancestors: re-defining the role of *os resectum* in rituals of purification and commemoration in Republican Rome. In M. Carroll and J. Rempel (eds.) *Living through the Dead: Burial and Commemoration in the Classical World*, 91–109. Oxford, Oxbow.

Graham, E-J. (2015) Introduction: embodying death in archaeology. In Z. Devlin and E.-J. Graham (eds) *Death embodied: archaeological approaches to the treatment of the corpse*, 41–62. Oxford: Oxbow.

Haüssler, R. (2010) From tomb to temple. On the rôle of hero cults in local religions in Gaul and Britain in the Iron Age and the Roman period. In J. Alberto Arenas-Estaban (ed.) *Celtic Religion across Space and Time: FERCAN IX*, 200–227. Toledo: Junta de Comunidades de Castilla-La Mancha.

Hay, S., Johnson, P, Keay, S. and Millett, M. (2010) Falerii Novi: further survey of the northern extramural area. *Papers of the British School at Rome* 78, 1–38.

Von Hesberg, H. and Zanker, P. (eds) (1987) *Römische Gräberstrasse - Selbstdarstellung, Status, Standard.* Munich: Bayerische Akademie der Wissenschaften.

Hiller, J. and Wilkinson, T. (2005) *Archaeology of the Jubilee Line extension: prehistoric and Roman activity at Stratford Market Depot, West Ham, London, 1991-1993.* London: MOLA.

Hope. V. M. (2007) *Death in Ancient Rome: A Sourcebook.* London: Routledge.

Hope. V. M. (2009) *Roman Death: Dying and the Dead in Ancient Rome.* London: Continuum.

Hope, V. M. and Huskinson, J. (eds) (2011) *Memory and Mourning: Studies on Roman Death.* Oxford: Oxbow.

Jiménez, A. (2008) A critical approach to the concept of resistance: new 'traditional rituals' and objects in funerary contexts of Roman Baetica. In C. Fenwick, M. Wiggins and D. Wythe (eds) *TRAC 2007: Proceedings of the 17th Theoretical Roman Archaeology Conference*, 15–30. Oxford: Oxbow.

Johanson, C. (2011) A walk with the dead. In B. Rawson (ed.) *A Companion to Families in the Greek and Roman Worlds*, 408–430. Chichester: Wiley-Blackwell.

Kaiser, M. (2006) *Das keltisch-römische Gräberfeld von Wederath-Belginum. 6. Die Aschengruben und Aschenflächen ausgegraben 1954-1985.* Trier: Rheinisches Landesmuseum.

Kilgrove, K. (2014) Bioarchaeology in the Roman Empire. In C. Smith (ed.) *Encyclopedia of Global Archaeology.* 876–82. New York: Springer. DOI: 10.1007/978-1-4419-0465-2.

Koster, A. (2013) *The cemetery of Noviomagus and the Wealthy Burials of the Municipal Elite.* Nijmegen: Museum Het Valkhof.

Kreuz, A. (2000) Functional and conceptual data from Roman cremations. In J. Pearce, M. Millett and M. Struck (eds) *Burial, Society and Context in the Roman World*, 45–51. Oxford: Oxbow.

Krier, J. and Henrich, P. (2011) Monumental funerary structures of the 1st to 3rd centuries associated with Roman villas in the area of the Treveri. In N. Roymans and T. Derks (eds) *Villa Landscapes in the Roman North*, 211–234. Amsterdam: Amsterdam University Press.

Le Bohec‚ Y and Buisson, A. (eds) (1991) *Le Testament du Lingon.* Lyon/Paris: de Boccard.

Lefebvre, A., Mondy, M., Cabart, H. Decanter, F. and Feller, M. (2013) Premières données sur l'archéologie funéraire de l'Antiquité tardive dans la cité des Médiomatriques: l'exemple d'Uckange (Moselle). *Revue Archéologique de L'Est* 62, 253–281.

Lindsay, H. (2000) Death pollution and funerals in the city of Rome. In V. M. Hope and E. Marshall (eds) *Death and Disease in the Ancient City*, 152–173. London: Routledge.

MacKinnon, M. (2007) Peopling the mortuary landscape of North Africa: an overview of the human osteological evidence. In D. Stone and L. Stirling (eds) *Mortuary landscapes of North Africa*, 204–240. Toronto: University of Toronto Press.

Martin-Kilcher, S. (2000) *Mors immatura* in the Roman world. A mirror of society and tradition. In J. Pearce, M. Millett and M. Struck (eds) *Burial, Society and Context in the Roman World*, 63–77. Oxford: Oxbow.

Maltby, M. (2010) *Feeding a Roman Town: Environmental Evidence from Excavations in Winchester, 1972-1985.* Winchester: Winchester Museums.

Mattingly, D. J. (2003) *Family values*: art and power at *Ghirza* in the Libyan pre-desert. In S. Scott and J. Webster (eds) *Roman Imperialism and Provincial Art*, 153–170. Cambridge: Cambridge University Press.

McKinley, J. (2013) Cremation: excavation, analysis and interpretation of material from cremation-related contexts. In S. Tarlow and L. Nilsson Stutz (eds) *Oxford Handbook on the Archaeology of Death and Burial*, 147–171. Oxford: Oxford University Press.

Meinecke, K. (2014) *Sarcophagum posuit: Römische Steinsarkophage im Kontext.* Ruhpolding: Verlag Franz Philipp Rutzen.

Metzler-Zens, N. and Metzler, J. (1999) *Lamadelaine: une nécropole de l'oppidum du Titelberg.* Luxembourg: Musée national d'histoire et d'art.

Metzler, J. (2009) *Goeblange-Nospelt: une nécropole aristocratique trévire.* Luxembourg: Musée national d'histoire et d'art.

Millett, M., Revell, L. and Moore, A. (eds) (forthcoming) *The Oxford Handbook of Roman Britain.* Oxford: Oxford University Press.

Moretti, J-C. and Tardy, D. (eds) (2006) *L'architecture funéraire monumentale: la Gaule dans l'Empire romain.* Paris: Édition du Comité des travaux historiques et scientifiques.

Morris, I. (1992) *Death Ritual and Social Structure in Classical Antiquity.* Cambridge: Cambridge University Press.

Müldner, G. (2013) Stable isotopes and diet: their contribution to Romano-British research. *Antiquity* 87, 137–149.

Niblett, R. (1999) *The Excavation of a Ceremonial Site at Folly Lane, Verulamium.* London: Society for the Promotion of Roman Studies.

Niblett, R. (2004) The native elite and their funerary practices from the first century BC to Nero. In M. Todd (ed.) *A Companion to Roman Britain*, 30–41. Oxford: Blackwell.

O'Connell, E. R. (2014) Settlements and cemeteries in Late Antique Egypt: an introduction. In E.R. O'Connell (ed.) *Egypt in the First Millennium AD: Perspectives from new fieldwork*, 1–19. Leuven: Peeters.

Orr, K. (2010) *Archaeological Excavations at 1 Queens Road, Colchester, Essex 2003 and 2004-2005.* Colchester: CAT.

Ortalli, J. (2008) Scavo stratigrafico e contesti sepolcrali: una questione aperta. In J. Scheid (ed.) *Pour une archéologie du rite. Nouvelles perspectives sur l'archéologie funéraire*, 137–159. Rome: Ecole française de Rome.

Ouzoulias, P. and Tranoy, L. (eds) (2010) *Comment les Gaules devinrent romaines.* Paris: Louvre/La Découverte.

Pearce, J. (2011) Marking the dead: tombs and topography in the Roman provinces. In M. Carroll and J. Rempel (eds) *Living with the Dead*, 134–158. Oxford: Oxbow.

Pearce, J. (2013a) *Contextual archaeology of burial practice: case studies from Roman Britain.* Oxford: Archaeopress.

Pearce, J. (2013b) Beyond the grave. Excavating the dead in the late Roman provinces. In L. Lavan and M. Mulryan (eds) *Field Methods and Post-Excavation Techniques in Late Antique Archaeology*, 441–482. Leiden: Brill.

Pearce, J. (2015a) A 'civilised' death? The interpretation of provincial Roman grave good assemblages. In J. R. Brandt, H. Ingvaldsen and M. Prusac (eds), *Death and Changing Rituals. Function and Meaning in Ancient Funerary Practices*, 223–248. Oxford: Oxbow.

Pearce, J. (2015b) Urban exits: commercial archaeology and the study of death rituals and the dead in the towns of Roman Britain. In M. Fulford and N. Holbrook (eds) *Commercial Archaeology and the Study of Romano-British towns*, 138–166. London, Society for the Promotion of Roman Studies.

Pirling, R. and Siepen, M. (2006) *Die Funde aus den römischen Gräbern von Krefeld-Gellep: Katalog der Gräber 6348-6361*. Stuttgart: Steiner.

Pitts, M. and Griffin, R. (2012) Exploring health and social well-being in Late Roman Britain: an intercemetery approach. *American Journal of Archaeology* 116, 253–276.

Polfer, M. (1996) *Das gallorömische Brandgräberfeld und der dazugehörige Verbrennungplatz von Septfontaines-Dëckt*. Luxembourg: Musée National d'Histoire et d'Art.

Quensel-von-Kalben, L. (2000) Putting late Roman burial practice (from Britain) in context. In J. Pearce, M. Millett and M. Struck (eds) *Burial, Society and Context in the Roman World*, 217–230. Oxford: Oxbow.

Raynaud C. (2006) Le monde des morts. *Gallia* 63, 137–170.

Rebillard, E. (2009) *Musarna 3: la nécropole impériale*. Rome: École française de Rome.

Redfern, R., DeWitte, S., Pearce, J., Hamlin, C. and Egging-Dinwiddy, K. (2015) Urban-rural differences in Roman Dorset, England: a bioarchaeological perspective on Roman settlements. *American Journal of Physical Anthropology* 157, 107–20.

Reece, R. (1977) Burial in Latin literature: two examples. In R. Reece (ed.) *Burial in the Roman World*, 44–45. London: CBA.

Revell, L. (2009) *Roman Imperialism and Local Identities*. Cambridge: Cambridge University Press.

Rife J. (2012) *Isthmia IX: The Roman and Byzantine Graves and Human Remains*. Athens: ACSA.

Ridgeway, V., Leary, K. and Sudds, B. (2013) *Roman Burials in Southwark, London*. London: PCA.

Riggs, C. (ed.) (2012) *The Oxford Handbook of Roman Egypt*. Oxford: Oxford University Press.

Rossi, D. (2012) *Sulla via Flaminia: il mausoleo di Marco Nonio Macrino*. Milan: Electa.

Scheid, J. (1984) *Contraria facere*: renversements et déplacements dans les rites funéraires. *Annali del Seminario di Studi del Mondo Classico. Istituto Universitario Orientale Napoli: Sezione di Archeologia e Storia Antica*, 6, 117–139.

Scheid, J. (2005) *Quand faire c'est croire: les rites sacrificiels des romains*. Paris: Aubier.

Scheid, J. (ed.) (2008) *Pour une archéologie du rite. Nouvelles perspectives sur l'archéologie funeraire*. Rome: Ecole Francaise de Rome.

Scheid, J. and Rüpke, J. (eds) (2010) *Bestattungsrituale und Totenkult in der römischen Kaiserzeit*. Stuttgart: Steiner.

Schrumpf, S. (2006) *Bestattung und Bestattungswesen im römischen Reich*. Göttingen: V & R Unipress.

Schucany, C. and Delage, R. (2006) *Die römische Villa von Biberist-Spitalhof/SO*. Remshalden: Greiner.

Scott, M. (2013) *Space and Society in the Greek and Roman Worlds*. Cambridge: Cambridge University Press.

Silvino, T., Blaizot, F., Maza, G., Argant, T., Carrara, S., Robin, L., Schaal, C. and Schenk, A. (2011) La *villa* des 'Vernes' à La Boisse (Ain): contribution des fouilles récentes à la compréhension de l'évolution d'un établissement rural antique et de son espace funéraire. *Revue Archéologique de l'Est* 60, 217–290.

Simmonds, A., Márquez-Grant, N. and Loe, L. (2008) *Life and Death in a Roman City*. Oxford: Oxford Archaeology.

Sissinger, B. *et al.* (2012) Fouille d'un quartier funéraire des Ier et IIe s. dans le suburbium de Soissons/Augusta Suessionum: aires de crémation et inhumations d'enfants. *Gallia* 69.1, 3–67.

Spanu, M. (2000) Burial in Asia Minor during the imperial period, with particular reference to Cilicia and Cappadocia. In J. Pearce, M. Millett and M. Struck (eds) *Burial, Society and Context in the Roman World*, 169–178. Oxford: Oxbow.

Šterbenc Erker, D. (2011) Gender and Roman funeral ritual. In V. Hope and J. Huskinson (eds) *Memory and Mourning in Ancient Rome*, 40–60. Oxford: Oxbow.

Struck, M. (1993) *Busta* in Britannien und ihre Verbindungen zum Kontinent. Allgemeine Überlegungen zur Herleitung der Bestattungssitte. In M. Struck (ed.) *Römerzeitliche Gräber als Quellen zu Religion, Bevölkerungsstruktur und Sozialgeschichte*, 81–94. Mainz: Johannes Gutenberg Institut für Vor- und Frühgeschichte.

Swift, D. (2003) *Roman burials, medieval tenements and suburban growth, 201 Bishopsgate, City of London*. London: MoLAS.

Tarlow, S. and Nilsson Stutz, L. (eds) (2013) *Oxford Handbook on the Archaeology of Death and Burial*. Oxford: Oxford University Press.

Taylor, A. (2008) Aspects of deviant burial in Roman Britain. In E. M. Murphy (ed.), *Deviant Burial in the Archaeological Record*, 91–114. Oxford: Oxbow.

Theuws, F. and M. Alkemade (2000) A kind of mirror for men: sword depositions in late Antique northern Gaul. In F. Theuws and J. Nelson (eds.) *Rituals of Power from Late Antiquity to the Early Middle Ages*, 401–476, Leiden: Brill.

Todd. M. (ed.) (2004) *A Companion to Roman Britain*. Oxford: Blackwell.

Thompson, T. J. (ed.) (2015) *The Archaeology of Cremation. Burned Human Remains in Funerary Studies*. Oxford: Oxbow.

Toynbee, J.M.C. (1971) *Death and Burial in the Roman World*. London: Thames and Hudson.

Trément, F. and Humbert, L. (2004) Une incinération spectaculaire au pied du Puy de Dôme. Le bûcher funéraire du col de Ceyssat (Saint-Genès-Champanelle). In M. Cebeillac-Gervasoni, L. Lamoine and F. Trément (eds) *Autocélébration des élites locales dans le monde romain: contextes, textes, images (IIe s. av. J.-C.–IIIe s. ap. J.-C.)*, 463–500. Clermont-Ferrand : Presses Universitaires Blaise Pascal.

Tucker, K. (2013) The osteology of decapitation burials from Roman Britain: a post-mortem burial rite? In M. J. Smith and C. Knüsel (eds) *The Routledge Handbook of the Bioarchaeology of Human Conflict*, 213–236. London: Routledge.

Van Andringa, W., Duday, H. and Lepetz, S. (2013) *Mourir à Pompéi – Fouille d'un quartier funéraire de la nécropole romaine de Porta Nocera (2003-2007)*. Rome: Ecole Française de Rome.

Vasić, M. and Tomović, M. (2005) Šarkamen (East Serbia): an imperial residence and memorial complex of the Tetrarchic period. *Germania* 83.2, 257–307.

Weekes, J. (2007) A specific problem? The detection, protection and exploration of Romano-British cremation cemeteries through competitive tendering. In B. Croxford *et al.* (eds) *TRAC 2006: Proceedings of the Sixteenth Annual Theoretical Roman Archaeology Conference*, 183–191. Oxford: Oxbow Books.

Weekes, J. (2008) Classification and analysis of archaeological contexts for the reconstruction of early Romano-British cremation funerals. *Britannia* 39, 145–160.

Weekes, J. (2014) Cemeteries and funerary practice. In M. Millett, L. Revell, and A. Moore (eds) *The Oxford Handbook of Roman Britain*. Oxford: Oxford University Press. DOI:10.1093/oxfordhb/9780199697713.013.025.

Wiethold, J. (2000) Die Pflanzenreste aus den *Aschengruben*. Ergebnisse archäobotanischer Analysen. In A. Miron (ed.) *Archäologische Untersuchungen im Trassenverlauf der Bundesautobahn A8 im Landkreis Merzig-Wadern*, 131–152. Saarbrücken: Staatliches Konservatoramt des Saarlandes.

Williams, H. (2004) Potted histories – cremation, ceramics and social memory in early Roman Britain. *Oxford Journal of Archaeology* 23.4, 417–427.

Willis, S. (2011) Samian ware and society in Britain and beyond. *Britannia* 42, 167–242.

Wilmott, T. (2010) Birdoswald Roman cemetery. *English Heritage Research News* 14, 16–19. https://content.historicengland.org.uk/images-books/publications/research-news-14/researchnews14.pdf/ (Accessed 10.13).

Winton, H. (2009) *Tar Barrow, Cirencester, Gloucestershire. A Roman or Iron Age ceremonial area. Aerial Photo Interpretation and Mapping*. London: English Heritage http://services.english-heritage.org.uk/ResearchReportsPdfs/051_2009WEB.pdf (Accessed 01.12).

Chapter 2

Space, object, and process in the Koutsongila Cemetery at Roman Kenchreai, Greece

Joseph L. Rife and Melissa Moore Morison

Introduction

Death signifies and compels many things in society. It is a biological event that brings an end to life, removing someone as a conscious participant from the activities of a family and a larger community. It evokes memories and feelings in those affected by the loss of a relative, friend, or associate. It is an abstraction that reflects a conception of the world and time, from an understanding of society to beliefs about an afterlife. Death also sets into motion a complicated series of behaviours among mourners. These behaviours activate various spaces and materials, all of which are intentionally chosen and meaningfully employed in rituals with a sequential order. In this sense, death is a process.

Many factors influence how people respond to death. The loss of a loved one can inject an element of randomness into the daily routine, a sense of timelessness or aporia. But even amid emotional turmoil, people turn to established practices. Members of a community are familiar with customary behaviours and their spatial and material components, and they can assess and negotiate the significance of local practice and meaning. They participate in the funerals of relatives and neighbours, they walk through cemeteries, and they see the burials of the dead. Such experiences frame their own choices and actions in the event of death. Within the framework of what is recognized as possible and acceptable, bereaved families employ specific rituals of burial and commemoration in order to communicate particular identities, such as socioeconomic status, ethnic affiliation, professional membership, age, and gender. Narratives of individual and corporate identity emerge through the interplay of personal and group experiences involving certain spaces and objects in a ritual process.

To examine processes of burial and commemoration in a past community and their role in the development of identity-narratives, we must rely on textual and archaeological sources that vary in quality and are often ambiguously determined (e.g. Morris 1992; Insoll 2004; Kyriakidis 2007). Literary descriptions of mourning and other funerary activities seldom afford the same focused or elaborate accounts as modern ethnographies. Moreover, the social historian must always consider authorial motives when evaluating ancient writings, as well as the perspective of the intended audience. On the other hand, the material remains of funerals are often either absent, fragmentary, or compromised by post-depositional factors. It can also be challenging to identify an affixed, special, and continuous range of activities – a ritual process – within a local mortuary record that overlaps spatially and materially with evidence for other spheres of public and private life. In the end, historians and archaeologists can sometimes trace particular events or episodes in longer processes. But a full picture of ancient mortuary behaviour and its relationship to self-presentation is often elusive, or at best open to reasoned speculation.

Rituals of death in the Roman Greece and Asia Minor

The evidence for death-rituals in the towns and cities of the eastern Mediterranean world under the Roman Empire, which is the setting for the present study, is plentiful but difficult to interpret. Numerous Greek authors who lived in the eastern provinces discussed or portrayed funerals and cemeteries. Their testimony, however, is often brief or oblique, and sometimes it is shaded by literary allusion, intellectual or stylistic proclivities, or fictional colour. Nonetheless, this literary record attests vividly to the central importance of civic topography, monumental space, the funerary procession, the burial rite, and commemorative visitation in the mortuary experience of local communities (e.g. Rife 1999; 2008; 2009). A vast body of material evidence for mortuary practices also survives: tombstones, graves, human bones, funerary artefacts, entire cemeteries. Yet burial in the Roman East remains an understudied field, and among the few comprehensive publications that have appeared, the chief concern remains sepulchral architecture, not mortuary behaviour. Several surveys of Greece and Asia Minor focus on architectural forms and to a lesser extent consider epitaphs, artefactual contents, the ritual function of structures, and their broader topographic context (e.g. Alföldi-Rosenbaum 1971; Haspels 1971, 164-199; Schneider Equini 1972; Fedak 1990; Goette 1994, 296–300; Cormack 1997; Spanu 2000; Berns 2003; Flämig 2007; cf. Laskaris 2000 for a similar propensity in the study of Late Roman to Early Byzantine Greek burial).

The present study investigates a broad range of material remains at one site to explore the ritual processes of death in the Greek world during the Roman era. The port of Kenchreai on the Isthmus of Corinth in Greece (Fig. 2.1) is an archaeological site with a well-preserved cemetery of chamber tombs of Early to Middle Roman date (c. first to third centuries CE). The chamber tombs, their surroundings, and their contents provide rich evidence for behaviours that mediated personal experience at the time of death and afterwards. Close examination of these spaces and objects helps us to understand the performance and experience of mourning in one provincial

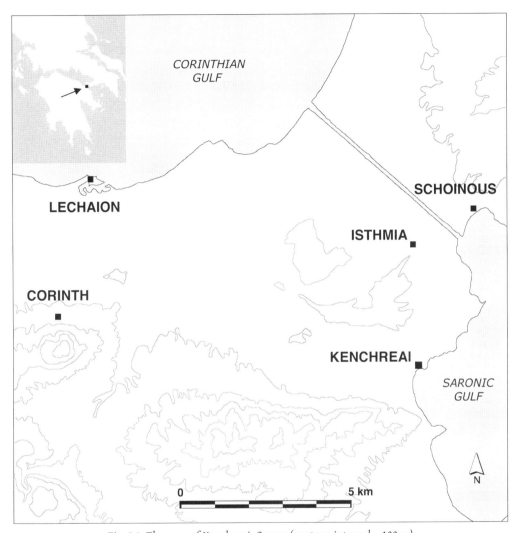

Fig. 2.1. The area of Kenchreai, Greece (contour interval = 100 m).

Greek community. On this basis, we can begin to explore relationships between the ritual process of death and the construction of identities for the deceased and the bereaved, both as individuals and as members of groups, at Roman Kenchreai.

Our examination can serve as a model for the exploration of comparable evidence from the Roman East. It makes an innovative contribution to the broader field of ancient mortuary studies in several respects. First, although the chamber tomb was a dominant form of monumental burial in Greece and Asia Minor during Hellenistic to Roman times, few reports provide sufficient evidence to reconstruct ritual behaviours associated with chamber tombs. Certain publications of chamber tombs at, for example, Knossos (Carington Smith 1982) and Corinth (Walbank 2005; Slane and

Walbank 2006; Slane 2012) have documented the funerary assemblages and sepulchral structures in great detail, but they have not fully correlated the physical remains with an ancient ritual process. The useful survey of Roman-era chamber tombs across Asia Minor by Sarah Cormack (2007) attempted to identify separate functional spaces and structural features with specific activities, such as dining, but it relied on only partly documented or unexcavated architecture, often without reference to depositional evidence. In contrast, we have investigated and recorded the Early to Middle Roman chamber tombs at Kenchreai with the express purpose of reading mortuary spaces and funerary artefacts as the material components of ritual behaviours that claim identity within the local community. A similar perspective has been applied to Bronze Age tombs in Greece, such as the disturbed chambers at Barnavos near Nemea (Wright *et al.* 2008), but it has not been tested on evidence of Roman date, which presents a very different range of historical and material conditions.

Second, our study imagines death at Roman Kenchreai to be a multi-layered experience involving a series of interconnected spaces and objects imbued with significance through ritual performance. Each space outside and inside the tomb, together with its artefactual associations, was a locus of behaviours that were meaningful to mourners. Our investigation of the process of death follows the experiential sequence of structural spaces and artefactual distributions, from the activities away from and outside the tomb down into the chamber and the burial compartments. This systematic approach will facilitate the comparison of the remains at Roman Kenchreai with those at other sites.

Finally, we understand the tombs at Kenchreai and their contents as tools that served an evolving set of social needs. This concept forms a useful interpretive bridge between the identification of activities and assemblages and the recovery of experiences and self-presentation. By tracing spatial and chronological variability among the tombs and their contents, and then examining the ways that the gradual accumulation of objects and reworking of spaces created an evolving bricolage of experiences and references for visitors, we can view artefacts and areas not only as elements in a behavioural process, but also as effective instruments of emotional regulation and identity-affirmation. Understanding the interplay between space, object, and process on the one hand and personal experience and social identity on the other, contributes to a richer picture of life in this one provincial community. While Kenchreai was surely distinguished by various local factors, from its history and topography to its commercial ties and demographic composition, it also shared certain essential qualities, such as economic and cultural affluence and internal complexity, with other vibrant port-towns in the Roman East.

Roman Kenchreai and its main cemetery

During the Roman era, Kenchreai on the Saronic Gulf flourished as the eastern harbor of Corinth, the largest administrative and commercial center in the province of

Achaea (Fig. 2.1). Although Kenchreai itself never obtained full political or economic autonomy from the urban hub located 8 km inland, it maintained its own cults, developed its own monumental landscape, and supported a stratified community that depended on frequent traffic from the east. The history and topography of the port-town are well attested in ancient literature and inscriptions, which portray a culturally and socially diverse community during the Early and Middle Roman periods (Rife 2010). Archaeological exploration over the past century, especially the large-scale excavations sponsored by the University of Chicago and Indiana University in the 1960s under the auspices of the American School of Classical Studies, focused on the central areas of the harbour. This program uncovered dense structures and artefacts along the waterfront (Fig. 2.2) that reflect the port's vitality from its rebirth around the reign of Augustus until at least the late sixth or seventh centuries CE (Hohlfelder 1976; Scranton, Shaw and Ibrahim 1979).

Recent research at Kenchreai has concentrated on the cemeteries that ringed the settlement on all sides. The main burial ground of the Roman era was situated on a prominent limestone ridge called Koutsongila along the coast immediately north of the harbour (Figs. 2.2 and 2.3). This cemetery seems to have stretched northward as far as 1 km, but its most prominent and elaborate burials were apparently situated on Koutsongila, just beyond the northeast periphery of the settlement and highly visible from both the sea and the harbour. The site has been frequently disturbed by illicit excavation in recent decades. Looters have sought valuable grave goods but have left intact numerous artefacts, bones, and deposits, as well as all architectural remains. American and Greek teams conducted brief salvage excavations of disturbed burials in 1969 and 1988–1990. The first systematic study of Koutsongila and its environment was the Kenchreai Cemetery Project (KCP), which was sponsored chiefly by Macalester College under the auspices of the American School and with the permission and oversight of the Hellenic Ministry of Culture. J. L. Rife directed the field research of KCP in 2002–2006 and has overseen analysis and publication thereafter, while M. M. Morison has served as the project ceramicist since 2003. The primary goal of KCP was to record the physical remains of burial in order to reconstruct funerary rituals and to trace their relationship with social structure (Rife *et al.* 2007).

The architectural, artefactual and skeletal remains in the Koutsongila cemetery illuminate mortuary practices at Roman Kenchreai (Rife 2007; Rife *et al.* 2007; Sarris *et al.* 2007). The cemetery was used until the late sixth or seventh century, but the most impressive burials were subterranean chamber tombs used during the Early to Middle Roman periods. So far, we have documented 30 tombs, most of which displayed a standardised plan (Figs. 2.3–2.7). At ground level the tombs were typically marked by a rectangular building of mortar and rubble with a massive threshold and door, probably a gabled roof and perhaps simple architectural decoration. These buildings not only ensured that the tomb was visible from a distance and provided a setting for the display of the epitaph, but also secured and enclosed the entrance. Immediately inside the building was the *dromos*, a stairway cut into the bedrock that descended

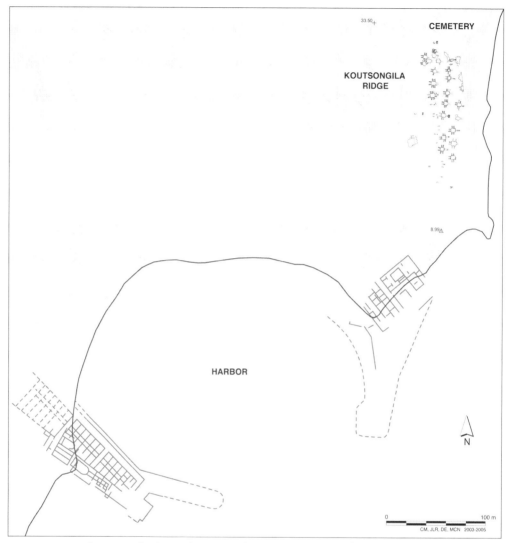

Fig. 2.2. The Roman harbor and Koutsongila Ridge at Kenchreai (contour interval = 1 m).

down to the chamber, which was entered through the *stomion*, a narrow space that could also be shut off by a substantial door. Inside the chamber was typically found a rock-cut bench on the front (east) wall and an altar against the back (west) wall, though a few tombs also possessed small movable altars. The walls and floors were finished in fine cement and white plaster, but some walls displayed rich painting with geometric, architectural and floral and faunal motifs. Numerous objects, particularly vessels and lamps, found on the floor of the chambers demonstrate that this was the main area for the enactment of burial and commemorative rituals. The dead were placed in compartments in the walls. In the upper register were niches designed to

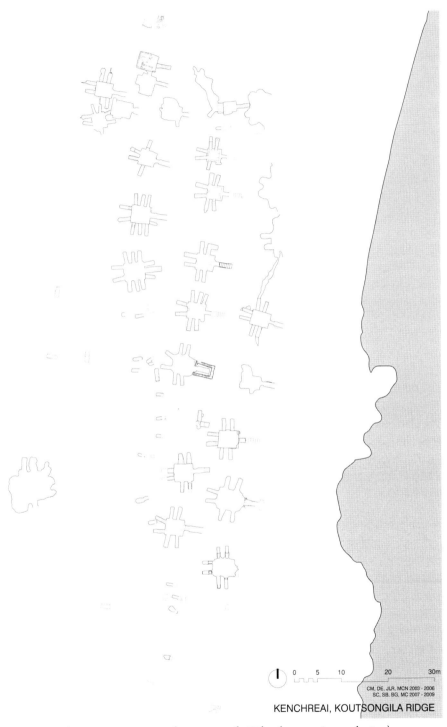

Fig. 2.3. The main cemetery on the Koutsongila Ridge (contour interval = 1 m).

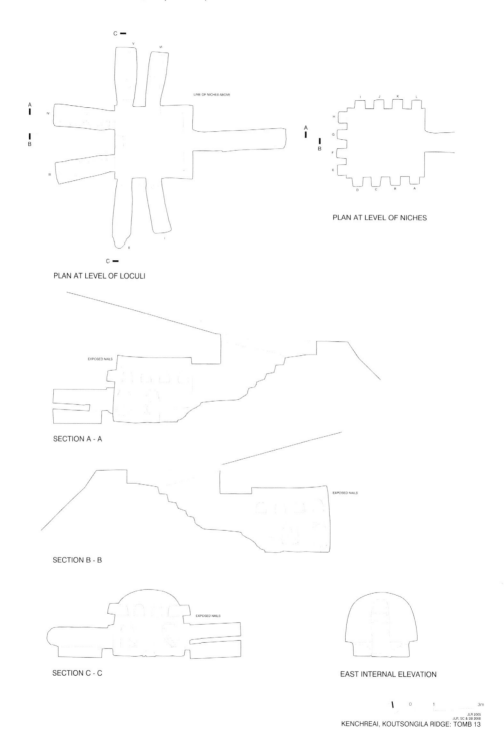

Fig. 2.4. Tomb 13, plans, sections, elevations.

Fig. 2.5. Tomb 13, views of interior.

receive heavy, cylindrical urns that contained the collected fragments of cremated bones and teeth. It is also possible that cremated remains were placed in niches without urns, in other vessels, or under lids. Apart from such containers, niches apparently held no objects, and they were not enclosed. Below the niches in the lower register of the walls were *loculi,* long compartments cut perpendicular to the plane of the wall that received intact bodies. *Loculi* typically held several bodies in a cist that was filled with shallow dirt and covered by stone slabs or terracotta tiles; the opening from the chamber was occasionally closed off. Sometimes the deceased were interred on a wooden structure, either a coffin or a bier, along with personal adornments, small vessels, and coins.

The form of the tombs and their contents reveal the length of use and certain aspects of their users' identities. The funerary assemblage shows that the tombs were built not later than the middle or third quarter of the first century CE. They were typically erected with a symmetrical arrangement of burial compartments in the walls, though a few tombs exhibit the addition of *loculi* or niches over time. Most tombs seem to have been used through the 3rd century. The development of the

Fig. 2.6. Tomb 6, tomb building at surface and dromos.

cemetery on Koutsongila would thus roughly coincide with the early emergence of the Roman port and its growth over subsequent centuries into a major commercial centre with numerous residents. Considering the monumentality of the tombs, the quality of their decoration and offerings, and the required cost of their construction and maintenance, we can reasonably conclude that they belonged to wealthy families. According to the few surviving epitaphs, which indicate the familial relations of the dead and not careers or offices, these tombs were erected by both men and women for themselves, their children, their descendants, and even their freedpersons. Over a use-life spanning several generations, as many as 50 or more persons could be buried in one chamber, some by cremation but most by inhumation. The difference in the meaning of these disposal modes within a burial group is obscure, but one explanation is that slaves or freed associates of a household were cremated for burial with their inhumed owners or patrons.

What is remarkable about the Koutsongila cemetery is the homogeneity of form and content among the tombs. Many are arrayed evenly in rows facing the sea, which reflects a conscientious placement or even careful planning (Figs. 2.2 and 2.3). The plans of the tombs are standardized, with minor variation in the number and location of the burial compartments and the quality of the parietal décor (Figs. 2.3–2.5). The funerary assemblage is also more or less consistent throughout the cemetery. This

Fig. 2.7. Reconstruction of a typical chamber tomb on Koutsongila.

overarching uniformity seems to indicate a concern among those who used the tombs to identify themselves as members of a distinct and cohesive group of wealthy families. This local élite claimed a place in the community that was significant, conspicuous, and lasting, just like their tombs. Furthermore, the specific form of the chamber tombs has almost no parallel in the broader region; it was apparently an innovation rooted in the sepulchral traditions of Asia Minor and the eastern Mediterranean more broadly. In devising and replicating this new form of burial for their families, powerful residents projected an image of themselves that was particular to Roman Kenchreai. They must have adopted many identities through professional and ethnic associations that stretched beyond the port's activities, connecting them variously to the urban centre of Corinth or ancestral homes overseas, as inscriptions attest. But in death they chose to commemorate their identity as local élites above others.

The evidence for a ritual process, and its limitations

We can expand and sharpen this basic picture of the tombs on Koutsongila as instruments of self-presentation by reconstructing the process of behaviours by which local residents activated objects and spaces around and inside the chambers.

Our study will concentrate on the four tombs that have been thoroughly explored by KCP (nos. 10, 13, 14, 22), though relevant evidence from tombs that have only been partly recorded by KCP and earlier campaigns (nos. 3, 6, 8, 9, 19, 23, 30) will also be considered (Fig. 2.3). We will examine the ritual significance of each successive space in the tombs, along with the associated artefacts and biological remains, from the above ground surroundings to the *dromos* and *stomion* to the chamber and its compartments.

The condition of the tombs and associated artefacts presents certain limitations to the interpretation of the ancient rituals, which should be based as fully as possible on the primary remains of mortuary behaviours. This is not an easy task, because those remains have been variously altered by taphonomic factors since deposition. Therefore, we have carefully reconstructed the depositional sequence in each chamber with an eye to environmental and anthropogenic disturbance (cf. Ubelaker and Rife 2008, 2011). The stratigraphy in the tombs consists predominantly of bedded sediment that had washed in from the surface. The lowest horizons typically encased the objects left by mourners on the chamber floors. Above this are massive deposits of colluvium that had collected in the tombs over the centuries from downslope erosion, intermixed with lenses of decomposed plaster and bedrock from the chamber walls. Filling the *dromoi* are dense deposits of disarticulated rubble and mortar, many apparently from the catastrophic collapse of the above-ground tomb buildings.

This depositional history shows that we have uncovered objects that were originally left in the chamber but were then sealed by the sediment that began to accumulate once the tomb fell out of use and was left open. There is no conceivable process for the past removal of burial artefacts from the chambers, because they were rapidly buried, and we have found little sure evidence that floor deposits were intentionally disturbed by looting in ancient, medieval or modern times. Environmental processes have impacted these primary remains of funerary activity, but only to a small degree. The introduction of water through the porous bedrock or in torrential floods, root infiltration, and burrowing, together with the initial arrival of coarse sediments into the chambers, have degraded or broken apart objects but have not dramatically displaced them. The areas where objects have most likely moved during structural collapse and colluvial deposition are those nearest the chamber's ingress, that is, throughout the *dromos*, inside the *stomion*, and on top of the benches. In general, it seems, objects located on chamber floors, in front of compartments, and around altars comprise a more or less total and undisturbed sample of the durable, non-organic materials that mourners left during the use of the tombs.

The contents of the burial compartments were found in a rather different state of preservation than deposits in the chamber and *dromos*. Burials interred in *loculi* have been frequently disturbed by graverobbing, a few apparently already in antiquity and all to some degree in recent times (Ubelaker and Rife 2008, 105–121, 2011, 6). Typically, looters have sought objects near the surface of the fill and then selectively dug deeper, especially at the front end of the cist. But they have been neither thorough nor systematic. Consequently, while looters have removed an unknown number of decorative objects from the burials, they overlooked several funerary artefacts and left behind all bones and teeth, in some places intact and in other places churned up, commingled, and scattered. In addition, natural factors such as moisture, pine roots, and animals have had a small but notable effect on the condition of artefacts, such as bronze objects, and particularly organic material, from human remains to wooden furniture. Despite these disturbances, careful documentation has in many cases permitted a detailed reconstruction of burial groups within *loculi*, even if datable finds have been relatively scarce. In contrast, only a few niches were found to contain cinerary urns or cremated remains, though that was clearly their intended purpose (Rife 2007, 108; Ubelaker and Rife 2011, 6, 12–14). There are several explanations: cinerary urns dropped out of some niches with coseismal earth-shaking, the gradual erosion of adjacent walls, or root invasion; urns were removed from niches (and sometimes dumped out in them) by local residents who wished to re-use the heavy, utilitarian vessels; urns were thrown from niches by vandals or looters rifling through; or some niches in fact never received urns. All these explanations are possible. In any case, the nature of the cremation burials in the niches can be recovered both from broken urns found on the floors and from burned bones and teeth found either inside niches or strewn across chambers.

It is also difficult to ascertain the exact duration and scale of rituals performed at the tombs. There are a few considerations that can help us to devise a most likely scenario. The existence of several individuals in both cremation and inhumation burials, and the fact that the tombs were designed for re-entry through accessible doorways rather than permanently sealed, imply that mourners participated in multiple ritual events. It is reasonable to assume that these participants were relatives and associates of the bereaved family. The tombs and their contents also seem to have attracted wider attention in the community. A magical prayer for justice inscribed on a lead tablet and placed above a cist grave in Tomb 22 called out a certain thief by name (Faraone and Rife 2007), while a prohibition painted on the front wall of Tomb 4 warned strangers not to ransack the place. The owners of the tombs knew that occasional interlopers and even vandals would enter the chambers, but such unpredictable visitors were not the target-audience for sepulchral design and funerary rituals.

The process of death in the chamber tombs on Koutsongila

With these limitations in mind, we will investigate the nature and distribution of funerary artefacts in different spaces in and around the chamber tombs on Koutsongila. This will help us to trace a picture of the ritual process enacted by mourners at Early to Middle Roman Kenchreai. It is clear that many activities must have occurred away from the tomb, such as the first phases of mourning, the cremation, the procession, and feasting. No activities are well attested in the immediate entrances to the tombs, though these areas must have been a locus for brief visitations and remembrance. Activity was concentrated inside the chambers, where the burial of the dead occurred in spaces built into the walls and the offering of food and personal objects occurred near the benches, altars, and burial compartments.

Activities away from the tombs

We cannot know exactly what happened away from the tombs, between the moment of death and the arrival at the burial site, but we can assume that local residents during the early centuries of the Roman era followed usual Greek practice for this period (Rife 1999, 54–76). The length of time probably varied, but most literary sources point to a customary two- or three-day interval between death and burial. During this time the bereaved family would have cleaned and dressed the body for display at home, privately mourned the deceased, entertained visitors, and generally prepared for the funeral and burial. The assembly of objects to be deposited with the body, from clothing and accessories to coins, lamps, and unguentaria, would have occurred at home before the procession. The only residential quarter that has been uncovered at Kenchreai is located between the base of the north mole and south end of the Koutsongila Ridge (Fig. 2.2). These lavish residences apparently belonged to the wealthy members of the community (Morgan 2009, 11, 14–15, figs. 14–17; Rife 2010, 400–401, fig. 13.3). If these were the homes of the tombs' owners, as seems plausible, we can imagine them moving in a procession northward up the ridge's seaward slope, a very conspicuous route that would have drawn the attention of the whole community as well as ships in or near the harbour.

One activity outside the tombs that we can partly reconstruct is cremation (Rife 2007, 106–108; Ubelaker and Rife 2007, 2011, 13–16). Since the burning of bodies must have been an extramural activity, and we have not found traces of pyre debris in the cemetery proper, we should conclude that the community used a cremation field. The most likely location for such an area would have been the upper slopes or summit of Koutsongila, which would have permitted not only access to and from the settlement and the main cemetery, but also strong wind currents to clear smoke and odours. Recent excavation has uncovered pyre debris of Roman date in an unoccupied area along the southern slope of the ridge (Morgan 2010, 22). Here was found an irregular depression or trench containing numerous but widely scattered human bone and occasional animal bone fragments and iron or bronze nails, all severely burned, mixed

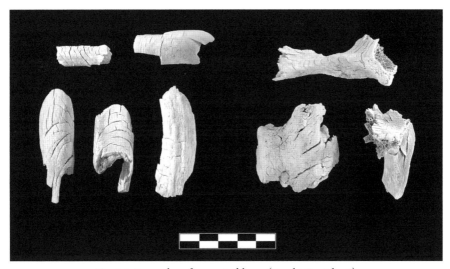

Fig. 2.8. Examples of cremated bone (Tomb 10, niche E).

with coarse fill and capped off with a dense stratum of rubble. This represents an attempt by local residents to dispose of, to enclose, and to conceal the remnants of pyres that burned several people over time. The actual site of cremation must not have been far away. For example, we might imagine that pyres were built during the Roman era on the southern promontory of the ridge's crest (Fig. 2.2). Unfortunately, this area was re-shaped and over-built by fortifications during the Second World War, and we cannot confirm whether it was an ancient cremation-site.

The burned bones and teeth found within the tombs attest to the process of cremation that preceded final interment. They are typically small fragments discoloured black to pale blue or white and exhibiting severe thermal alterations, such as calcination, warping, fissures, and transverse splitting (Fig. 2.8). Such deformation points to the thorough and intense burning of the body over a long period of time at temperatures exceeding 700° Celsius. The burned fragments left in the tombs only represent a fraction of the elements in the skeleton. This evidence implies the existence of a sizable pyre with plentiful fuel that was closely attended. There is slight evidence that animals or objects were also burned with the body on the pyre, including the few animal bones already noted, as well as pieces of molten glass from small vessels, perhaps unguentaria (Tombs 9, 13). Once the pyre had burned out, participants must have worked hard to sift through the ash pit and to extract small bones and teeth, which they then carried to the tombs. In this reconstruction, cremation at Kenchreai would have been a highly visible event that required a significant investment of time, energy and resources as well as coordinated effort by the bereaved family. The ability to perform a cremation expressed to the community a family's prosperity and solidarity. Furthermore, if cremations belonged to the slaves or freedpersons of a wealthy family, perhaps they reflected the family's devotion to patronage, and to the coherence of their household and its wider social and economic network.

Activities around the tomb building and in the entrance

Once the funeral procession reached the tomb, mourners must have gathered outside before entering. The tomb building at the surface was a solid construction with thick walls, a prominent façade, and a massive threshold that would have been clearly visible from roads and shipping lanes in the vicinity (Figs. 2.6 and 2.7). The structure itself would have served to advertise the prosperity of the family that erected and used it. It also distinctly divided the open exterior from the burial space within. The walls and the threshold seem to have been the primary delineation of ritual space, because exploration around the tombs has not turned up secure evidence for a fence or any other physical limit tracing a wider precinct beyond the building. Furthermore, we have found no facilities outside the tombs to accommodate burial or commemoration, such as accessory chambers for dining, wells or water pipes, or altars, and no artefacts that are commonly associated with rituals. There is one exception to this pattern on Koutsongila. Outside Tomb 6, along the southeast corner of the tomb building, was located a dense deposit of large amphora sherds that had been broken on the spot. This material seems to represent either debris from a funerary meal immediately outside the tomb, or a dump of vessels removed from the chamber during a cleaning operation. Even though the exterior of the tomb was not usually a locale for formal activities involving ritual implements, it was also not completely detached from the experience of burial. Participants in the procession who did not enter the chamber must have waited here mindful of the deceased, while later visitors, whether familiar with the deceased or not, would have stood on the tomb's periphery and contemplated the eternal rest of those members of the community named in the epitaph.

In most tombs, the *dromos* and *stomion* served simply to connect the surface with the underground chamber. There is little secure evidence that any rituals were enacted in this area, and indeed the narrowness and steepness of the space would not have allowed elaborate or prolonged activity. A few typical funerary artefacts have been found in certain *dromoi,* such as lamps (Tombs 6, 10, 22), glass vessels (Tombs 6, 13, 14, 22), and pottery (Tomb 14). But these are small fragments and they come from the upper horizons of fill, in which case they were most likely dropped or tossed here by recent looters rather than deliberately deposited here by ancient mourners. Only one object found in a *dromos,* the smashed base of a cup near the lower stairs of Tomb 10, might have been used in antiquity for commemorative drinking just outside the chamber. Glass vessel fragments found near the floor of the *stomion* and *dromos* in Tomb 22 were probably either dropped there by early looters or were deposited there with in-washed sediment. A collection of sherds typical of the local funerary assemblage has been found in the lower to middle horizons of fill in another *dromos* (Tomb 14), but this pottery was scattered and fragmentary, and it might represent a general cleaning effort late in the tomb's use. In one exceptional instance (Tomb 30), a *loculus* had been cut outside the chamber at the north base of the *dromos* for the burial of an old woman and two children (Fig. 2.3). Opposite the *loculus* the *dromos* had been widened with a shallow recess that was apparently used as a receptacle for offerings, namely glass vessels. The anomalous placement of inhumation burials in this *dromos* is hard to explain. Perhaps the chamber, which has collapsed since antiquity, had been gradually filled with bodies, and the owners were simply forced to extend burial out into the tomb's entrance.

Besides these unusual cases, the *dromos* served predominantly as a corridor through which mourners passed from the tomb's exterior to its interior, a liminal space entered through one door and exited through another. The upper and lower portals are evident from sill cuttings and nails, perhaps from doorframes, clustered between the *dromos* and the *stomion* in certain tombs (Tombs 6, 22, 30). This passage must have made a potent sensory and emotional impression on ritual participants, who were crowded together with the corpse, feeling their way down a steep, tight descent with increasingly still air and dim light. The transitional space of the *dromos* effectively defined two stages in the process of burial: the public mourning and the funeral procession above ground, and the deposition of the body and its offerings below ground.

The use of the burial compartments

When activity moved into the chamber, the number of participants undoubtedly decreased as the funeral became more private and strictly focused on burial, and as the experience of mourning intensified. While we cannot ascertain the number of people directly involved in the deposition of the body, we can estimate generally from the minimum number required to complete certain tasks and from the holding capacity of the chamber itself, which measured on average 3.73 m long by 3.27 m wide by 2.53 m high. The transport of a body and its offerings down a steep descent and their placement in narrow spaces would have probably called for at least three or four persons. Even if we assume that a heightened emotional state would have compelled mourners to endure the discomforts of a subterranean cavity, we must acknowledge that the interior of a chamber could not accommodate more than around six or seven persons.

Upon arrival in the underground chamber, the physical impact of the space on the mourners must have been strong. In the moment of passage from the *dromos* and *stomion*, which were still linked directly to the world above, into the tomb proper, which was completely detached from the surface, mourners would have been struck by the dim, stagnant, dank surroundings. As their senses adjusted to this otherworldly space, they gazed upon the walls. In certain chambers (Tombs 2, 4, 9, 20) these were colourfully painted, sometimes with great complexity and skill (Barbet and Rife 2007; Rife *et al.* 2007, 163–166). Pictorial details such as *aediculae*, delicate birds, flowering plants, and garlands created the atmosphere of a garden, beautiful and special, a sacred place (Fig. 2.9). The mural schemes were aligned with the burial compartments:

Fig. 2.9. Wall-painting: detail of garland, ribbons and bird (Tomb 4, south wall).

architecture framed the *loculi* and niches, between which were hanging garlands. In chambers without painting (e.g. Tombs 6, 7, 10–14, 21, 22, 23), the walls were finished with fine, white plaster, against which the dark compartments would have stood out. The treatment of the walls, whether painted or not, made the *loculi* and niches visual focal points for the entire chamber. For the viewer, the walls seemed to be not so much architectural elements, or structures of enclosure, but rather solid matrices into which bodies were inserted.

The burials themselves provide the best evidence for the next stage in the ritual process. The tombs furnished two contexts for interment: several niches for cremations and several *loculi* for inhumations. Both spaces were used over time for the deposition of several individuals. As mourners entered, they would have recognized the compartments and chosen one for burial. This crucial decision must have been based chiefly on the identity of the deceased, because ample space was available for multiple interments over many generations. Even without textual indices after the epitaph above ground, mourners presumably recalled how the burial compartments were organised, and familial or associative relationships between persons both alive and dead would have influenced where the deceased was placed. The act of interment was an important time for contemplating these relationships. They were physically constructed through the design of the tombs, where the seriated compartments represented a model of the family or household, and they were conscientiously enacted, as mourners worked together to transport the body of a loved one and then to deposit it among others of dear memory. These well-choreographed choices and movements, combined with the use of objects carefully selected beforehand and carried in the procession, helped to regulate emotion and to reify claims of identity. The performance of burial involving spaces and objects in the chamber tombs thus reinforced an image of familial relationships and solidarity within the local community.

In the absence of nameplates or inscribed or painted names, we cannot know who was buried where inside the chambers. The basic discrepancy in the forms of niches and *loculi* supports the conclusion that individuals of different status were buried in these two distinct spaces. A degree of priority among the *loculi* is reflected in the greater complexity of the painted decoration around the central compartments in the back wall. Moreover, the discovery of nails or pins at even intervals in the wall-plaster immediately alongside or above *loculi* suggests that mourners either affixed grills or screens over them, or hung garlands or fillets around them (Figs. 2.4 and 2.5). In contrast, the shallow niches lack all elaboration. They do not seem to have been adorned with enclosures, swags or ribbons, and in the pictorial tableau they are embedded within a continuous series of faux marble panelling. In the perception of the viewer, the structural and decorative features of the chambers expressed an overt priority of *loculi* over niches that also connoted a differential importance among those buried therein. As has been noted, one explanation based on the epitaphs is that members of the founding family and their descendants occupied the *loculi* and their slaves or freedpersons occupied the niches. In this scenario, the largest and most prominent compartments on the back wall would have been reserved for the most important members of the household, such as the originator of the tomb and his or her closest relatives. The other *loculi* around the walls might then have contained other members of the same lineage, constituting over time one synchronous vision of a unified ancestry. If slaves or freedpersons were cremated and placed in the smaller compartments above, their burials became secondary and supplementary to those of their overseers, just as in life they had been in slavery the possessions of the household or in freedom the ornaments of patronage.

The central object in the ritual process of burial was of course not architectural or artefactual but biological: the dead body. The few mourners who carried the deceased down the narrow *dromos* and into the dark chamber placed it inside the chosen compartment. Those who were leaving cremated bones and teeth either brought the fragments extracted from the pyre for insertion into an urn already in the tomb, or they brought an urn filled with fragments at the pyre and placed it in a niche. In this case, the final face-to-face encounter between mourners and the deceased occurred at the site of cremation, not in the chamber. The cremation was a public event away from the tomb that highlighted in spectacular fashion the family's obligation to care for the dead, even at great expense. But once the cremated remains reached the tomb chamber, the act of burial itself, typically in plain urns without accompanying objects, left little scope for individual expression. The burial of cremated remains in niches promoted an image of anonymity.

In the case of inhumations, however, mourners interacted with the dead up until the final moment of burial. The limited quantity of burned remains and urn sherds found in the tombs reveals that inhumation was much commoner than cremation in the Koutsongila cemetery. The activity of burial in *loculi* would have brought the living and the dead into close contact. Mourners opened up the shut-off compartment, saw the remains of those previously interred, and remembered the personalities and the lives behind the decaying tissue, even if it was hard to differentiate individuals from a mass of flesh and bone. Then they introduced the newest member of the buried family by adding the body on top, inserting the feet first and arranging the torso and limbs for an attractive appearance (Fig. 2.10). The presence of nails in rows inside some compartments, many exhibiting massive corrosion product that preserves the

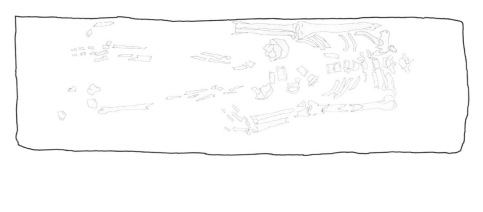

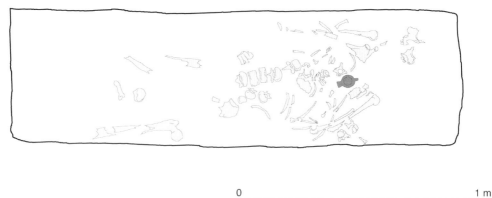

0 _____ 1 m

JLR 2006/2009

Fig. 2.10. Upper and lower skeletons in Tomb 10, loculus VI.

grain of wood, shows that biers or coffins were sometimes interred with the body. Before re-sealing the *loculus*, mourners had a last chance to look at, to talk to, and to touch the deceased. The sense of finality must have been potent, even wrenching, but the confirmation of familial bonds through such an interactive burial might have afforded a measure of solace.

The objects that mourners brought with the dead body into the tomb had various purposes and meanings. Some objects, which we can call personal, were considered appropriate for burial with the dead, while others, which we can call communal, were intended for activities elsewhere in the tomb. Communal objects, which will be discussed below, were used outside the burial compartments but inside the chamber, where people gathered on numerous occasions. In contrast, personal objects were used specifically for the adornment, protection, accommodation, and burial of the body. Most have been found inside the *loculi* and niches, though some were dispersed by looters into the chambers. Mourners selected these objects before the funeral, carried them in the procession, and then left them with the body inside the tomb. This process mediated the emotional experience of burial; it also drew a special personal connection between an individual survivor and the individual deceased.

The standard suite of personal objects from the *loculi* included articles of bodily adornment, unguentaria, and coins (Figs. 2.11 and 2.12). We have uncovered a well-defined burial assemblage that was remarkably consistent over the course of several generations across the Koutsongila cemetery (Rife *et al.* 2007, 166–175). The use of unguentaria, for example, was highly conventional. These small bottles with minimal decoration were mass-produced to hold ointments or aromatic substances. Mourners placed them around the head and feet of the deceased, probably after anointing the body during the procession and burial. The early unguentaria from Kenchreai belong to the standard bulbous type common in Greece and elsewhere in the eastern Mediterranean during the first century (Anderson-Stojanović 1987; Fig. 2.11a). The bulbous form was gradually replaced by glass types, examples of which have also been found in the tombs. Unguentaria were tools of interaction between the living and the dead, because the act of anointment substantiated a fleeting bond. At the same time, unguentaria show no patterning by age or gender in the tombs on Koutsongila. Rather than the expression of specific personae through typological variability, it was the statement of behavioural conformity and group identity made by the presence of these simple, readily available objects that mattered to mourners. While it is difficult to say whether the use of unguentaria was emblematic of social status, this form of funerary vessel was so ubiquitous in Greek burials of the era that it seems unlikely. Nonetheless, the deposition of unguentaria in tombs throughout the Koutsongila cemetery was a practice that identified mourners with one another, whether as residents of a single community or as individuals who shared in the common experience of bereavement.

Coins and decorative articles in burials carried different meanings (Fig. 2.12). Such objects may well be systematically under-represented in the tombs, because

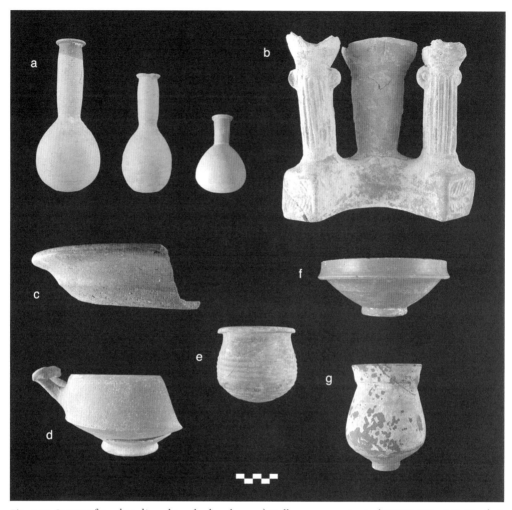

Fig. 2.11. Pottery from loculi and tomb chambers: a) Bulbous unguentaria (KP1989-040, -041, Tomb 3; KP1988-002; Tomb 19); b) incense-burner (KM017; Tomb 22); c) Imitation Pompeian Red Ware pan (KP1989-014; Tomb 3); d) Knidian bowl (KP051; Tomb 14); e) regional stewpot (KP061; Tomb 22); f) Çandarlı Ware bowl (KP031; Tomb 14); g) Thin-Walled Ware cup (KP1989-043; Tomb 3).

looters have undoubtedly removed many from loculi over the years. Nonetheless, the evidence we have documented points to certain patterns. Decorative pieces such as hairpins, cosmetic spatulae, and ear- and finger rings varied considerably in design and material. These were probably personal possessions of the deceased that mourners selected and attached to the body for attractive display during the funeral and burial. Formal variation in jewellery must have reflected the personal taste or preferred style of the dead to some degree. Certain precious items, such as gold objects or rings with gemstones (Fig. 2.12b), must have been valued possessions

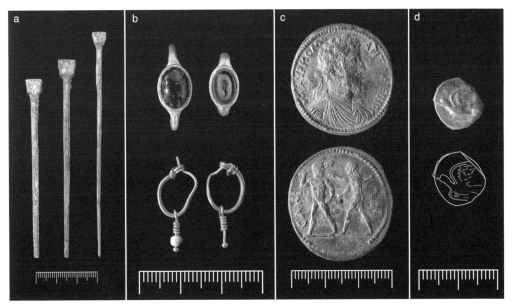

Fig. 2.12. Personal items from loculi: *a) silver* spatulae *(KM008-10; Tomb 13); b) gold finger rings with cornelian (KM109; Tomb 10), gold finger ring with nicolo intaglio (KM012; Tomb 13), and pair of gold earrings with glass bead (KM013; Tomb 13); c) bronze* sestertius(?) *of Commodus from Samos, AD 180-192 (KC016; Tomb 22); d) gold bracteate of Early Hellenistic issue from Sikyon (KM014; Tomb 14).*

that connoted prestige and wealth, a characteristic of the tombs' owners that is also apparent in the monumentality and decoration of the tombs themselves. Other objects, such as *spatulae* and pins (Fig. 2.12a), would seem to conform to an image of beauty that was specifically gendered, particularly because they are usually found with female skeletons.

Mourners often placed coins over the face, mouth, or chest of the deceased at the time of burial. The intimate gesture of setting a coin on the dead body was an opportunity for a mourner to communicate with the deceased. Perhaps it also betokened the personal loss suffered by the bereaved, or a personal sacrifice through offering an object of stated value. Coins were also selected for what they might communicate. The coins from *loculi* represent a surprising range of origins, dates, and forms. Most of them are Late Classical or Hellenistic issues that predate the burials with which they are associated by several centuries. The one trait shared by these coins, which were long out of circulation and which displayed varied imagery, was their antiquity. While mourners presumably preferred to inter coins without monetary value, the choice of personal objects for burial was deliberate and meaningful, and thus we must consider whether these coins were chosen for their ancient date. An antique coin might have sent several messages to participants in the funeral: a respectful connection to the past; the longevity of a lineage, household, or group; a sense of rootedness or stability. The special character of these coins as antiques might have also connoted the special status or prestige of the dead. We should not, however, overlook the fact that the choice of special objects as funerary offerings, even among non-elite residents, could have simply expressed affection or personal attachment.

In tombs across the Koutsongila cemetery, relatively few coins of Roman date were found in or near burial compartments, apart from Late Antique issues that might well have been introduced during early episodes of looting. Noteworthy among these are two very rare issues from Samos under Nero and Commodus (Tomb 22). The latter in particular was a large piece with a mythological scene depicting an archaizing figure of Herakles on the reverse (Fig. 2.12c). In this case, we must wonder whether the deceased were Samian immigrants, whose mourners chose coins to recall their homeland.

Another numismatic practice was the burial of bracteates, or gold foil impressions of coins, with the dead (Fig. 2.12d). Several of these have been found at chamber tombs, associated either with *loculi* or with disturbed contexts (e.g. Tombs 10, 14), but many more have presumably been removed by graverobbers over the centuries. This was a widespread custom: examples of bracteates have been found in Roman-era burials at Corinth (Wiseman 1969, 87; Rife 1999, 281–282; Walbank 2005, 276–277) and at other sites of various dates on the Greek mainland, the islands, and the colonies (e.g. Kurtz and Boardman 1971, 211, 363; Stefanakis 2002). The residents of Roman Kenchreai typically selected a common Hellenistic coin type – in this case a bronze issue of the city of Sikyon that showed a flying dove on the reverse – and they wrapped a thin disk of gold foil over it to create a raised image of the dove. These bracteates, which mourners left with the dead in the same manner as actual coins, might have carried layered levels of meaning. The consistent preference for the image of the dove, combined with the medium of gold foil, could indicate a talismanic or apotropaic significance. On the other hand, the bracteate was a special object due to the precious quality of its material and the ancient date of the host coin. The placement of bracteates with certain bodies in *loculi* could have both protected and distinguished the dead.

Personal adornments and coins thus had multiple referents in the ritual process. Certain objects were private possessions that drew a connection between the living and the dead, some connoted social status or gendered identity, while others linked the deceased to a particular household or situated them within a community with shared mortuary behaviours. These objects reflect a desire on the part of mourners to conform with group expectations and to assume a corporate identity on the one hand, while negotiating a space for familial and individual identity on the other. By burying a mixture of objects representing multiple roles and relationships, mourners could simultaneously celebrate the individuality of the deceased and reify connections with a larger community. These objects, moreover, could have helped to regulate emotion as mourners placed them directly on or adjacent to the dead, often in the confined space of the *loculi*, before covering the body and shutting the compartment.

Certain other funerary artefacts found in the tombs, in particular lamps, served as both personal and communal objects, thereby bridging the interests of the individual and the community at Roman Kenchreai (Fig. 2.13). These small terracotta vessels with decorative embellishments were carried in the funeral procession and left in the tomb at the time of burial and probably later, during visits. Lamps provided illumination in the dark subterranean chamber, but they also created a visual and even olfactory focus, with their single flame and the familiar odour of burning oil, during a time of reflection and remembrance.

The lamps found in the tombs have a typological distribution that reflects, with few exceptions, the normal trading patterns and general availability of lamps in the Corinthia and at the harbour of Kenchreai itself, as we have discussed elsewhere (Rife *et al.* 2007, 168, 171–172). The earliest lamps in the tombs, which date from the middle to late first century, have been found both in and around *loculi* and in various parts of the chambers, sometimes near benches and altars or in front of *loculi*. However, from the second through third centuries, lamps were no longer deposited with the dead but were left most frequently around the chamber outside the burial compartments. This diachronic pattern suggests that the importance of lamps as personal objects declined, though they remained popular as communal objects until the tombs fell out of use.

It seems that, during the early phase of the Koutsongila cemetery, the bridging effect of lamps as both personal and communal objects was important for local residents. Lamps from the initial use of the tombs, including various imports from Italy and Asia Minor as well as regional products, display broader typological range than later lamps. This might show a conscious effort on the part of mourners to choose lamps by origin or appearance that would celebrate particular personae of the deceased, such as ethnic background or familial history, beyond their elite status in the local community. Over time, however, as lamps were used more consistently as communal objects inside the chamber, such personae were expressed through other channels, such as coins and jewellery. It will be seen that a growing investment in communal rituals over the life of the tombs is reflected also in the pottery assemblage.

The chamber as ritual space

Outside the burial compartments, the main context for ritual activity was the chamber itself. This area would have been rather confined and dark, but the presence of rock-cut or movable furnishings demonstrates that mourners gathered here for activities in small numbers. Certainly, they did so at the time of burial and presumably also during periodic visits to the chamber after burial, a common commemorative practice throughout Greek and Roman antiquity. The main artefactual evidence for mortuary behaviour in the chamber includes ceramic and glass vessels of many forms, incense-burners (*thymiateria*), and lamps (Figs. 2.11 and 2.13). Mourners employed these objects in communal activities that played an important role in self-presentation.

The chambers were planned in such a way as to provide distinct areas for ritual performance which employed objects brought by mourners. These areas were the

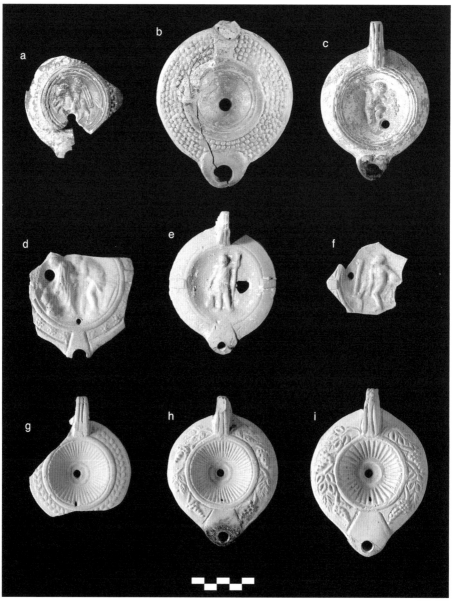

Fig. 2.13. Lamps from tomb chambers: a) lamp with Celtic warrior on discus and ovules on rim (KL030; Tomb 14); b) lamp with plain, concave discus and rows of globules on rim (KL033; Tomb 13); c) red-on-white lamp with dancing child holding grapes on discus and plain rim (KL1989-004; Tomb 3); d) lamp with erotic scene on discus and dotted rosettes and panels on rim (KL035; Tomb 22); e) lamp with Hephaistos on discus and plain rim with panels (KL024; Tomb 14); f) lamp with dancing grotesque(?) and colonette on discus and ovules and panels on rim (KL042; Tomb 22); g) lamp with rays on discus and wreath on rim (KL1989-001; Tomb 3); h) lamp with ray-and-vine design (KL1988-001; Tomb 23); i) lamp with ray-and-vine design (KL1988-003; Tomb 23).

walls, altars, benches, and the floor immediately in front of *loculi*. The content of the wall-paintings, the placement of the *loculi* and niches, and the location of altars and benches in symmetrical array around a central visual axis from the *stomion* to the back wall together created a strong visual imprint of spatial differentiation and hierarchy. When mourners entered the chamber, their attention fell first on the walls and the burial compartments, but then on the surrounding furniture, laden with lamps and vessels, and the floor, scattered with objects and food debris. As they comprehended separate functional spaces within the chamber, mourners were enveloped in a cloud of sensory impressions. Flickering torch-light enhanced the vibrancy of painted images and colourful garlands or textiles; the sight of the prominently displayed gifts of previous mourners reminded visitors of familial ties that the tomb represented and maintained; the quiet of the underground chamber reinforced a sense of awe and purpose; the rich textures of cool smooth plaster, a carved bier or coffin, or a hefty ceramic urn gave comfort or even pleasure; the competing smells of unguents, flowers, smoke, and putrefaction in a confined space no doubt aroused both positive and negative emotions; the taste of a ritual meal evoked memories and sympathies with loved ones past and present. These potent sensory experiences, combined with the deployment of special objects in specific areas, drew a connection between emotion and ritual that became a key element in the creation of identity. The use of personal objects in the deposition of the body in the niches and *loculi* was an important stage in this process; the use of communal objects in other areas of the tomb played an equally important role in burial ritual.

The communal objects found in the chambers were lamps, rare *thymiateria*, glass vessels, and various ceramic utensils, including stewpots, basins, casseroles, lids in cooking ware, frying pans, pitchers, jars, amphoras, bowls, cups, and an occasional plate or grill (Fig. 2.11). While the numerous glass fragments are often hard to reconstruct, most or all represent various forms of bowls, cups, beakers, or pitchers. These artefacts show a discernible spatial patterning that reflects their intentional placement during funerary rituals. Some had been moved or crushed by the massive accumulation of coarse sediments in the chambers or the gradual erosion of the walls and ceilings since the Roman era, but many were found in a well-preserved state approximately where they had been left by mourners.

In the largely intact chamber deposits of Tombs 10 and 14, lamps were found in three locations: most often on the benches that wrapped around the front walls, rarely in front of side *loculi*, and sometimes along the western wall, either in front of the central *loculi* or alongside (once on top of?) altars. One ornate incense-burner flanked by lampstands on a moulded plinth (Fig. 2.11b) was found in the front of Tomb 22, where it might have been situated on a central altar. The vessels display a similar situation in Tombs 10, 13 and 14, with distinct concentrations around altars and benches and in front of *loculi*, though they were also found dispersed across the floors in these tombs and others (nos. 3, 4, 8, 9, 23). In Tomb 22, which had an unusual plan lacking benches and exhibiting three cists built into the floor, numerous vessels had been piled up on the front floor to the left and right of the *stomion*. It is difficult to trace the original location of glass vessels, but in Tomb 14 they had clearly been placed on the sills in front of *loculi*. Sparse organic material has also been found in association with funerary artifacts in primary chamber deposits. While all tombs have produced small quantities of animal bone from such contexts, the best-preserved evidence comes from Tombs 10 and 14, where small fragments of sheep-goat, pig, and bird bones, many displaying burning or cutmarks, have been found either alongside benches or altars or spread across the floor.

These remains provide direct evidence for the communal rituals of mourners in the chambers. The lamps and incense-burners served a practical purpose by giving light in a dark place and by generating smells that would mask the strong odours of dankness and decay. These tools of burning also provided a visual focus for mourners in an intense emotional state as they contemplated their loss and remembered the deceased. The various vessels in the chambers were utensils for the preparation, service, and consumption of small meals, and the animal bones were food debris. While most vessels reflect consumption of food and drink, limited preparation of food on site is suggested by frying pans and a grill with traces of burning, as well as small pockets of wood charcoal in floor deposits that most likely represent fires. Apparently, mourners congregated in the front of the tomb near the benches for eating and drinking, but the conscientious placement of cups and bowls at altars and before *loculi* also indicates that they shared portions of their meals with the dead. The relatively limited number of vessels and animal bones, representing several generations of activity, indicates that single meals were small. While some meat was evidently cooked (or at least warmed) in the chambers, most food and drink was brought from the bereaved home in procession, along with the serving utensils. In any event, the size and design of the chambers would not have allowed for large-scale dining, only eating and drinking by a few persons; if mourners participated in a full funerary banquet, it must have occurred outside the tomb. We cannot differentiate vessels with such depositional and chronological precision as to determine whether eating, drinking, and offering food occurred at the time of burial or during visits after burial, or whether such practices changed over time. We can only conclude that the debris from meals slowly accumulated in the chamber over the use-life of the tomb. Mourners generally respected the remains of earlier rituals, leaving cups, bowls, and pitchers from previous funerals more or less in place. In at least one tomb (no. 13), however, mourners pushed vessels to the north side of the chamber floor to create space for a new interment in the tomb's south wall.

It is interesting to observe the strong continuity in the occurrence of these objects in chamber tombs over at least two centuries. Evidently the communal activities they served were essential stages in the process of burial and commemoration at Early to Middle Roman Kenchreai, indicating a desire by the mourners to claim and to maintain a strong and active connection with multiple generations of their families. The differential use of certain artefact classes (e.g. lamps and cuisine-related pottery) in constructing this connection, however, shows some chronological variability.

As noted above, imported lamps from Italy and Asia Minor were found in both the chambers and the *loculi* during the early phase in the use of the tombs, when they carried both personal and communal connotations (Fig. 2.13). By the late 2nd century, lamps were used almost exclusively in communal contexts, where they were typically left on benches or in front of *loculi*. The vast majority of these lamps were Corinthian products of the common Broneer XXVII type, which is not surprising given the frequency of these items on the eastern Mediterranean market at the time (Fig. 2.13e–2.13i). But study of discus imagery reveals little preference in the selection of sub-types for use in the tombs. In the later phase of the tombs, the presence and display of a lamp or lamps was a significant means of claiming group identity, while the specific imagery and even type were matters of individual discretion.

This pattern contrasts strongly with the use of communal, cuisine-related pottery over the life of the tombs, illustrating the importance of the technology of cuisine in the construction and negotiation of identity (Moore Morison 2000, 2006; Morison forthcoming). In all periods the pottery assemblage was highly standardized, but its internal composition varied across time (Rife *et al* 2007, 166–173). In the early phase of the tombs, ceramic variability was very limited. Food preparation was denoted by frying pans, many of which directly imitated Pompeian Red Ware pans (Fig. 2.11c). A single imported drinking-cup type, the so-called Knidian bowl, represented consumption (Fig. 2.11d). Thus, the ritual meal eaten with the dead combined the consumption of liquid with a particular type of cooking associated with Italian patterns of food preparation. The lack of serving platters and dishes in this context suggests that the consumption (and perhaps even cooking) of solid food was in this period largely symbolic in the communal area of the tombs.

Visitors to the tombs in subsequent periods would have taken note of these older items and their position on the benches, altars, and floors, reminded by such antiques of the longevity of their family group and of the fundamental physical and emotional connections between themselves, their ancestors, and the recently deceased. These later mourners, however, selected a different suite of cuisine-related items with which to express these connections, preferring a still standardised but more elaborate assemblage of vessels. From the late 2nd century onward, standard regional stewpots, casseroles, mortaria, and grills represented food preparation (Fig. 2.11e); pitchers of regional form represented liquid service; a variety of terra sigillata vessels, predominantly imported Çandarlı ware, signified food service and consumption (Fig. 2.11f); and two types of eastern Thin-Walled Ware mugs denoted liquid consumption (Fig. 2.11g). This growing complexity of cuisine-related vessels in the life of the tombs might reflect the increasing importance of dining rituals, as local residents during the second to third centuries were more concerned than before to construct and propagate identity-narratives about the enduring stability of their families.

Thus, the use and placement of lamps and cuisine-related vessels in the communal area of the tomb, as opposed to a niche or *loculus*, reflects a desire by mourners to claim and maintain a strong and active connection with multiple generations of their family. The utensils themselves were not inherently valuable or emblematic of social or ethnic status; it was the behavioural process of deploying these objects that claimed an identity for the deceased within the community at Roman Kenchreai. The collection of ceramic and glass vessels with recognizable functions, and their transport in plain view by procession up into the Koutsongila cemetery, signalled to viewers that the mourners were planning a communal experience at the tomb. Once mourners had entered the chamber, the activities of preparing, serving, and consuming food and drink promoted solidarity among participants both living and dead, inasmuch as these activities involved offering food to the deceased buried in the walls. Since mourning was an obligation of relatives and close friends or associates, and since the tombs were intended for use by lineages or households, the process of dining, even on a circumscribed or symbolic scale, reinforced the image of a unified family. The ritual of commensality reconstituted an ancestral group that cross-cut generations, in one timeless moment conjuring a living memory of the dead and engaging them in the life of the community.

Conclusion

Our conception of the process of funerary ritual at Roman Kenchreai is one in which the tombs themselves were loci of sequential and overlapping performances that together formed a composite image of the living and the dead. Public and private behaviours with spatial, chronological, and experiential dimensions intersected at the tombs over the course of centuries, employing a range of physical objects in the regulation of emotion, the display of status, and the negotiation of identity. In each of the tombs' spaces – the exterior, the *dromos*, the burial compartments, and the chamber – the physical constraints of movement and the expectations of spectators stimulated the emotions of the participants in funerary ritual. The mourners' desire to construct and to display particular identities both for the deceased and for themselves also influenced their activities in each space. Chief among these were the wealth, power, and stability of the family and its household. The long duration of the tombs' use was in itself a form of display and social negotiation, reifying familial and community relationships through accumulation of, and frequent interaction with, an ever-expanding body of physical and experiential elements. Indeed, the increasing complexity of food-related activities within the chambers over time points to a deepening concern among mourners to connect with their past and to preserve ancestral memory through commensality.

The activities involving these spaces and objects in the chamber tombs on Koutsongila thus reinforced an image of familial prosperity and solidarity within the local community. The marked consistency in the ritual deployment of spaces and objects across the cemetery also indicates the overarching significance of group identity. For the group of leading families who built and used these tombs, an identity as elite members of that small but cosmopolitan local community appears to have mattered at least as much as an ethnic, religious, or professional affiliation. The significance of this pattern with regard to larger questions of the construction of provincial identities across the Empire should not be underestimated. As we have observed elsewhere (Rife 2007; Rife *et al.* 2007), the processes of death at Kenchreai differed from those in other nearby communities, including Corinth, as well as in other more distant communities in Achaea, such as Athens. The future of Roman provincial archaeology – in explorations of burial, identity-narration, and other areas of social history – may well lie in fine-grained reconstructions of intraregional variability and close attention to the choices made by individuals who lived in a multitude of small but important towns like Kenchreai.

Acknowledgements

All photographs and drawings are used courtesy of the American Excavations at Kenchreai.

Bibliography

Alföldi-Rosenbaum, E. (1971) *Anamur Nekropolü; The Necropolis of Anemurium*. Türk Tarih Kurumu Yayımlarıdan seri 6.12, Ankara: Türk Tarih Kurumu Basimevi.

Anderson-Stojanović, V. (1987) The chronology and function of ceramic unguentaria. *American Journal of Archaeology* 91, 105–122.

Barbet, A. and Rife, J. L. (2007) Un tombeau peint de la nécropole de Cenchrées-Kenchreai, près de Corinthe. In C. Guiral Pelegrín (ed.) *Circulación de temas y sistemas decorativos en la pintura mural antigua: Actas del IX Congreso internacional de la "Association Internationale pour la Peinture Murale Antique"*, 395–399. Calatayud Universidad Nacional de Educación a Distancia.

Berns, C. (2003) *Untersuchungen zu den Grabbauten der frühen Kaiserzeit in Kleinasien*. Asia Minor Studien 51, Bonn: R. Habelt.

Carington Smith, J. (1982) A Roman chamber tomb on the south-east slopes of Monasteriaki Kephala, Knossos. *The Annual of the British School at Athens* 77, 255–293.

Cormack, S. H. (1997) Funerary monuments and mortuary practices in Roman Asia Minor. In S.E. Alcock (ed.) *The Early Roman Empire in the East*, 137–156. Oxford: Oxbow.

Cormack, S. H. (2004) *The Space of Death in Roman Asia Minor*. Wiener Forschungen zur Archäologie 6, Vienna: Phoibos.

Faraone, C. A. and Rife, J. L. (2007) A Greek curse against a thief from the North Cemetery at Roman Kenchreai. *Zeitschrift für Papyrologie und Epigraphik* 160, 141–157.

Fedak, J. (1990) *Monumental Tombs of the Hellenistic Age. A Study of Selected Tombs from the Pre-Classical to the Early Imperial Era*. Toronto: Toronto University Press.

Flämig, C. (2007) *Grabarchitektur der römischen Kaiserzeit in Griechenland*. Internationale Archäologie 97, Rahden: Marie Leidorf.

Goette, H. R. (1994) Der sog. römische Tempel von Karystos: ein Mausoleum der Kaiserzeit. *Mitteilungen des Deutschen Archäologischen Instituts, Athenische Abteilung* 109, 259–300.

Haspels, C. H. E. (1971) *The Highlands of Phrygia: Sites and Monuments*. Princeton: Princeton University Press.

Hohlfelder, R. L. (1976) Kenchreai on the Saronic Gulf: aspects of its imperial history. *Classical Journal* 71, 217–226.

Insoll, T. (2004) *Archaeology, Ritual, Religion*. London/New York: Routledge.

Kurtz, D. C. and Boardman, J. (1971) *Greek Burial Customs*. Ithaca: Cornell University Press.

Kyriakidis, E. (2007) Archaeologies of ritual. In E. Kyriakidis (ed.) *The Archaeology of Ritual*, 289–308. Los Angeles: The Cotsen Institute of Archaeology.

Laskaris, N. G. (2000) *Monuments funéraires paléochrétiens (et byzantins) de Grèce*. Athens: Stefanos D. Vasilopoulos.

Moore Morison, M. (2000) Ceramics, Cuisine and Social Class in Southern Epirus, Greece. Unpublished thesis, Boston University.

Moore Morison, M. (2006) Romanization in Southern Epirus: a ceramic perspective. In B. Croxford (ed.) *TRAC 2005*, 12–24. Oxford: Oxbow Books.

Morgan, C. (2009) Archaeology in Greece 2008–2009: Kenchreai. *Archaeological Reports* 55, 11–15.

Morgan, C. (2010) Archaeology in Greece 2009–2010: Kenchreai. *Archaeological Reports* 56, 12–23.

Morison, M. (forthcoming) *Pots for the Ancestors: The Role of Ceramics in Roman Funerary Ritual.*

Morris, I. (1992) *Death-Ritual and Social Structure in Classical Antiquity*. Cambridge: Cambridge University Press.

Rife, J. L. (1999) Death, Ritual and Memory in the Greek World during the Early and Middle Roman Periods. Unpublished thesis. University of Michigan.

Rife, J. L. (2007) Inhumation and cremation at Early Roman Kenchreai (Corinthia) in local and regional context. In A. Faber, P. Fasold, M. Struck, and M. Witteyer (eds.), *Körpergräber des 1.-3. Jh. in der römischen Welt: Internationales Kolloquium Frankfurt am Main 19.-20. November 2004*, 99–120, Schriften des Archäologischen Museums Frankfurt 21, Frankfurt: Archäologisches Museum Frankfurt.

Rife, J. L. (2008) The burial of Herodes Atticus: elite identity, urban society, and public memory in Roman Greece. *Journal of Hellenic Studies* 128, 92–127.

Rife, J. L. (2009) The deaths of the sophists: Philostratean biography, mortuary behavior, and elite identity. In E. Bowie and J. Elsner (eds.) *Philostratus*, 100–129. Cambridge: Cambridge University Press.

Rife, J. L. (2010) Religion and society at Roman Kenchreai. In S. J. Friesen, D. N. Schowalter, and J. C. Walters (eds.) *Corinth in Context: Comparative Studies on Religion and Society, Novum Testamentum* suppl. 134, 391–432. Leiden: Brill.

Rife, J. L., Morison, M., Barbet, A., Dunn, R. K., Ubelaker, D. H. and Monier, F. (2007) Life and death at a port in Roman Greece: The Kenchreai Cemetery Project 2002–2006. *Hesperia* 76, 143–181.

Sarris, A., Dunn, R. K., Rife, J. L., Papadopoulos, N., Kokkinou, E. and Mundigler, C. (2007) Geological and geophysical investigations at Kenchreai (Korinthia), Greece. *Archaeological Prospection* 14, 1–23.

Schneider Equini, E. (1972) La necropoli di Hierapolis di Frigia: contributi allo studio dell'architettura funeraria de età romana in Asia Minore. *Monumenti antichi* 48, 93–142.

Scranton, R. L., Shaw, J. W. and Ibrahim, L. (1979) *Kenchreai* I: Topography and Architecture. Leiden: Brill.

Slane, K. W. (2012) Remaining Roman in death at an eastern colony. *Journal of Roman Archaeology* 25, 441–455.

Slane, K. W. and Walbank, M. E. H. (2006) Anointing and commemorating the dead: Funerary rituals of Roman Corinthians. In D. Malfitana, J. Poblome and J. Lund (eds.) *Old Pottery in a New Century: Innovating Perspectives on Roman Pottery Studies*, 377–387. Catania-Rome: L'Erma di Bretschneider.

Spanu, M. (2000) Burial in Asia Minor during the Imperial period, with a particular reference to Cilicia and Cappadocia. In J. Pearce, M. Millett and M. Struck (eds.) *Burial, Society and Context in the Roman World*, 169–177. Oxford: Oxbow.

Stefanakis, M. (2002) An inexpensive ride? A contribution to death-coin rites in Hellenistic Crete. *Numismatica e antichitá classiche* 31, 171–189.

Ubelaker, D. H. and Rife, J. L. (2007) The practice of cremation in the Roman-era cemetery at Kenchreai, Greece: the perspective from archaeology and forensic science. *Bioarchaeology of the Near East* 1, 35–57.

Ubelaker, D. H. and Rife, J. L. (2008) Approaches to commingling issues in archaeological samples: A case study from Roman era tombs in Greece. In B. J. Adams and J. E. Byrd (eds.) *Recovery, Analysis and Identification of Commingled Human Remains*, 97–122. Totawa: Humana.

Ubelaker, D. H. and Rife, J. L. (2011) Skeletal analysis and mortuary practice in an Early Roman chamber tomb at Kenchreai, Greece. *International Journal of Osteoarchaeology* 21, 1–18.

Walbank, M. E. H. (2005) Unquiet graves: burial practices of the Roman Corinthians. In D. N. Schowalter and S. J. Friesen (eds.) *Urban Religion in Roman Corinth: Interdisciplinary Approaches* Harvard Theological Studies 53, 249–280. Cambridge: Harvard Divinity School.

Wiseman, J. (1969) Excavations in Corinth, the Gymnasium Area, 1967–1968. *Hesperia* 38, 64–106.

Wright, J. C., Pappi, E., Triantaphyllou, S., Dabney, M. K., Karkanas, P., Kotzamani, G. and Livarda, A. (2008) Nemea Valley Archaeological Project, excavations at Barnavos: final report. *Hesperia* 77, 607–654.

Chapter 3

Archaeology and funerary cult: The stratigraphy of soils in the cemeteries of Emilia Romagna (northern Italy)

Jacopo Ortalli

The Roman cemetery as material and ideological context

Roman cemeteries were complex settings in which there existed a close relationship between material structure and ideological expression: these two different types of factors fused to create a human environment that was highly diverse but substantially organic and coherent. Today, however, it can be difficult to understand well the original unity of these contexts, since research is often directed at context-specific problems or conditioned by specialized interests.

For a long time, scholars have concerned themselves in a particularistic way with architectural typologies, with their associated epigraphic and figurative apparatus, or with the tombs themselves, the remains of inhumation and cremation burials, these also being analysed from an anthropological and taphonomic perspective, and finally with the composition of grave goods. In this respect, we have achieved a good understanding of both the monuments which stood above the ground surface and of what was found buried beneath. Thus, we are able to reconstruct many aspects of cemetery organization and their cultural and social significance (Toynbee 1971; von Hesberg and Zanker 1987; von Hesberg 1992; Fasold *et al.* 1998; Pearce *et al.* 2000; Heinzelmann *et al.* 2001; Moretti and Tardy 2006).

In most cases the data at our disposal are, nonetheless, incomplete and do not allow us fully to understand all aspects of the problem. In excavation reports, for example, plans are generally published with an undefined background from which only the forms of various monuments and the position of tombs stand out, sometimes in relation to a road. This type of documentation, which summarises fieldwork results, does not allow us fully to understand the links between the archaeologically attested elements of the cemetery 'organism'. Its real 'connective tissue' was represented not so much by these conspicuous elements as by the seemingly empty ground surfaces which extended around them, to which it is appropriate to draw further attention. The areas situated between one tomb and another did not have a neutral role and cannot be considered in the abstract and in general, as topographic voids within which to locate the main funerary structures. On the contrary, these really were the free spaces within the cemetery which link all its various components and define the functional unity of the setting in which the dead lay and where, at the same time, the living circulated who came there to commemorate them.

These spaces can therefore preserve many material indicators which are of limited archaeological visibility but are very important for their documentary and ideological value, in so far as they attest the way in which the cemetery was used. This concerns traces of the original natural environment, the remains of ephemeral structures, and objects whose presence is due to the continued frequenting of the cemetery and, above all, to repetitive ritual acts and funerary cult through which the memory of the dead was preserved (Toynbee 1971, 43–64; De Filippis Cappai 1997, 49–123; Scheid 2005, 161–188).

Minor traces of functional and ritual elements

To gain a clearer idea of these minor and sometimes near invisible components which can be investigated in cemeteries (Ortalli 2008, 137–145), it is first necessary to refer to those elements which, despite their modest archaeological presence, were in the past perceived by the whole community as reference points in the articulation and use of cemetery space. Such elements also include those natural factors able to condition the human environment, for example vegetation. The presence of wooded areas, recognizable in the soil by the traces left by tree roots, must have sometimes limited the development of tombs. We should remember too the natural or artificial water courses documented by palaeochannels and by cuts for ditches which could border cemetery plots or regulate surface drainage.

Of greater importance were other infrastructural elements of public character, artificially created in order to enable the frequenting and use of funerary areas. Small roadways and internal paths were commonly created to allow passage through cemeteries and easier access to burial areas which were further from the main road. These also provided spatial markers for subdividing the various areas of the cemetery and for the orientation of burials. In many cases, such routes are marked on the palaeosoils by linear strips of beaten earth, rich in rubbish. These are flanked by regular series of tombs.

Enclosures which marked plots for family or social groups with cadastral precision were also distributed in cemetery complexes. In addition to walled structures, which are easily recognised and well documented in many cemeteries, the limits of these were in many cases marked by small ditches or in perishable materials. The latter include wooden posts fixed in the ground and fences, or even simple hedges, traces of which appear in negative on ancient ground surfaces.

Other utilitarian facilities scattered among the tombs were more directly linked to funerary ritual. We may note here the wells and stores of water necessary for various ceremonial procedures, or the areas which accommodated pyres for cremating the dead. For fixed *ustrina*, i.e. those pyres not used for a single individual but intended for repeated burning, traces including charcoal patches, ash and baked earth mixed with cremated bone and fragments of burnt objects can still be identified on the ground.

From the perspective of our research, evidence linked to individual behaviour is more deserving of attention than aspects related to the community. This is not very conspicuous but certainly very widespread and of considerable importance for its ideological implications. For funerary ceremonies, numerous textual references are undoubtedly at our disposal, yet, as has been noted (Scheid 2008), it is essential to check the effective validity of what is recorded in written sources through comparison with the reality of the facts which we can only establish through archaeological excavation and the analysis of material evidence.

From what we know, for the Romans the relationship between the world of the living and that beyond the tomb was expressed above all within the cemetery, with its twofold static and dynamic character. It was not only the space which received the remains of the dead in their final and lasting home, but was also the meeting place with the living who gathered there to honour the *Di Manes* and to institute a dialogue of a sort intended to satisfy the main requirement of funerary cult, i.e. to maintain the remembrance of the dead and celebrate their memory, so as to guarantee them an ideal and consolatory form of survival.

In order to achieve this end, it was essential to carry out a series of rites which in some way, established a relationship between the *Di Manes*, the deceased and the relatives, while maintaining a clear distinction in their roles. These cult actions generally took place under precise circumstances, codified by tradition and the festival calendar. These include those at the time of the funeral, the burial of the body, the leave-taking, the meals and liturgical libations poured in the *circumpotatio*, at the *silicernium* and the *cena novendialis* and those that took place later, during the periodic commemorations in the public honouring of the dead and the private anniversaries related to the day of birth or death, i.e. the *Parentalia, Feralia, Lemuria, Rosalia*, and *dies natalis*. At these times flowers, food and especially wine were brought to the tomb in the form of *inferiae, edulia, libationes* and *profusiones* (Toynbee 1971, 50–54, 61–64; De Filippis Cappai 1997, 70–73, 95–104; Scheid 2005, 167–182).

With this in mind, it is therefore important when a cemetery is excavated to identify the remains of all these actions which were addressed to the dead several times a year. Yet this objective is more easily met only when the material remains of funerary cult practice are represented by fixed installations connected to the burials. In some cases, as in the ruins of the *triclinia* at Ostia with walls for couches and tables (Baldassarre *et al.* 1996, 35–41), the record of banquets and funerary libations presents itself in the form of solid architectural evidence. More often however for such rites structures were employed which were much more modest but equally indicative from a conceptual perspective. In this connection we can note, for example, the libation tubes in terracotta, lead or wood fixed vertically in the ground, which put in direct contact the base of the burials with the outside world, allowing wine and other liquids to reach the remains of the dead. We can also note the tiles or bricks placed against the base of the funerary marker or set on the ground surface near the tomb, used as *mensae* for placing flowers, portions of food and other gifts.

In comparison to these fixed structural elements, anchored durably in the earth, a completely different argument must be made for the objects and portable materials occasionally used in the many commemorative acts which took place outside the tomb not only at the funeral but also afterwards at periodic anniversaries, i.e. by those tombs that had been closed for some time. In many cases, such gestures of *pietas* included the offering of organic foodstuffs which have almost completely perished with the passage of time and are now only detectable through palaeobotanical and palaeozoological analysis. Beyond this, nonetheless, many durable items were certainly used in funerary cult and deposited or abandoned on the ground. It is very important to identify the material traces of these which are still preserved within the soil.

In reality the archaeological evidence normally at our disposal for such remains survives as rare chance finds or not at all, given the difficulty of finding a context for small objects which are highly fragmented and dispersed in the stratification of the cemetery because of prolonged exposure on the ground surface. Nevertheless, the interest of these indices for ritual activity outside the tombs must again be underlined: they have a fundamental importance for a full understanding of the functional and cultural character of the cemetery.

To fill the lacunae in our knowledge it is therefore necessary to devote the maximum attention to these significant minor elements. First of all, we must be aware of their existence as an integral part of the funerary context; secondly, we must not concentrate only on tombs but also systematically analyse the formation processes of the soil deposits between the tombs and the meaning of the finds which these deposits contain.

Stratigraphic problems

Such objectives are not easy to realise, above all because of the nature of burial areas, the particular configuration of which has always conditioned and constrained fieldwork. If we consider the ways in which cemeteries are normally investigated, as has been said the principal attention is focused on individual structures which emerge from the soil and on individual burials with their grave goods. These are immediately identifiable elements which can in some ways be evaluated as distinct units, examined and excavated individually and then re-integrated in a synthetic interpretative framework only later in analyzing the data. In reality this procedure obstructs a sense of the linkage and connections between the multiple material and cultural factors of which the cemetery was an expression. It is however important already to consider these aspects during excavation through a holistic approach to research and a comprehensive view of the context.

The problem is essentially a stratigraphic one and lies in the essential difference between cemetery deposits and those normally found in settlement contexts, where depositional sequences and physical relationships between structures and stratigraphic units are clearly legible. In contrast to settlement excavations, which find their fullest expression in urban archaeology, excavations of cemetery areas concern sites which served a single, repeated function. They are characterised by variable soil thicknesses which tend to be homogeneous and seemingly without internal distinctions. From the 'natural' and the levels related to their first creation to the uppermost abandonment layers there do not seem to be clear discontinuities in the deposits or series of extensive interfaces. In sum, they appear to comprise an undifferentiated single layer.

In reality, if we consider the composition and evolution of the funerary 'organism', we should recognise that this so-called 'monostratum' cannot be uniform. It forms progressively, sometimes over several centuries, and must necessarily incorporate within it the remains of all human actions, structural, utilitarian and ritual which successively took place on multiple ground surfaces. If the burials and the markers are excluded, the analytical difficulty lies in the fact that generally such actions are attested by modest and restricted traces or limited material scattered through the soils. Such indicators we normally do not succeed in clearly understanding or connecting.

This problematic stratigraphic situation doubtless arises from the specific environment and development of cemetery contexts. Roman cemeteries were normally located in open areas at some distance from the locations in which human settlement and activity were concentrated. Moreover, their frequenting was rather occasional and desultory, linked to days of funerals and to the ceremonies which periodically took place by burials, or to the period of construction of the tombs and monuments. All these activities were intermittent and from time to time involved limited separate areas which were distant from one another.

In the process of formation and accumulation of cemetery deposits, natural factors predominated in practice over human. This circumstance is demonstrated by the pedogenesis of the strata, in which the soil matrix is analogous if not identical to that of the local natural, its internal composition is substantially homogeneous and poor in material and its formation continuous and slow, with a regular raising of the ground level.

Reflecting on what has been said so far, in spite of these real difficulties it is nevertheless appropriate in many cases to aim at integrated analysis of stratigraphic funerary complexes, obliging ourselves to follow less usual methods of investigation. Only research aimed at recording every small detail will permit us to understand the way in which these environments were really experienced by the collectivity and the individuals who celebrated their rites at the tombs.

A different approach to the excavation of cemetery deposits

As we have tried to explain, traditional excavation techniques do not allow accurate extensive analysis of the deposits and materials present in cemetery stratification, as much in larger cemeteries as in small groups of burials. Here we therefore offer a series of theoretical considerations, supported by some practical experiments in the field, which can indicate a procedure better suited to this research problem. In this regard, we need to adopt a fundamental premise, related to the fact that only contexts with good preservation of use deposits can offer information useful for reconstructing past activity on cemetery ground surfaces above and outside the tomb. For example, an area with a constant build-up of soil and covered or protected by alluvial sediments guarantees the best preservation of the original cemetery surfaces and, as a consequence, the greatest opportunities for study. On the other hand, stable ground eroded by atmospheric agents or damaged by agriculture, as on a mainly rocky substrate, may have undergone degenerative processes such as to impede the survival of anthropically generated soils. In these cases, the disappearance of deposits corresponding to phases of use of the cemetery leaves the deepest tomb cuts and most substantial monuments, without any direct connection between them, as the only archaeological evidence. Any effective stratigraphic observation is then impossible.

Even where cemetery deposits still remain, and therefore offer good opportunities for investigation, the reliability of contexts is not always equal, being dependent on the different composition and consistency of the base soils. The optimum soils are those with a compact matrix, clays or silts, which allow the preservation of the activity layers and provide stability for the material scattered on exposed surfaces in the past. By contrast, soft, unstable or sandy soils are more at risk of stratigraphic instability, since they can easily lose their shape, even through human movement across them, altering the original surfaces and causing the displacement or mixing of the objects associated with them.

When taking into consideration a cemetery area appropriate for the analysis of layers and deposits related to use, we should first define the field of interest for the specific questions which we are discussing. As has been said, from our point of view the greater importance lies in the seemingly empty spaces which are free of structures but nonetheless contain scattered traces of the funerary behaviour repeated during the progressive accumulation of soils. Rather than a single cemetery stratum we should think of a heterogeneous deposit which has within it a progressive sequence of surface layers. From an archaeological point of view these generally appear as undifferentiated, except those particular areas and levels which accommodated a more intense human activity so as to cause the dispersal on the ground of inorganic material and organic residues linked to burial rites in some way.

The task of the archaeologist is therefore to find, document and integrate in a coherent picture the traces which survive on the subsoil outside the tombs, at different levels and often clustered in small areas without, however, it being possible to identify clear interfaces or distinctions between layers. We have defined a method which we consider effective for achieving this result as 'targeted microstratigraphy', which aims to bring to light every trace of human activity, even the slightest, which can possibly be related to a real activity layer.

If this procedure is used it is appropriate to work slowly across reasonably extensive areas and progressively to remove layers of a few centimetres in thickness. The major difference with the old planum method of excavation, which involved the removal of regular spits of predetermined depth, lies in the fact that in our case the depth of soil excavated is variable. In general, it can be adapted to the natural morphology of the excavation area and above all should be guided by archaeological indicators which give some orientation for calibrating the thickness and direction of the spits.

Generally speaking, the first archaeological layer which can thus be detected is situated at the top of the cemetery deposit sequence, corresponding to the abandonment phase. Beneath this further exploration should continue with a sequence of micro-spits, each extended until the appearance of elements which reveal actual, if circumscribed, surface levels to be carefully delimited and documented analytically.

Usually the most indicative traces which appear during excavation comprise the cuts which mark the burial pits, holes for wooden posts or roots, rubble and pieces of brick, and above all zones of denser rubbish and fragmentary objects. The latter are disposed in a particular horizontal configuration which distinguishes them from erratic residual material which is usually randomly mixed in the soil and is typical of ground surfaces directly exposed to human passage and action.

In general, these comprise finds characteristic of funerary contexts, linked to the creation of tombs, to burial procedures and to funerary cult in which the same types of objects used as grave goods could also be offered to the dead and placed on the ground outside the tombs. Among this evidence, we note, for example, the carbonised remains of funerary pyres, the leftovers of meat dishes comprising animal bones and carbonised remains of vegetable foodstuffs, as well as artefacts including lamps, balsamaria, coins, ceramic incense burners and trays for offerings, vessels for serving and consuming wine, cups, bowls and beakers used in libations. All these objects may be found complete, sometimes even in series composed of several examples, or in groups of small fragments which can be reconstructed, the witnesses to intentional breakage for ritual ends or accidental breakage through trampling. Calibrating the depth of excavation on the basis of these indicators, which need recording in three dimensions, it is therefore possible to recognize what remains of the more or lesser extensive areas of activity with some consistency.

As work progresses the excavation of targeted micro-spits and the documentation of archaeological evidence should be repeated as long as layers are encountered which contain significant traces of human activity. The number of times depends on the duration and intensity of the funerary activity that took place in a particular area. Tombs identified during the removal of these spits should be excavated before the next spit is begun, since this will reveal layers preceding these burials.

With its particular attention to the ground surfaces between tombs, this procedure allows some significant aspects of the cemetery complex to be focused on during excavation. Still more important are the results which emerge later, thanks to the holistic examination and interpretation of the documentation and the finds excavated in the field. Their integrated study can provide new information on activities and cult practices in the cemeteries, considerably enriching our usual framework of understanding of these contexts.

The working method which has been set out here is not simply a statement of theoretical principles. It has already been tried out in various excavations at different times by the Soprintendenza Archeologica for Emilia Romagna. Here in particular we note as examples the experience derived from two important sites, characterized by different environmental conditions, in which the validity of this approach to funerary stratification has been demonstrated.

The Pian di Bezzo cemetery at Sarsina

In antiquity Sarsina was the most important centre of the northern Apennines. Today it is known to scholars above all for the Roman cemetery on the floodplain at Pian di Bezzo, which differs from the normal archaeology of the region in its exceptional state of preservation, due to having been flooded and protected by a riverine lake that formed through natural causes in c. AD 200 (Ortalli 1989). The alluvial sediments, some metres thick, sealed for centuries the grand mausolea of the Augustan period and the tomb markers which had been raised alongside a road. These elements were brought to light in the successful excavation campaigns undertaken in the first half of the 20th century (Aurigemma 1963; Ortalli 1987). Not only have the funerary structures kept their integrity but so too has the extensive archaeological base deposit, in a compact clay matrix, containing interesting evidence for human activity.

Given the particularly favourable conditions, between 1981 and 1984 it was decided to experiment for the first time here with analytical excavation of the original layers related to cemetery use following the method of targeted microstratigraphy (Ortalli 1988; Ortalli 2008, 145–149). The investigation concerned more than 300 m^2 near to the zone of monuments which had been previously explored and revealed a length of gravelled street flanked by 25 tombs of modest scale, mainly *bustum* burials made between the first and mid-second century AD (Ortalli, Baldoni and Pellicioni 2008).

Having removed the alluvial silts which covered the Roman period deposits we sought first of all to discover in its entirety the abandonment layer from the period at which the cemetery was flooded. This came to light in its original form, slightly undulating and sloping towards the river Savio, with the road demarcated by a ditch and bank. The fields beside the road were only partly occupied by tombs. A wide strip of earth had not been used for burial as it was covered by thick vegetation, as the clearly recognized scattered traces of roots have shown. Essentially the only real interface found during the whole excavation comprised this layer at the top of the anthropic stratification, which corresponded with the final phase of use of the cemetery.

Heterogeneous material was distributed across the surface of the soil, without any clear structure, giving an impression of neglect. This can be explained by the disuse of this area for funerary activity for several decades before it was flooded. By the end of the second century AD, to which the final soil is dated, the spatial progression of the cemetery along the road had led to burial in other zones at some distance from the area under excavation. From this comes the impression of an area that was frequented little if at all, across which were scattered rubbish, rubble and organic remains, a few vessel fragments, two upturned stelae, and the necks of some amphorae which had partly disintegrated. The latter had originally been fixed above tombs as markers and receptacles for offerings (Fig. 3.1).

Gradually, as this abandonment layer was brought to light its profile was drawn in detail, recording its level, all the traces of activity present in it and the various materials which lay on the surface (Fig. 3.2). At the end of the excavation an integrated plan of the context was created which was not limited to the structural remains, such as the road and the visible parts of the tombs, but also included the portable minor elements observed on the activity surface. It was thus possible to note some interesting aspects of function by space, differentiating zones reserved for burial, areas covered with vegetation and the paths which allowed access to the burial plots from the main road.

Fig. 3.1. Sarsina, Pian di Bezzo: layer corresponding to the period of abandonment of the cemetery.

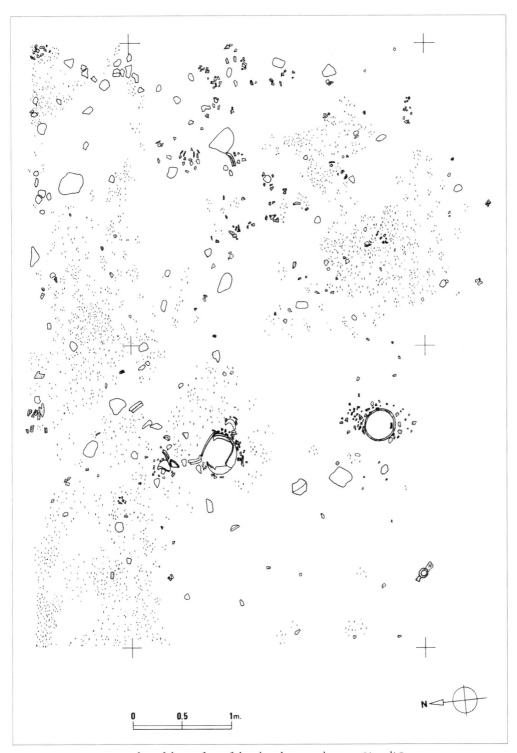

Fig. 3.2. Plan of the surface of the abandonment layer at Pian di Bezzo.

Fig. 3.3. The depositional sequence in the cemetery Pian di Bezzo: left, the upper deposit corresponding to the abandonment phase (c. AD 170-200); centre, the intermediate deposit with the activity surface corresponding to the phase of burial in this specific sector (c. AD 80-170); right, the lower deposit corresponding to the first phase of activity (c. 30 BC-AD 80).

After this first stage, gradual further excavation revealed the homogeneous composition and regular natural accumulation of this stratum, which contained a few scattered finds until, between 10 and 20 cm from the abandonment palaeosoil, the level corresponding to the period of most intensive activity in this area of the cemetery was reached. Here the situation changed radically. The considerable number of finds in the clay soil were used as reference points for discovering the extent of what could with some confidence be interpreted as the surface which for more than 50 years between the first and second centuries AD was destined for burial of the dead (Fig. 3.3).

Following the many traces of activity connected to funerary practices brought to a light a second palaeosoil with features which could be archaeologically well characterized, above all by the material found within it. As well as numerous grave cuts and charcoal patches the surface revealed zones rich in organic remains and ceramic fragments which often joined. Among these were recognized objects which had either been deliberately placed on the ground near the burials or used for ritual offerings and then abandoned at that spot (Fig. 3.4).

After also documenting analytically this zone (Fig. 3.5), the various tombs belonging to the same chronological phase were dug and a further spit was excavated of the surrounding deposit. As foreseen this showed little evidence for human activity, as if it was related to an earlier period when this sector was not yet used for burials or monuments (Fig. 3.3).

Fig. 3.4. Layer corresponding to the phase of most intense funerary use in the sector excavated at Pian di Bezzo.

In substance, this microstratigraphic method applied to the deposit within the Sarsina cemetery, from the upper interface with the abandonment and flood layer to the 'sterile' substrate without archaeological remains, allowed us to identify directly in the field the changes which occurred to this area through the progressive shifting of the burial zone. As well as this general character, we should also note the detailed and more significant results which were obtained by identifying the levels corresponding to the principal period of burial.

The graphical reconstruction and interpretation of the documentation gathered in the sequence of spits gives particular prominence to the soil related to the years of most intensive funerary use of this area. Around the *bustum* tombs, in simple pits or with a tile cover *a cappuccina*, numerous traces of burial practices and ceremonies carried out during the funeral and successively afterwards were observed. These include obvious spreads of ash and charcoal scattered during cremation and *ossilegium* [collection of the cremated bone], items deposited outside the tomb such as dishes placed on the edge of grave pits, remains of food offerings comprising animal bones and carbonized plant remains, many objects of clearly ritual use, identical to those found among the grave goods, including coins, lamps, balsamaria and above all, series of cups and beakers in thin-walled wares, amongst which can be recognized drinking vessels used for collective libations in honour of the dead (Fig. 3.6).

Among the various data obtained, those which have allowed reconstruction of rituals belonging to a cultural sphere which is otherwise little archaeologically known are undoubtedly the most important. This experiment confirms that minor traces of limited visibility on ancient activity surfaces, when studied systematically, can effectively contribute to integrated and properly contextual research.

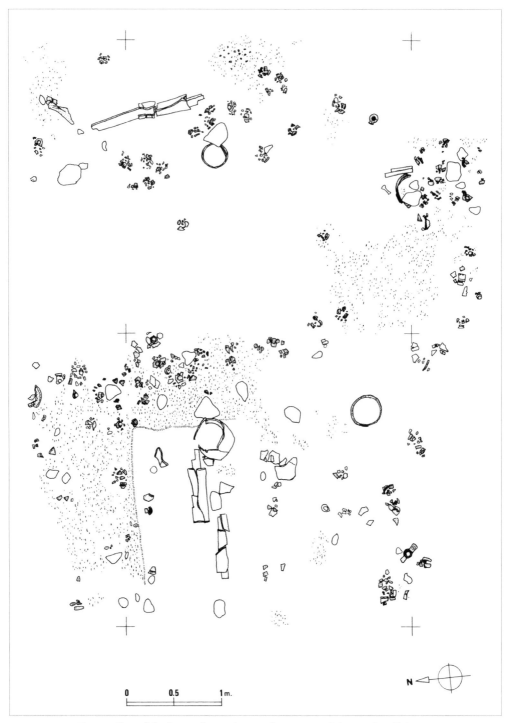

Fig. 3.5. Plan of the layer of most intense funerary activity at Pian di Bezzo.

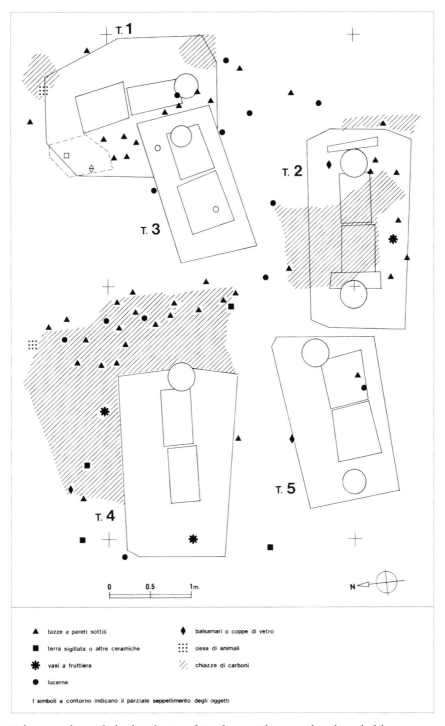

Fig. 3.6. Schematic plan with the distribution of ritual material scattered on the soil of the same excavation sector at Pian di Bezzo.

The cemetery at Podere Minghetti, Classe (Ravenna)

The most recent systematic application of microstratigraphic excavation in Emilia Romagna took place between 2003 and 2005 in a cemetery area of the former Roman port at Classe, near Ravenna (Maioli, Ortalli and Montevecchi 2008). The investigation was initiated within the framework of a research project which brought together various European archaeologists, the aim of which was to discuss methods of excavation and study in order to improve the understanding of evidence for funerary cult and of acculturation in the Roman period (Scheid 2004; Scheid (ed.) 2008). For the field investigation, an area was chosen at Podere Minghetti, known for some time for the discovery of stelae and grave goods indicating a long funerary use for the area. In antiquity, it was situated on a dune a short distance from the Adriatic Sea (Maioli 1990, 390–412).

The characteristics of the Ravenna context were substantially different from those at Sarsina. First, because of their high density and the many centuries over which deposits were made, the tombs were not at a single level but were superimposed in various layers, so as to produce a very complex stratigraphic sequence. Furthermore, the soil matrix essentially comprised marine sand, so unstable and soft that it did not allow for good preservation of the original activity surfaces. To these problems, we should also add the proximity of the archaeological deposits to the modern ground surface: the highest, corresponding to the final phases of use of the cemetery in late antiquity, had been damaged by agriculture.

Although these conditions were not optimal for this particular kind of investigation, it also seemed appropriate in this setting to excavate by applying targeted microstratigraphy, so as to verify the effectiveness of the technique by putting it to the test in a context where the conditions were undoubtedly problematic.

The area concerned extended over 200 m², of which only a third was explored down to the natural. In this complex were found 16 brick bases of small funerary monuments, arrayed in lines on paths through the cemetery, and 188 burials datable from the Augustan period to the 5th century AD, comprising *bustum* burials, cremation burials in cinerary containers and inhumations in pits or amphorae (Fasold *et al.* 2004; Maioli, Ortalli, Montevecchi 2008; see papers by the following in the volume edited by John Scheid (ed.) 2008: V. Bel, C. Demangeot, H. Duday, C. Leoni, M.G. Maioli, G. Montevecchi, J. Ortalli, J. Scheid, V. Zech-Matterne) (Fig. 3.7).

Like other cemeteries, here too the continuous and undifferentiated build-up of the deposit was evident. The funerary deposits had accumulated so rapidly that they caused the gradual burial of the bases of the oldest monuments until they reached a depth of one and a half metres.

Taking as a reference point either the foundation level of the brick monuments or the slightly sloping profile of the substrate dune, excavation proceeded with the removal of spits of limited depth in the matrix into which the burial pits were dug; these were excavated at a second stage. Thus, at different points and varying depths surfaces were identified which could be related to environmental and human factors

Fig. 3.7. Classe (Ravenna), Podere Minghetti: part of the cemetery area with the bases of funerary monuments in the foreground.

of interest, including roots, animal bones, remains of burial processes, debris from the collapse of the structures which had once stood above the ground and, above all, objects deposited near the burials to commemorate the dead (Ortalli 2008, 149–156). Each element was registered and documented through a series of detailed levelled plans (Fig. 3.8).

The limited compaction of the sandy soil and the many cavities from the burial pits created at different times and levels did not generally allow preservation (and therefore observation) of extensive activity layers rich in material in the horizontal plane. Only in some limited areas could the remains of palaeosoils be securely identified in which a significant artefact concentration could be noted arrayed on the surface, mostly of whole or reconstructable items. Among these traces of ritual activity, we should note the many fragments of cups in thin-walled wares and terra sigillata scattered at the base of a stele, to be interpreted as the remnants of a collective libation, or the three whole lamps, one of which was covered with an upturned tazza, placed on the ground above the pit of a *bustum*, next to the mouth of a wooden tube for *profusiones* (Fig. 3.9).

Given the homogeneity of the sandy matrix, variation in the depositional sequence was limited and only perceptible in macroscopic terms. For example, in the higher levels, which could be attributed to the final occasional frequenting of the cemetery, finds were abundant but minute, scattered and residual, while at the lower levels, attributed to the early and middle empire, they were rarer but less fragmented and more easily identifiable as types. Furthermore, in the intermediate levels dated towards the second century AD, in which the *bustum* rite was prevalent, the earth had taken on a slightly dark colouring because of the charcoal remnants dispersed in the soils.

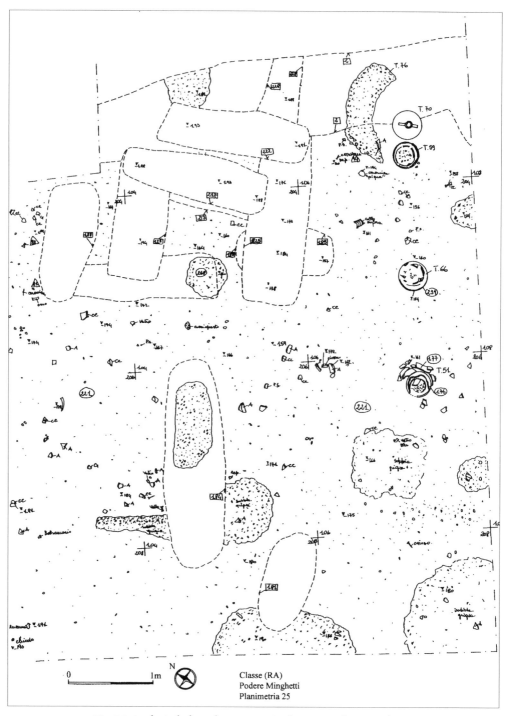

Fig. 3.8. Analytical plan of an excavation layer at Podere Minghetti.

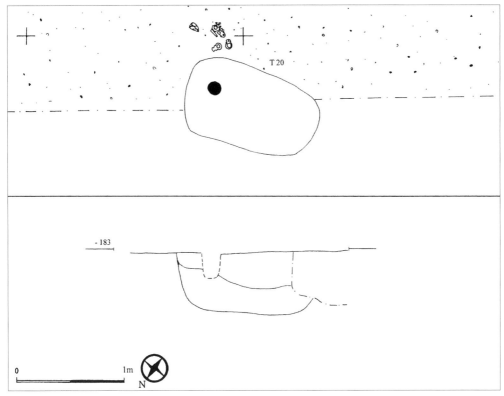

Fig. 3.9. Plan and section of tomb 20 with material deposited on the activity surface outside the tomb.

Generally speaking, at Classe too the excavation of micro-spits led to good results, providing interesting information on the development of the burial complex and on some ritual actions which were periodically conducted outside the tomb. In this case, we should nonetheless admit that unfortunately the particular ground conditions limited the possibility for confident identification of extensive and consistent traces of ancient activity surfaces. As has been said, in the one-and-a-half-metre deep cemetery deposit only generic differences could be seen.

Fortunately, the limits to the recognition of strata within the deposit and of the detail of its formation were balanced by the positive results achieved in the general understanding of the context, thanks to the integration and analysis of all the data recorded in the field. As an example, we can offer a first attempt at synthesis of the distribution of all the finds made outside the tombs, considered separately by type and findspot. Essentially a series of cumulative stratigraphic projections are worked out which summarise the topographic location and the level of each example of the classes of material normally employed in funerary rituals, in order to provide a statistical assessment of such objects in relation to the phases of use of the cemetery area (Fig. 3.10). In this way, some diverse trends in funerary cult have been identified

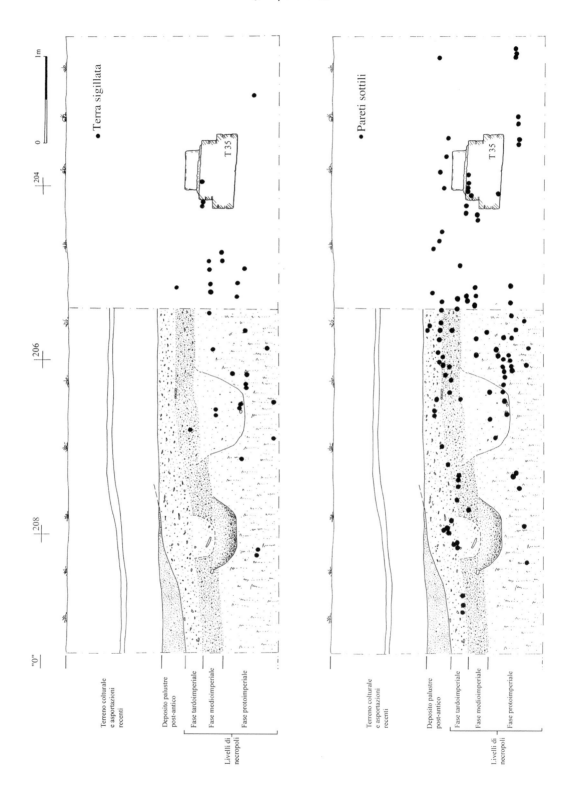

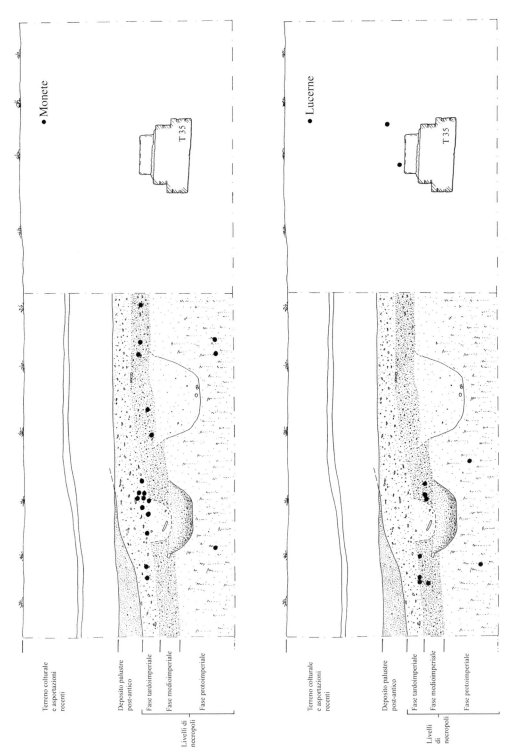

Fig. 3.10. Stratigraphic section with cumulative projections of the objects found in the phases of use of the cemetery at Podere Minghetti (from top to bottom: terra sigillata; thin-walled wares; coins; lamps).

from the changing preference over time for some types of offerings over others (Ortalli 2008, 155–157).

In the early imperial period the offering of terra sigillata vessels was frequent. These were associated with and then substituted by thin-walled drinking vessels, long used in considerable quantities for libation ceremonies. The deposition of balsamaria near the tomb was rare but consistent over time; by contrast lamps and coins occur in large numbers above all in levels relating to the mid-imperial phase, when *bustum* burials dominated.

Yet it has been observed that although such materials belonged to the same classes which were habitually deposited as contemporary grave goods, in comparison with these their numerical relationship changed significantly. The percentage differences among the offerings placed outside the tomb in relation to those within it testify to the selective and variable use which was made of the same objects in relation to different ritual ends. They may have been regularly dedicated to the dead but were occasionally in use among the living.

* * *

In short, the excavation method which has been illustrated here and tested in various different situations has been able to offer interesting data on certain uncommon aspects of funerary archaeology, in particular basic elements of context such as the meaning in human terms of cemetery surfaces and the traces of the rituals which were conducted there in honour of the dead. For this problem, more refined and effective techniques can certainly be elaborated which should go beyond the fundamentally empirical nature of this type of microstratigraphic targeted excavation. This technique however has the advantage of being an addition rather than an alternative to traditional techniques for investigating cemeteries and small groups of burials, requiring only some greater attention in the approach to and execution of research on the ground. Finally, we note that the application of the methodological principles we have set out should not be considered exclusively relevant to contexts of Roman date; the same procedures apply to cemetery areas of earlier and later date, in which the living visited tombs to fulfil their cult obligations in honour of the dead.

Editor's note

The paper was translated by John Pearce, who records his gratitude to Grazia and Roberto Piva for their help in preparing the translation.

Bibliography

Aurigemma, S. (1963) I monumenti della necropoli romana di Sarsina. *Bollettino del Centro di Studi per la Storia dell'Architettura* 19, 1–107.

Baldassarre, I., Bragantini, I., Morselli, C. and Taglietti, F. (1996) *Necropoli di Porto. Isola Sacra.* Rome: Istituto Poligrafico e Zecca dello Stato Libreria dello Stato.

De Filippis Cappai, C. (1997) *Imago mortis.* Naples, Loffredo Editore.

Fasold, P., Fischer, T., von Hesberg, H. and Witteyer M. (eds.) (1998) *Bestattungssitte und kulturelle Identität.* Kolloquium Xanten 1995. Cologne: Rheinland-Verlag GmbH.

Fasold, P., Maioli, M. G., Ortalli, J. and Scheid, J. (eds.) 2004. *La necropoli sulla duna. Scavi a Classe romana.* Catalogo della Mostra. Frankfurt-Ravenna, Museo Archeologico di Francoforte.

Heinzelmann, M., Ortalli, J., Fasold, P. and Witteyer, M. (eds.) (2001) *Römischer Bestattungsbrauch und Beigabensitten.* Colloquio Internazionale Roma 1998, Wiesbaden: Dr. Ludwig Reichert Verlag.

Von Hesberg, H. (1992) *Römische Grabbauten.* Darmstadt: Wissenschaftliche Buchgesellschaft.

Von Hesberg, H. and Zanker, P. (eds.) (1987) *Römische Gräberstrassen, Selbstdarstellung- Status- Standard München.* Kolloquium München 1985, Munich: Bayerischen Akademie der Wissenschaften.

Maioli, M. G. (1990) Topografia della zona di Classe. In G. Susini (ed.) *Storia di Ravenna 1, L'evo antico,* 375–414. Venice: Marsilio Editori.

Maioli, M. G., Ortalli, J. and Montevecchi, G. (2008) La nécropole du port militaire de Classe. *Les Dossiers d'archéologie* 330, 74–81.

Moretti, J.-Ch. and Tardy, D. (eds.) (2006) *L'architecture funéraire monumentale. La Gaule dans l'empire romain. Colloquio Lattes 2001.* Paris: Édition du Comité des travaux historiques et scientifiques.

Ortalli, J. (1987) La via dei sepolcri di Sarsina. Aspetti funzionali, formali e sociali. In von Hesberg, H. and Zanker, P. (eds.), *Römische Gräberstrassen, Selbstdarstellung-Status- Standard München. Kolloquium München 1985,* 155–182. Munich: Bayerischen Akademie der Wissenschaften.

Ortalli, J. (1988) Proposte metodologiche per lo scavo di necropoli romane. *Archeologia Stratigrafica dell'Italia Settentrionale* 1, 165–195.

Ortalli, J. (1989) Sarsina. La fine di una necropoli romana nel II-III sec. d.C. In E. Guidoboni (ed.) *I terremoti prima del mille in Italia e nell'area mediterranea,* 474–483. Bologna: SGA-Storia Geofisica Ambiente.

Ortalli, J. (2008) Scavo stratigrafico e contesti sepolcrali: una questione aperta. In Scheid (ed.) (2008) *Pour une archéologie du rite. Nouvelles perspectives de l'archéologie funéraire,* 136–159. Collection de l'École Française de Rome, 407, Rome: École Française de Rome.

Ortalli, J., Baldoni, D. and Pellicioni, M. T. (2008) Pian di Bezzo di Sarsina. La necropoli Romana. In A. Donati (ed.) *Storia di Sarsina 1. L'età antica,* 431–663. Cesena: Editrice Stilgraf.

Ortalli, J. (2011) Culto e riti funerari dei Romani: la documentazione archeologica. In *Thesaurus cultus et rituum antiquorum (ThesCRA), VI, Stages and circumstances of life,* 198–215. Basel, Los Angeles: Getty Museum.

Pearce, J., Millett, M. and Struck, M. (eds.) (2000) *Burial, Society and Context in the Roman World.* Oxford: Oxbow.

Scheid, J. (2004) Analyse de l'évolution des rites funéraires au début de notre ère. *La Lettre du Collège de France* 10, 12–13.

Scheid, J. (2005) *Quand faire, c'est croire. Les rites sacrificiels des Romains.* Paris: Éditions Flammarion.

Scheid, J. (2008) En guise de prologue. De l'utilisation correcte des sources écrites dans l'étude des rites funéraires. In J. Scheid (ed.) (2008) *Pour une archéologie du rite. Nouvelles perspectives de l'archéologie funéraire*, 5–8. Collection de l'École Française de Rome 407, Rome: École Française de Rome.

Scheid, J. (ed.) (2008) *Pour une archéologie du rite. Nouvelles perspectives de l'archéologie funéraire*. Collection de l'École Française de Rome 407, Rome: École Française de Rome.

Toynbee, J. M. C. (1971) *Death and Burial in the Roman World*. London: Thames and Hudson.

Van Andringa, W., Duday, H. and Lepetz, S. (2013) *Mourir à Pompéi: fouille d'un quartier funéraire de la nécropole romaine de Porta Nocera (2003-2007)*. Collection de l'École française de Rome, 468, Rome.

Chapter 4

Funerary archaeology at St Dunstan's Terrace, Canterbury

Jake Weekes

This paper presents an analysis of the Romano-British funerals represented by part of a cemetery exposed within a housing development in Canterbury in 2001. Apart from disseminating the insights from a significant site more widely, the methodological and interpretive aims of this paper are twofold. The first and more straightforward intention is to match classes of archaeological evidence more closely and with more clarity with those aspects of the funeral that they can particularly help to elucidate. Secondly, and partly as an experiment, I intend here to develop a more integrated method of reconstructing and comparing entire funerary sequences, in direct relation to the dead individuals with which they were associated.[1]

I suggest that this disparate evidence is often fragmented still further by typical methods of post-excavation analysis undertaken by separate specialists, and then even more by comparative studies that focus on individual facets of funerals, such as the frequency of certain grave goods, rather than the whole assemblage. Such divisive approaches can alienate the material from its original context: the particular funeral of a person (or people) as experienced and structured by participants, which surely detracts from a proper understanding.

In effect, I prefer to treat *the funerals themselves* as the shattered 'artefacts' under study, which first need to be tentatively reconstructed using the understandably selective and compromised evidence available to us. As will be seen, particular strands of evidence are often missing on certain sites, or impossibly equivocal, or in fact eloquent of more than one funerary aspect. Such evidential difficulties and possibilities need to be teased out before comparisons are made and conclusions drawn that relate to the entire funerary sequence.

In the following I therefore begin with giving the background to and a reappraisal of the St Dunstan's Terrace evidence before moving on to a treatment of it as a case study in funerary analysis and interpretation.[2]

St Dunstan's Terrace 2001

Background

The cemetery at St Dunstan's Terrace first came to the attention of archaeologists in 1925 and 1926, when cremation burials were noted during construction work on a telephone facility on the site (Whiting 1927). Evaluation prior to the construction of new housing in 2000 uncovered further cremation burials (Rady 2000), some of which form part of this analysis. Open area excavation followed in the winter of 2000–2001, and the results of this work form the majority of the evidence considered below (Diack 2003).

The cemetery dates from the late Iron Age to the late Roman period, although its *floruit* lay in the first two centuries AD. It was located next to what is still the London Road (Fig. 4.1), probably the first Roman road in *Britannia*, which, having followed the Stour Valley from Richborough (*Rutupiae*) and the Wansum Channel, would have crossed the Stour at *Durovernon*. It was this road around which Canterbury's early *forum* area,[3] and, we suspect, the street grid as a whole, was planned, taking in what may well have been a pre-Roman ritual focus to the south-east of the river, later given monumental form by a suite of public buildings comprising a temple precinct and associated theatre and baths. Having crossed the Stour, the road climbed the gravel terraces forming the northern approaches to the river, avoiding another late Iron Age

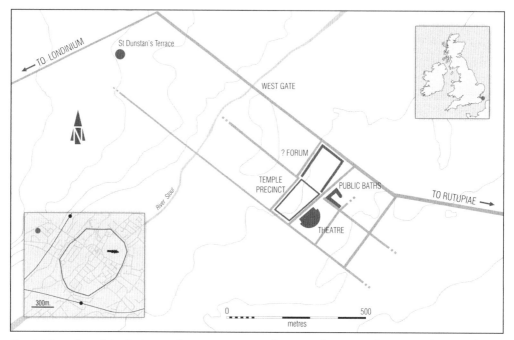

Fig. 4.1. Location of the St Dunstan's Terrace site in relation to the UK, modern Canterbury and early Durovernum.

settlement focus in St Dunstan's (Frere *et al.* 1982, 45–54) before turning west once again, past the St Dunstan's Terrace cemetery area, and towards an already familiar gap between the North Downs and the Blean overlooked by Bigberry hillfort, heading for *Londinium*.

The St Dunstan's cemetery in this area has remained largely undisturbed by post-depositional processes since the Roman period, and has the potential therefore to provide a particularly interesting case study of the development of early and developing Romano-British funerary styles; a careful consideration of the evidence it presents is called for.

A reappraisal

The first step in the process of interpreting cemetery evidence is an understanding of archaeological contexts, and the St Dunstan's Terrace site required reappraisal in light of theoretical and analytical advances that have moved towards the centre of our study in recent years, as well as some more basic re-reading of the archaeology, with the acknowledged benefit of considerable hindsight. Of the cremation burials recorded for example, at least three are, on reflection, more likely to have been partially seen inhumation burials ('Cr86', 'Cr92', and 'Cr94'), and a number of 'cenotaphs' reported in fact relate either to features that had undoubtedly been compromised, or again could only be partly excavated at the edge of site.

More significantly, however, a number of contexts excavated or later re-interpreted as cremation burials might actually be funerary deposits of a rather different and interesting type (see below). There were typical problems around discernment of un-urned burials, other types of burials that include pyre material, and pyre related deposits in non-burial contexts, and this especially required some re-thinking on firmer theoretical grounds. A number of these features actually represented large spreads of material, while others certainly suggest a possible practice of commemorative deposition within the cemetery of material other than human remains (see McKinley this volume).[4]

The development-driven machine excavation used to 'find' the cemetery horizon on the site is also an issue, leading inevitably to truncation of the upper contexts of burials, and loose or bagged deposits of cremated remains within burials in some cases (Weekes 2007). I mention this here because it has analytical implications, but would reiterate that machine excavation of Roman cemetery sites is a methodological compromise; certainly at St Dunstan's Terrace many of the burials were only located once some of the contents were destroyed. Nonetheless, the assemblage is considered to be largely intact for the purposes of the comparative analyses that follow.

Following a reappraisal of all the evidential factors, then, we are actually left with 86 burials for comparison, representing 93 dead: 21 more or less certain late Iron Age and early Romano-British inhumation burials, broadly contemporary in the Roman period with 62 cremation burials (representing 69 dead),[5] and three late Romano-British inhumations. A number of more or less probable alternative funerary features of various types were also uncovered, some potentially very significant for understanding commemorative ritual, for example.

Cemetery development

The first visible phase of cemetery activity on the site came in the form of Late Iron Age inhumation burials, apparently aligned with a ditch (Ditch 1; Fig. 4.2), or some earlier marking of this boundary. Certainly the ditch was also dug around this time, and went on to form the southern limit of the Romano-British cemetery. Despite difficulties of dating (relying on residual material and stratigraphy in most cases, or a qualitative assessment of spatial association in some) this earliest phase of activity seemed to be largely focussed in the south-west corner of the site,[6] although two potential further burials to the east could also be tentatively placed in this early phase based on the terminus ante quem of abraded prehistoric pottery sherds from their backfill deposits.[7]

Cremation burial at St Dunstan's began before c. AD 70 (but apparently post-conquest), with two definite burials,[8] also aligned along Ditch 1 (Fig. 4.3) and a placed deposit cut into the upper backfills of late Iron Age inhumation burial Ih6 (Cr5) comprising a single butt beaker but very little cremated bone (less than 1g). The quantity suggests it may have resulted from commemorative acts of a different sort from burial (considered further below). By c. AD 100 (Fig 4.4) a further two placed deposits[9] seemed to focus on the upper backfills of late Iron Age inhumation burials, comprising single vessels (beakers and a miniature vessel) but no cremated bone.

Fig. 4.2. Development of the cemetery plot by c. AD 50.

Fig. 4.3. Development of the cemetery plot by c. AD 70.

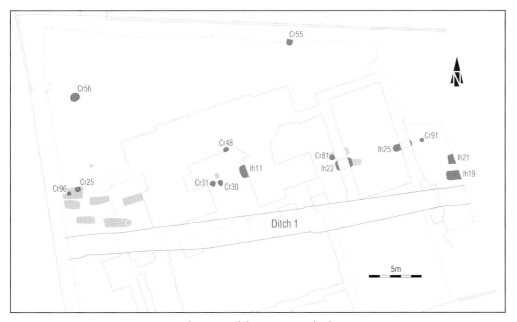

Fig. 4.4. Development of the cemetery plot by c. AD 100.

Further and more convincing cremation burials continued to be focussed along the boundary,[10] however, with some notable clusters forming; burials Cr55 and Cr56 appear to mark the beginnings of burial foci nearer the road which may have been built at around this time, to the north. Inhumation aligned on the boundary ditch also apparently continued as a tradition into the early Roman period, noticeably towards the east.[11] Further inhumations were added here at least as late as AD 150.[12] Some of these potential burials were clearly but only partially identified in the difficult soil conditions, however, and may equally have represented alternative activities (see below).

Northerly cremation burial groups had clearly developed by c. AD 150 (Fig 4.5).[13] Meanwhile further burials[14] were added to the southern clusters discernible near the boundary ditch which was evidently silting up during this period. Three further early to mid-Romano-British inhumations, again only partially observed,[15] were not focused on the 'west–east' ditch but placed in the northern half of the site (and on a north-south alignment). They may represent a subgroup different from those focussed on the boundary to the south.

According to the typological evidence of assemblages, both northern and southern clusters of cremation burials were added to by c. AD 175, potentially providing an interesting snapshot of the organisation of the cemetery at its height, with a clear suggestion of linear arrangements, and alignments both broadly north-south as well as west-east, and a southward shift of burial in the northern part of the site (Fig 4.6).[16] This is in keeping with the local topography whereby such burials would have been placed at an increasing distance from the road to the north, but also appears to

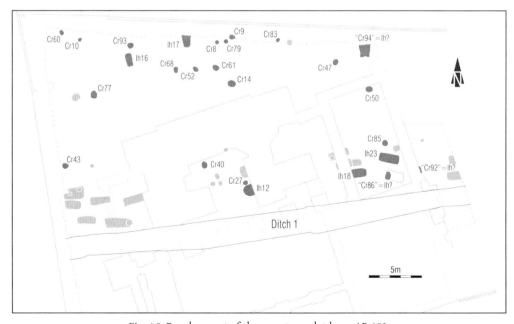

Fig. 4.5. Development of the cemetery plot by c. AD 150.

Fig. 4.6. *Development of the cemetery plot by c. AD 175.*

emphasise a west/east linear zone across the site; the restriction of burial to the areas north and south of a pathway through the cemetery could be suggested, although the limitations to fieldwork and the disturbance caused by the demolished buildings should be borne in mind.

By c. AD 200 the zenith of use of the cemetery would appear to have been reached, with a number of burials added to existing clusters across the site (Fig 4.7).[17] In keeping with increasing levels of burial activity, the silting of the cemetery boundary ditch (down slope) apparently became pronounced enough around the end of the second century to require an extensive re-cutting. The period c. AD 150–200 also marked the early chronological boundary of the distinct area of disturbance/cemetery detritus recorded in the northern part of the site: it is suggested that the material represents a soil build-up resulting from increased cemetery activity at this time, being both the *floruit* of the cemetery for deposition of the dead and of offerings to them. 'Cr11', for example, was in fact a layer covering 1.5 m, with potsherds lying flat within it, and 'Cr12' was actually a number given to a thin buried surface layer at least 5 m² in extent, containing various materials, including at least one upright truncated pot base (see below). The majority of the discrete deposits recognised within this material dated to around AD 200, although some contained material as late as c. AD 250.[18]

A number of cremation burials date to c. AD 250, again with relatively even distribution, contributing to existing linear patterns and clusters (Fig 4.8).[19] Cr28, initially considered a disturbed burial or, and Cr90, again with very little bone, date to the late third century and could perhaps be commemorative placed deposits near

Fig. 4.7. Development of the cemetery plot by c. AD 200.

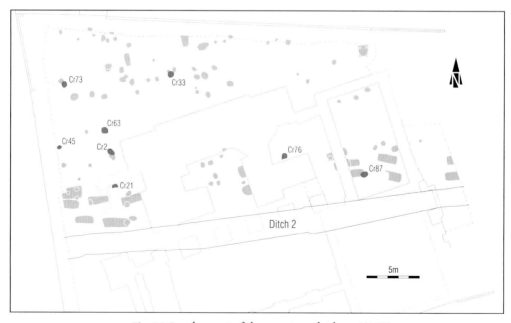

Fig. 4.8. Development of the cemetery plot by c. AD 250.

Fig. 4.9. Development of the cemetery plot by the late third to fourth century.

the cemetery boundary (Fig 4.9). The final convincing cremation burial recorded (Cr65), again relatively near the boundary ditch, was probably in place before the end of the third century.

A cluster of late Roman inhumations focussed in the south-west corner of the site, again on the by now heavily silted ditch. Inhumation Ih2 was dated by a small flagon to c. AD 260–400, while inhumations Ih8 and Ih15 contained pottery of c. AD 270–370 and 270–400 respectively. It is conceivable that remnants of a cremation burial, represented by a broken jar and associated cremated bone (Cr4) found nearby in the upper silts of Ditch 1, were disturbed during the cutting of one of these late graves.[20]

Funerary Analysis

Methodology

Funerary evidence can be seen as falling into three general classes: human remains, associated material, and cut features and layers. In each case we will consider how the evidence types attest to the different aspects of the funerary sequences that produced them at St Dunstan's Terrace. I term these funerary 'phases': selection, preparation, modification, location, deposition and commemoration (Weekes 2014). It will be noted that only certain types of funerary evidence are identifiable within the St Dunstan's Terrace data. More troublesome, but certainly not fatal for the analysis, is that certain evidence is basically equivocal, eloquent of multiple stages of the funerary sequence in different ways. In such cases, rather than considering all the interpretive possibilities in one place, I have returned to the evidence as and when it has a potential bearing on a particular aspect of the funerary sequence, in order to maintain the integrity of considering each 'phase' of the sequence in turn.

Selection

Human remains

The first consideration in funerary archaeology must be exactly who, through time, were the subjects of the funerary treatments represented in the archaeological record. The primary evidence for discerning the type of person who 'goes with' a funeral type is osteological, along with other archaeological science techniques and environmental evidence (including sampling of the stomach area for parasites, for example), which aid in reconstructing who the dead in question were in every sense possible, in both life and, sometimes, in the nature of their deaths. However post-depositional processes, and in particular acidic clay soil conditions, had considerable implications for the study of those selected for inhumation at St Dunstan's Terrace, with just a few fragments of tooth enamel and skull the only survivals.[21] Study even of the minimum number of individuals (MNI), let alone any other osteologically-established characteristics, was deeply compromised.

In treating with the founding tradition of Iron Age inhumation at St Dunstan's Terrace, which continued into the early to mid-Romano-British period, establishing funerary selection is largely restricted to what can be deduced from body 'casts', carefully revealed by excavators, although grave features also provided clues (see below). From the casts, it is certainly possible to say that several burials had clearly contained adults of unknown sex.[22] Of the more likely Romano-British inhumations to the east, only Ih19 produced a body stain, again of an adult.

Cremated bone, as should be expected, had fared better in the corrosive soil, with additional protection having been afforded by the frequent provision of urns (McKinley 2008, 3). It has already been noted that three of the early 'burials' (Cr5; Cr81 and Cr96) contained little or no cremated bone. It would be acceptable these days to label these as 'cenotaphs', burials in all respects save for the lack of human remains, but the fact that they cut the backfills of late Iron Age inhumations might suggest alternative functions. Table 4.1 demonstrates that there would certainly seem to be an increase in the excavated area in the numbers of individuals given cremation funerals culminating in burials in each phase, apparently marking a growth of the cemetery during the second century: up to c. AD 100, 10 burials; c. 100–150, 20 burials; c. 150–200, 24 burials; c. 200–250, eight burials.[23]

Taking the apparently broadly contemporary inhumation burials into account, the area of the cemetery revealed at St Dunstan's Terrace, *may* have received on average about one burial every two years during the second century.[24] If upper crude death rates in the 'developing world' of 20 deaths per 1000 annually are considered in comparison, this would mean that the total number of burials in the second century

Table 4.1. Cremated individuals represented by burials, and age ranges by phase

Before c AD:	Total	Infant	Child	Young adult	Adult	Older adult	Unknown
70	2	0	0	0	2	0	0
100	8	2	0	0	6	0	0
150	21	2	1	2	15	0	1
175	13	2	0	1	6	4	0
200	13	0	1	1	8	3	0
250	9	1	0	0	6	2	0
300	3	0	0	0	2	0	1
Totals	69	7	2	4	45	9	2

Table 4.2. Sexed burials by phase represented by cremation burials

Before c AD:	Total	Female	Possible female	Male	Possible male	Unknown
70	2	1	1	0	0	0
100	8	3	0	0	1	4
150	21	3	3	0	3	12
175	13	1	3	0	1	8
200	13	0	2	1	2	8
250	9	0	1	0	0	8
300	3	0	0	0	1	2
Totals	69	8	10	1	8	42

at St Dunstan's Terrace could have been produced from a relatively small 'pool' among the living of, say, 25 people: an extended family, perhaps? There is a spread of age ranges that could be considered commensurate at each phase with such a group, but the higher proportion of adults is very noticeable. Especially given the fact that remains of three of the infants and one of the children were found mixed with adult remains, it seems clear that cremation funerals culminating in burial at St Dunstan's Terrace were primarily for adults. More women than men were apparently buried in this area in most periods, although the number of only possible identifications (and cases where burials were not sexed) means this observation must be treated with caution (Table 4.2).

Due to the vicissitudes of the process and selective collection, cremated bone from burials typically held very little or no potential for the study of geographical origins of the population, familial groups, diet, disease or congenital factors, occupation, class and lifestyle, injury, or cause of death. Some ante-mortem tooth loss and dental abscesses were noted in a number of individuals throughout the phases[25] and a possible infection of the mandible (Cr21). It is worth noting that Cr48 presented a healed infection, while Cr64 and at least one of the deceased represented by Cr77 present active infections; Jackie McKinley (2008, 6) points out that it was most likely that these derive from infection in the overlying soft tissue, possibly linked to trauma: could this have had implications for what is in fact an untypical burial practice?

Of the three late Roman adult inhumations in the south-west corner of the site, deep burials Ih2 and Ih8 produced body stains of adult corpses, but no further information could be discerned from human remains concerning selection of the dead for a given type of funeral.

Associated material

Apparent coffin nails at the corners of Ih18 indicated a coffin site of just 1 m × 0.35 m, which must have housed either an extremely diminutive adult or a child. Associated materials other than coffins etc., i.e. objects deliberately brought into physical association with the dead during the funerary sequence, particularly grave goods, were in the past taken at face value by archaeologists as speaking plainly of who the dead were in life and as manifesting their afterlife locations and activities. We now consider the special context of the burial and its contents more carefully (after Ucko 1969, for example; Pearce 2000). Rather than ignore a potential connection between an object and the deceased in life, however, it is surely of value to consider potential clues pertaining to funerary selection, which might, albeit through a special funerary lens, tell us something about the people at the heart of the ritual actions. This could include objects suggesting cause of death, or indeed culturally-specific markers of social status, gender, age, occupation etc., or mass produced, curated, broken or defective objects, or potential personal objects.

I must emphasise that the principle here is to consider objects that may have had a resonance in the lived context; the specialised *funereal treatment* of objects and aspects, such as inclusion of pyre material in certain burials, or selection of objects for the grave, are the subject of the analyses that follow. In fact objects associated with the human remains in funereal contexts have the potential to indicate both the deceased in life *and* the funerary representation of them in death. At such an early stage in the analysis it is right to reserve judgements on attendant meanings of objects, but it also does no harm to indicate that some items have perhaps greater potential for manifesting a specific association with the deceased *in their lifetime*, which has implications for selection of the dead for certain types of funeral.

The fact that so many of the cremated dead may have worn copper alloy, iron or bone dress accessories on the pyre, for example, is worthy of note for discussing the cremated and cremators. On the other hand, I would not consider common items like accessory vessels in burials as being evidentially eloquent of a particular life, except perhaps in those cases where there is something distinctive about otherwise typical objects (e.g. graffiti),. I would instead focus in this area on rarer types of accessories at various stages of the funeral (Weekes forthcoming). At St Dunstan's Terrace an inventory of such objects, in chronological sequence, might include: late Iron Age/conquest period beads from possible inhumation burial Ih3/4, a glass beaker and an ornate wooden cover from early Romano-British burial Cr55, two brooches accompanying Cr91 as well as a small 'honey jar', and a miniature 'pinch pot' in Cr30 and another miniature vessel in Cr62, two glass beads and a pipe clay Venus figurine associated with Cr61, a brooch and a glass vessel in Cr40, a casket in Cr50, a mirror in Cr14 (before c. AD 175), a probable gaming board and four counters in Cr76 and a possible wooden box in Cr63; there is also a minor tradition of some cremation burials throughout the life of the cemetery being furnished with hobnailed shoes. That glass vessels and brooches were associated with females while the glass beads and figurine in Cr61 and the mirror in Cr14 accompanied probable male remains is of note, as is the fact that the miniature vessels in this list were associated with infant remains, and, by the same token, the shoes with adults.

Features and layers

The main feature type to contribute to an understanding of who was selected for certain types of funeral is the grave itself, although other features containing human remains, if indeed they can be defined as not being 'graves', have much to offer in terms of comparison (neonates under thresholds, for example?). At St Dunstan's Terrace the inhumation burial cuts actually provided necessary proxy evidence where actual remains were largely unavailable (see below). The best clue to perhaps all the late Iron Age and the majority of the early Romano-British inhumations being those of adults came in the form of the size of grave cuts, although of course this cannot be taken as proof of the size of the occupant. At least one of the more convincing inhumations was noted as being child-sized (Ih16), and Ih18, one of the partially seen burials aligned with Ditch 1, also appears suspiciously small (see below).

While we would tend to recognise 'burials' as placements of human remains within discrete features ('graves'), interment of remains in an existing feature or a developing layer, perhaps meant for another purpose, is another alternative to consider. If we wish to diagnose such practices at St Dunstan's Terrace we are limited to cremated bone survival, in particular deposits of this material not in obvious graves. The prime candidate is an area of mixed cemetery soil in the northern part of the site,[26] although it might just as easily represent alternative deposition of pyre material.[27] Cr6, Cr42 and Cr74 may represent disturbed or redeposited burials, but again their location among the former deposits is noteworthy; the location of Cr4 within Ditch 1 is more interesting, given the date of the vessel (before c. AD 200), its fragmented nature and late Roman stratigraphic position, most likely a redepositing of a burial disturbed by cutting of one of the nearby late inhumation graves. Further small, and likely disturbed, deposits of cremated bone (Cr19 and Cr 22) were also found un-stratified in the same area.

Preparation

Human remains

The limited preservation of unburnt bone at St Dunstan's Terrace, and the selective nature of the cremated human bone deposits meant that there was little evidence to be obtained from the human remains themselves for the preparation phase of the funerals, such as laying out, shrouding or wrapping the dead, hair styling, cosmetic application etc., or evidence for preliminary burial or storage. In some cases it was possible to say, from the location of the skull cast and scant dental remains, that late Iron Age and early Romano-British inhumations lay supine in the grave, and coffins are evidenced (see below), suggesting this may have been the posture for formal laying out, at least for some adults (noting taphonomic caveats regarding the reconstruction of pre-interment positions).

Associated material

Objects associated with inhumed and cremated bone would seem to have considerable potential for reconstruction and comparison of 'laying out' rituals, again as long as certain inferences are acceptable. With inhumations this chiefly relies on considerations of *in situ* dress accessories (etc.) which are likely to have been placed on the body prior to transportation to the burial site, but any objects within the area of a reconstructed coffin, which may well have been closed before leaving the preparatory ritual context, are also admissible (Weekes 2014, 5–6). Laying out display may also be hinted at within the cemetery itself. At St Dunstan's Terrace, for example, 'Cr94'[28] is the best candidate for this. In this case some of the 'associated material' took the form of staining resulting from decayed wood, suggesting a carefully prepared grave with structural features that was probably meant to be appreciated by the living, if perhaps for a short time.

A possible late Iron Age/conquest period inhumation burial (Ih3/4) produced beads that may have adorned the deceased during laying out. Early Romano-British inhumations Ih12 and Ih18, both with evidence for coffins in the form of nails, also produced hobnails from the presumed feet end of the graves, indicating that shoes were apparently put on the deceased. Whether shoes were worn or placed in the south-east corner of likely inhumation 'Cr94' is unknown, although the heels seemed to be touching; this burial also yielded a small glass unguentarium, an object that may be especially associated with laying out (Weekes 2008: 148; Lepetz *et al.* 2011). Another likely inhumation burial thought at first to be a cremation burial ('Cr86', of similar date), may have contained a ceramic flask within a line of coffin nails, again suggesting it may have been placed there during preparatory rites.

Coffins themselves perhaps reveal a certain style of laying out and preparatory ritual; as well as those already mentioned, late Iron Age/Conquest period inhumations (Ih1, Ih10 and Ih14) and early Romano-British burials Ih11, Ih23, and Ih22 all yielded nails whose position suggested coffins: an iron nail fused to a fragment of cremated *femur* shaft in Cr68 (au; by c. AD 150) could also relate to a coffin, or a bier.[29]

At St Dunstan's we have several interesting examples of burnt objects found mixed with cremated bone and therefore probably survivals of pre-pyre preparatory rites: these include bone hairpin fragments.[30] Hobnails, apparently mixed with pyre material, in Cr49 also indicate either wearing or depositing of shoes on the pyre.

Table 4.3. Preparation as evidenced by staining in cremation deposits. Age/sex: af=adult female, afp=adult, possibly female, am=adult male, amp=adult, possibly male, o=older adult, ya=young adult, c=child, i=infant and u=unknown. Staining: 'b/g', i.e. described by McKinley as 'b/g'=blue-green

Cr number	TAQ	Age/sex	Stain	Bones
14	150	amp	b/g	Humerus
85	150	au	Fe	Femur
77	150	afp + amp	Fe	Vault
60	150	yau	b/g	Mandible
40	150	af	b/g	Vault
62	175	ofp + i	b/g	Radius
46	175	omp	b/g	Distal femur
34	175	au	b/g	Humerus
59	175	au	b/g	Vault
49	200	amp	Fe and b/g	Vault and metacarpal
2	250	au	b/g	Tibia and humerus
33	250	au	b/g	Femur

Furthermore, osteological analyses have identified staining on a considerable number of cremation deposits from objects that may have been 'worn' by the deceased on the pyre, and therefore beforehand (Table 4.3).

Given the vicissitudes undergone by the corpse on an open pyre (and post-depositional processes), the location of staining cannot be simply taken as the original location of an object such as a dress accessory while the body was displayed before burning or on the pyre. Nonetheless these data suggest the importance of adornment of the body, at least from the second century, and of manifesting gender and/or other aspects of identity, as well as individual biography.

The occupants of late inhumation burials Ih2 and Ih8 both wore hobnailed shoes, and a small beaker at the western end of Ih15 certainly lay within the coffin in this burial. Ih2 yielded a small flagon placed next to the feet and parts of a small knife from the backfill, both of which could also feasibly have been placed within a possible coffin during laying out. A small iron hook from Ih15 perhaps formed part of now disappeared coffin furniture.

Features and layers

No mortuary structures or other features relating to laying out of the dead were recorded at St Dunstans Terrace. At least in the case of inhumations, the digging of graves most likely represents a preparatory act as part of a recognisable funerary sequence. None of the Iron Age or early Romano-British burials were particularly deep or ornate, and none of the cremation burial cuts were elaborated. However we have already noted the case of 'Cr94', where a preparatory wood lining of a grave cut is a real possibility, along with posts to secure the structure. Late inhumations Ih2 and Ih8 were also especially deep in comparison with other graves in the cemetery, respectively at 1.1 m and 1.5 m (archeologically), denoting considerable effort in pre-burial preparation.

Modification

Human remains

Many (not necessarily mutually exclusive) ritualised impacts on the body's integrity can be found in mortuary rites the world over in archaeological and ethnographic records (e.g. Metcalf and Huntington 1992, ch. 1). At St Dunstan's Terrace, as for many Romano-British funerals, cremated material makes up a considerable proportion of the evidence.

It is important to acknowledge that inhumation and cremation were practised contemporaneously in this cemetery. This is particularly the case in the early Roman period, with some inhumations occurring among cremation burials, showing conservation of a late Iron Age tradition, but also later; the latest cremation burials (Cr65 and ?Cr90) may overlap chronologically with the more typical late inhumation group (Ih2, Ih8 or Ih15). Cremation was however the coming fashion in the early Romano-British funerals represented at St Dunstan's Terrace and McKinley's analysis of the cremated bone from the site (2008) produced key insights into the process.

McKinley describes minor variations in bone oxidation across the entire St Dunstan's Terrace assemblage, reflecting a general 'shortfall' in 'pyre technology' across all phases, probably in the amount of pyre wood used, and apparently not qualified by the age or sex of the deceased. McKinley also suggests practical reasons for the higher incidence of skull and femur, and in some cases humeri, among this less mineralized material, including a location peripheral to the core of the pyre, extra fleshiness, or covering with clothing/pyre goods and resultant oxygen deprivation. Some variation in skeletal elements affected across chronological phases suggests to McKinley the varying practices of different groups of professional cremators. Of course, as a primarily symbolic act, the practical concerns of cremation 'efficiency' are contingent on many non-practical concerns; the osteological analysis appears rather to evince a style of cremation which only really subtly changed over some considerable time. Exactly how localised this practice was must await more comparative evidence, but the evidence from St Dunstan's Terrace could suggest typically relatively small pyres, rather than an emphasis on carefully 'complete' cremation of all bone.

It is also interesting in this light to note that a few of the best protected cremation deposits (urned and intact burials Cr7, Cr23, Cr53 and Cr79) yielded some of the most fragmented bone. One of these (Cr79) represented the remains of an infant, where smaller fragment sizes might be expected, and another (Cr53) was in fact a truncated loose or bagged deposit, here considered a cremation-related deposit but not certainly a burial, and certainly not immune from post-depositional processes. McKinley informs us that there 'is rarely evidence to suggest deliberate fragmentation of the bone occurred prior to burial at any period in which the rite was practiced, but in these three instances at least, there is evidence to suggest a variation in the mortuary rite which resulted in heavier than normal fragmentation of the cremated bone' (2008, 9–10); elsewhere however she reiterates her argument that it is less likely that 'there could have been some deliberate breakage prior to burial' (ibid, 10). I would again suggest, however, that rigorous pyre maintenance (perhaps with a particular anatomical focus) could well be a factor (Weekes 2005a; Jolicoeur 2014), albeit perhaps focussed in the centre of a small pyre (see below).

Cremated bone deposits derived from Romano-British burials are frequently 'token' deposits, not representing the full complement that might be expected from full corpses (McKinley 1993), but the average weight of bone in St Dunstan's deposits was somewhat less than that of contemporary local Kent sites (McKinley 2008, 9). Of special note is the very small amount of bone in first-century Cr30, the remains of a child aged 1–3 years; this burial apparently emphasized the diminutive in other ways (see below). The bone weights collected/deposited also do not appear to have matched container type (McKinley 2008, 8), although intact loose/bagged deposits produced much more than the average for the site. For example, the 1190 g and 880 g respectively recorded in Cr55 and Cr80 exceeded the quantities documented in any of the intact deposits in ceramic containers.

Turning to the collection of skeletal elements, despite a commonly observed deficiency in elements from the axial skeleton in most of the deposits, some interesting variants were observed. One of the earliest burials, Cr26, which also produced a bone weight above average, contained a relatively high count of axial skeleton elements; this suggested an especial effort on completeness in this case to McKinley (2008, 10). Interestingly, other deposits with more axial material tended to be from contexts unlikely to represent 'burials' per se, again hinting at the selective recovery by hand of individual fragments from the pyre inferred by McKinley in most burial deposits (2008, 12). Were the bones of the several dual cremation burials, i.e. deposits containing the remains of more than one individual, also picked out? The bone in adjacent first-century burials Cr30 and Cr31 may, intriguingly, have derived from the same pyre but have been divided between two adjacent burials (see below), while most cases, dating to the second and early third century, suggest mixing, post-cremation, from shared pyres or pyre facilities (the deposit in burial Cr77 was mixed and shared out between two primary containers in the same burial). In Cr58, most of the bone was in a jar, but a small amount was in an accompanying flagon, either as a result of post-depositional processes, or, conceivably, of careful sorting and sharing of the remains of one or two individuals between two vessels.

There may seem to be contradictions in the foregoing evidence between extreme measures meted out in pyre maintenance and a prevalence of only partially cremated long bones and skull fragments in the collected material. Yet we can perhaps discern a more detailed narrative of the pyre here, separating the actual cremation process from post-pyre collection of cremated remains. Vigorous interventions in the cremation process would not necessarily bring about, on a small pyre, the equal oxidation of the limbs and skull, which may have been pushed towards the centre of the pyre later in the process, and therefore have been less fully cremated.[31] Nonetheless, these same long bones and skull fragments would be the most visually 'obvious' and easily selected remains for those collecting the remains for burial: hence a bias to less well cremated limbs and skull fragments?

Associated material

We have already noted several items that were apparently carbonised on pyres along with their 'hosts', namely hairpins with Cr23, Cr62 and Cr63. It could be significant that two of these were associated with possible females, and two with apparently elderly deceased. As is now well recognised, other items found mixed with cremated bone deposits would seem to be associated with 'pyre-side rituals', in particular possible food items. The likely 'accidental' qualities of these inclusions within the deposited remains should be remembered, so that a general picture of such items placed or thrown on the pyre is probably the best we can hope for.

What emerges from the St Dunstan's corpus suggests some homogeneity of pyre offerings in terms of who was being cremated (Table 4.4). It will be noted that, across the chronological spectrum, pig was burnt alongside adult and young adult females and probable males, and infants. The presence of other species (birds, dogs, chickens, mandibles, and immature animals) is more likely to be a chance function of the deposit formation process, but may indicate how 'colourful' the cremations may have been, as do other items reported from environmental analyses. These included possible walnut shell fragments (Cr13), and, in adjacent and contemporary burials Cr57 and Cr49 'at least 14 large, round pulse seeds that were almost certainly peas (cf. *Pisum sativum*) as well as a few other food items such as hazelnut shell fragments' (Curruthers 2014, 2), the former an urned burial of at least partly-sorted cremated bone, the latter a deposit of mixed pyre material with accessory vessels. There is surely the possibility here once again, therefore, of twin burials that derive from the same pyre, the cremation of an adult (possibly female) and a child (see below).[32] In addition, Cr49 yielded 'charred fragments of irregularly vesicular material with traces of plant tissue embedded within them... [which] ...have the appearance of charred bread' (Curruthers 2014, 2).

It is notable that Cr13, C49 and Cr59 were all possible *Brandgrubengräber*,[33] hence the 'accidental' inclusion of a variety of pyre materials in these cases and a glimpse into what may well have been more typical pyre practice. Associated materials can relate to pyre method, and Wendy Curruthers notes that Cr59 'contained an unusual mixture of charred tubers. Onion couch (*Arrhenatherum elatius* var. *bulbosum*), lesser celandine (*Ranunculus ficaria*) and possibly pignut (*Conopodium majus*) could represent

Table 4.4. Cremated animal bone compared with phase and age and sex of human remains (for key to age/sex categories see Table 4.3)

Cr number	TAQ	Age/Sex	Type	Weight (g)
56	100	af	Bird (partridge?)	1
25	100	amp	Pig	25.9
31	100	au + i	Pig	3.8
8	150	af	Pig	5.2
60	150	yau	Pig	12.5
68	150	au	Pig	6.1
52	150	au + c	Unknown mammal	2.3
79	150	i	Pig	2.3
83	150	au	Pig	2.6
43	150	afp	Unknown mammal	4
40	150	af	Pig	2.5
50	150	af	Pig and dog	21.6
34	175	au	Bird?	0.3
13	175	ou	Unknown mammal	4.2
59	175	au	Chicken and unknown mammal	6.5
7	175	au	Pig (mandible with teeth)	2.2
1	175	afp	Unknown mammal	0.4
80	200	am	Pig	7.7
36	200	amp	Pig	21.8
32	200	au	Pig	4
15	200	au	Pig (immature)	8.3
21	250	au	Unknown mammal (immature)	1
33	250	au	Pig	17.7

a deliberate burning of common grassland tubers, the gathering of tubers for kindling or possibly the burning of turf beneath the pyre...'[34]

Wood fuel was also testable using charcoal samples. Dana Challinor's analysis (2014) demonstrates the primary use of 'sizable' oak (*quercus*) for pyre structure and fuel, throughout the use of the cemetery for cremation burial. Other taxa were generally found in smaller quantities and likely mainly used for kindling. Intriguingly, the two interesting cases stand out from this pattern are again adjacent burials Cr49 and Cr57. Cr49 contained an 'unusually diverse charcoal assemblage, with a range of taxa including cherry/blackthorn, willow/poplar, alder and hazel. Oak was also present (including sapwood and small roundwood fragments) but no single taxon was dominant...' (ibid). Cr57, on the other hand, was 'the only burial to be dominated by a large quantity of ash', a fuel that burns well even if unseasoned (ibid). Again, there

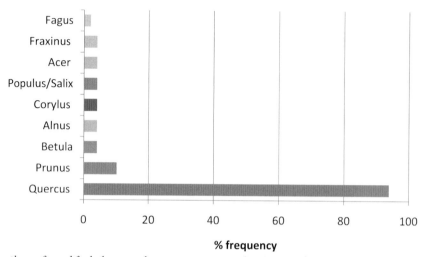

Fig. 4.10. Chart of wood fuel ubiquity, showing % presence of each taxon (49 samples) (data from Challinor 2014) Note that much of the sample was not derived from burial contexts but deposits of pyre material (including disturbed material).

is a strong suggestion of particularity about these two burials. The remains may have been collected from separate but concurrent pyres, but the evidence is surely better explained by different areas of the same pyre being represented in adjacent cremation burials. Could the unique use of ash (in this assemblage) suggest other contingencies, perhaps related to circumstances of the deaths?

The nail fused to a cremated femur fragment in Cr68, if not from a coffin, bier (see below) or pyre offering of some sort, could suggest that at least some fuel wood was re-used. Further nails were recorded from cremation deposits.[35]

Features and layers

No cremation sites or other (convincing) modification areas were recorded at St Dunstan's Terrace, although a focus of pyre material in the north of the area suggests that these may have lain to the north, nearer the road (see below).

Location

Human remains

Location is conceptual, relational and again heavily symbolic, and may be used to denote the movement and positioning of human remains during funerals in relation to domestic space, settlements, modification sites, other burials, local topography, cosmological space and so on. The first thing to consider is the location of a 'place for the dead' in relation to the settlements of the living.

We know that the St Dunstan's Terrace site uncovered part of the south-eastern (and rear?) boundary of a cemetery plot which very likely fronted onto the early

Roman road to London. In the late Iron Age, Ditch 1 may indeed already have been a boundary, a suitable location for inhumation of the dead (see above). The Romano-British cemetery plot, which seems to have been associated with at least another two plots separated by a further ditch at the adjacent Cranmer House site (Frere *et al.* 1982, 56–74), seems therefore to have developed in an already liminal late Iron Age setting, perhaps relating to Stour side settlement to the south-west, that was further marked out as suitable for funerary use following the construction of the road to London. The suburbs of the early Roman town were not defined by a wall or otherwise as far as we know before c. AD 270–90, and were more likely characterised by ribbon development along main roads out of the centre.[36]

Turning to the burials themselves, evidence of binding of the inhumed body could suggest a desire to restrain the dead in a 'correct place' or restrict movement of the corpse during transport, but no evidence of this survived at St Dunstan's Terrace. However, other locational aspects were clear and many have already been referred to, in particular the alignment of a group of late Iron Age and early Romano-British inhumation burials with the ditch and other local topography, and alternative orientations among the inhumations in the northern part of the site ('Cr94', Ih16 and Ih17).

As for the cremation burials, it should first be noted that, while no cremation sites were recorded at St Dunstans Terrace, deposits of pyre material are indeed plentifully present, especially in the northern part of the site, both in non-burial (e.g. Cr74) and burial contexts. This suggests that pyres were constructed nearby, perhaps nearer the road frontage to the north. There would also seem to be a convincing clustering of *Brandschuttgräber*[37] and dual burials[38] in the north-west corner of the site, and we have already noted, the potential for spatial grouping of related burials Cr30 and Cr31 on the basis of human remains. General phasing of clusters and alignments of cremation burials have been noted above during the site narrative. In burials containing sorted remains the primary containers could well have been filled at the pyre side and then transferred to the burial site.

This brings us to the position of human remains within each burial context, the ultimate *location*. Within the late Iron Age/early Romano-British inhumation tradition it is possible to consider here posture and orientation of the interred bodies to some degree. Late Iron Age inhumations Ih1 and Ih6 apparently lay extended and supine, and these and Ih10 were laid to rest with the head at the eastern end of the grave. Ih13, probably Iron Age, is recorded as having the legs towards the east. Of the early Romano-British inhumations aligned along Ditch 1, only Ih19 produced a body stain, being extended and supine with the head again to the west. It has already been noted that north-south alignments were restricted to the northern part of the site at this period, but the bodies were archaeologically invisible. In the late Romano-British inhumations, it is notable that two aligned with Ditch 1 (Ih2 and Ih15), while the third (Ih8) lay on a north-south axis. The corpse in Ih2 had apparently lain supine and extended within a probable coffin, the head to the east, while that in Ih8 lay with the head at the northern end of the grave.

There was no evidence of cremation deposits being placed in particular quadrants of the burial pits, with all but twelve of the ceramic primary containers being centrally placed, and the others in a mixture of locations with no apparent patterning.

Associated material

Equipment and (implied) modes of transport are clues to the necessity for the dead to travel. Late Iron Age/Conquest period (Ih1, Ih10 and Ih14) and Romano-British inhumations Ih11, Ih23, and Ih22 and all three late inhumations (Ih2, Ih8 and Ih15) all presented with nails and fittings suggesting coffins. The nail fused to a cremated femur fragment in Cr68, and those from cremation deposits in Cr50, Cr79, Cr85 (by c. AD 150), Cr49 (by c. AD 200) and Cr73 (by c. AD 250) could also derive from a coffin or bier used for transportation of the deceased to the pyre (see below cf. Cool 2004).

There was no evidence for certain types of objects being recurringly placed in particular quadrants of the cremation burial pits, or of parts of objects being oriented in a particular direction beyond the burial. A certain spatial emphasis is observable, however, in the placement of the two pieces of the pipe clay Venus in Cr61 either side of the jar containing the cremated remains, for example. Also one of the facets of the minor tradition of hobnailed shoes in cremation burials (as at the neighbouring Cranmer House site; Weekes 2005b) was specialised placement, overlapping the ends of the shoes and sometimes placing them either side of the vessel containing the human remains, as in burial Cr47. Some 'accessory vessels' were also found to be inverted or within other vessels; however, for reasons that will become apparent, these are considered below in view of commemorative rites.

Features and layers

A key feature locating the funerary activity at St Dunstan's Terrace is clearly Ditch 1, and while it is quite possible that this ditch formed a late Iron Age boundary, its funerary purpose was clearly also established then, as a focus for inhumations. It is clear that the ditch bounded the Romano-British cemetery throughout its use, with no encroachment, even when it was probably almost completely silted, in the late Roman period: it had been re-cut on at least one previous occasion, at the height of the use of the cemetery. Parallel with both ditch and road, a potential linear zone apparently largely avoided by burial or other features, perhaps a track or path providing access to burials, was again noted within the site narrative (above). While two (Ih2 and Ih15) of the late inhumations aligned on the by then silted ditch, the other (Ih8) is on a north/south alignment.

Note should also be made of the apparently focussed location of potential cenotaphs/placed deposits within the upper backfills of several of the early inhumations,[39] and of the placement of a disturbed second-century cremation vessel (Cr4) in Ditch 1, also near the late burials, one of which may in fact have disturbed it.

Deposition

Human remains

All of the inhumation burials at St Dunstan's Terrace seem to have contained single individuals. We can also generally characterise the cremated remains, which form the primary deposit of cremation burials as well as suggesting traditions of post-pyre collection considered above, as token deposits. This tokenism was apparently a local emphasis (with smaller tokens than nearby sites, see above), and particularly interesting if the apparent predilection for small pyres as inferred from colour variability is accepted. Exceptions to the 'rule' included the first-century loose or bagged deposit Cr55 (1190g) and Cr56 within an amphora, and the second-century urned burial Cr36 (1026g), while Cr77, at 1142 g, was derived from two adults.

The deposition of material deriving from more than one individual is again part of a widespread phenomenon, but interestingly took several forms at St Dunstan's. Several of the dual burials were the more 'typical' combination and ratio of an adult and an infant/child (McKinley 2008, 12), with the majority of the bone relating to the older individual, but it is interesting that in dual burials Cr52 and Cr73 the majority of the bone derived from the immature individual. Bones of two adults were mixed within two jars in burial Cr77 and also deposited between them, somewhat in the form of a *Brandschüttungsgrab*.[40] The choice of combining sorted human remains with unsorted pyre material within the burial, is elsewhere represented at St Dunstans Terrace, with two similarly dated examples containing lenses of pyre material as well as sorted bone within ceramic containers.[41] Three possible burials contained unsorted pyre material.[42] Finally, there are possible dual burials at St Dunstans where the same pyre may have furnished the primary deposits for separate but adjacent burials.[43]

Associated material

It is in the funerary archaeology of *burial* practice that placed deposits have always come to the fore. The fact that some objects placed separately within a burial or cremation pit might also be tied to earlier events in the funerary process, such as incense or other materials associated with laying out, has already been considered. If these objects were contained in a now enclosed container or coffin at the laying out stage, it could be argued that the ritual emphasis of such objects belongs there. Objects placed outside such a container or coffin more definitely form part of burial rite *per se*. In light of this, it could be suggested that the real significance of the latter in fact lies in their combination and collective inclusion in the grave, to the exclusion of other objects, and indeed their spatial arrangement.

A small amount of unburnt pig bone was also recovered from the grave backfill of late Iron Age Ih6. The only other unburnt pig remnant recovered from a burial was a fragment of tooth from Romano-British cremation burial Cr30, which could either be intrusive or another interesting quality of this burial. A small jar and Gallo-Belgic platter placed by the feet of the Iron Age adult buried in Ih6 in fact mark this burial out as unique within this phase of inhumation within the cemetery in containing *bona fide* 'grave goods'. The early to mid-Romano-British vessel in probable inhumation 'Cr86', part of the continuing inhumation tradition aligned on Ditch 1, and objects within the late inhumations, are all likely to have lain within the coffin, and therefore suggest display at the preparation stage, rather than the grave. This would seem to be in stark contrast to the majority of the cremation burials.

In terms of the types of primary containers of cremated remains at St Dunstan's Terrace, there would appear to be no pattern related to age or sex of the deceased. It could be said that container types in the earliest (up to c. AD 100) and latest (after c. AD 200) phases of the cremation burials are slightly more diverse; certainly cremated remains of various ages and sexes were most commonly deposited in a jar form at the height of the cemetery's use.[44] We might wonder what the use of a heavily distorted waster jar in Cr27 or the misfired jar in Cr9, or the fact that the jar in Cr65 had a hole drilled in the base, might have said about attitudes to the respective dead (Weekes 2008, 155).[45] We can also point out that the later second century produced at least some variant burial types in the form of *Brandschüttungsgraber* and possible *Brandgrubengräber*.

Turning to accessories in the cremation burials, comparative analyses in 2005 for both accessory vessels and other accessories from south-east England produced the clear and widely applicable finding that, although general trends in certain types of object being deposited can be adduced, *combinations* of objects clearly demonstrate very considerable diversity between burials (Weekes 2008, 156ff). This, coupled with the rarity or even uniqueness of many objects found in cremation burials, has led me to the increasing analytical and interpretive focus on the individual funeral (Weekes 2014; forthcoming; this volume) that I present and augment here. Enhanced osteological and scientific analyses now available allow for a consideration of how such items related to the subjects of the funerals. These reveal that no general pattern of age or sex of the deceased and numbers or types of accessory vessels or other accessory types can be discerned (see appendix). The number of burials in the sample suffices to show that socially determined age or gender are unlikely to have directly governed this aspect of the funeral. Instead, a diversity of *combinations* persists, suggesting that burials were governed by particularised treatment throughout the use of the cemetery.

Over and above such an observation, certain burials stand out. Some are distinguished by 'secondary' containers like the amphora (Cr56), the certain casket (Cr50), and a possible box containing just a jar with cremated bone (Cr63). Others contain especially large numbers and types of placed deposits compared to others (Cr50, where the casket contained the equal largest number of objects; cf. third-century

Cr87), or rare or unique objects, at least within this assemblage, for example a glass beaker in Cr55, the pinch pot in infant burial Cr30, a pipe clay pseudo-Venus figurine in Cr61, a mirror in Cr14 and a possible gaming board with counters in Cr76.

Other burials in the excavated area were particularised in ways that formed part of wider traditions, such as the first-century amphora burial Cr56 (Philpott 1991; Weekes 2005b; 2008). This also applied to some minor traditions within the cemetery area, discernible through the distinctiveness of the objects concerned. Samian accessory vessels were recorded in single burials from different phases.[46] There was also a recurring incidence of speciality miniature vessels,[47] and, as noted above, a tradition of placing footwear in some burials that appeared to remain roughly constant as a proportion of the total number of burials being deposited.[48] Footwear seems to be associated with the graves of adults, although of both sexes.

The depositional context is also the one where we can directly observe ritual modification of objects. The removal of the head of the pipe clay Venus in Cr61 provides a good example, and was the hole drilled in the base of the jar in burial Cr65 similarly a 'ritual killing' of this vessel to set it beyond use by the living?

Features and layers

The design of inhumation grave cuts demonstrates a preponderance for extended burial postures for the interred throughout the late Iron Age and Romano-British periods. Grave cuts for cremation burials varied, with some apparently more or less clearly carefully designed, such as the pits for amphora burial Cr56, and Cr55 with its wooden cover, or those containing 'caskets' (Cr50) or 'boxes' (Cr63). Most other burial pits were ovoid and either snugly accommodated or sometimes barely contained their contents, which could no doubt suggest rather 'ad hoc' pit digging, perhaps even by parties other than those directly associated with the deceased, implying a limited significance for the size and shape of the pit.

Commemoration

Human remains

An alternative explanation for mixing of human remains within a particular burial might be continued access and use of the context for addition or manipulation of skeletal material. Certain micro-stratigraphic examples where cists have clearly been the focus for more than one deposit can be cited (Duday 2009, 148–149), but the St Dunstan's cases are not so clear cut in this regard. None of the burials with possibly accessible secondary containers or lids (see below) contained evidence of more than one individual, and the dual burials all seem to represent single deposits, or at least deposits relating to a single pyre.

Associated material

Placed objects and materials can also represent closure deposits, completing the cremation or burial phase, or, in fact, *entirely separate* deposits, denoting acts of commemoration focussed on the location of the burial, for example.

So, did the apparent post and plank shoring of probable inhumation burial 'Cr94' allow continued access following deposition? Moreover, lids and covers for cremated remains and other items, often apocryphally considered protective in perpetuity, in fact would have made continued access to and manipulation of remains easier. Various types of lids and possible lids are recorded for a considerable number of cremation burials at St Dunstan's Terrace (truncation through post-depositional processes and/ or machine stripping of the site almost undoubtedly played a part in reducing this number), with diverse objects being used. The wood cover, with a number of copper alloy ring fittings in burial Cr55, for example, may well have functioned as a 'lid', visible at the surface, providing access to the cremated bone within; the associated glass beaker, footwear and samian vessel lay outside the area of bone, and no trace of the replaced wood was found beneath the bone deposit: perhaps this was a modified and re-used lid from a casket? The beaker in this burial seems to have been damaged, perhaps through lifting and replacement of the putative wooden cover after initial deposition: while the beaker lay at the north-east of the grave cut, the broken shard was found beneath the south-western edge of the putative cover, as though moved there. The amphora cist of Cr56, if broken transversally (as many were) so as to allow access, may also have provided continued access to the jar within containing cremated bone.

Actual lids were used in a number of cases at St Dunstan's[49] while various types of vessels were also used including jars,[50] dishes,[51] and an inverted flagon in Cr7. In some cases these vessels must have required modification for the purpose. In Cr45, a modified tile was used. In fact, the mirror in Cr14, if indeed collapsed into the vessel, may also have served this purpose, as could the probable gaming board in the case of third-century Cr76.[52] Finally, we might also wonder if the copper alloy ring from Cr87 might at some point acted as a handle for a lid covering a loose or originally bagged deposit. It is interesting to note how many dual burials are represented here.

We should also note that the dish in 'casket burial' Cr50 would seem to be surplus to requirements as a lid if the casket was indeed closed. Were this and other vessels used for other forms of secondary ritual actions, commemorative in nature? Miniature vessels placed within primary containers might certainly be considered in this light.[53] The early 'cenotaphs' Cr81 and Cr96, in the tops of Iron Age inhumation burials, both contained miniature vessels, perhaps reinforcing the idea that these vessels were especially linked with commemoration. Again, later possible cenotaphs Cr28 and Cr90 could equally represent ongoing, perhaps commemorative, placed deposits.

Features and layers

The fact that so few cremation burials were found to inter-cut on this site would seem to testify to marking of burial locations, which would also provide some opportunity for continued access to particular burial sites.

Finally, the mixed material which was not entirely understood in the northern part of the site may represent a remnant cemetery soil. The fact that this material contained occasional large potsherds lying flat (Cr11, Cr12), including at least one foot ring apparently in situ, along with mixed lenses of pyre material, could suggest remnants of a cemetery surface which at some stage formed the context for commemorative offerings to be placed. The vertical stratigraphy of such a surface was indeed partially recorded at nearby Cranmer House in the 1980s (Weekes 2007). Further investigation will be required in order to analyse this potentially very important (but likely compromised) resource in finer detail, but the evidence is suggestive.

Reconstructed funerary profiles

The foregoing analytical framework demonstrates some of the possibilities of properly reassembling the evidence of mortuary rituals in order to compare entire funerary sequences. At St Dunstan's Terrace, for example, apart from the general pattern of late Iron Age inhumation, early to mid-Romano-British mixed rites and late inhumations, we have various new dimensions to consider. Perhaps what comes across most of all is that the deposition stage, far from producing 'typical' burial assemblages (as has been asserted in the past through quantitative analyses), is actually *the* paramount context for funerary diversity.

The late Iron Age inhumations, part of an increasingly recognise regional tradition (Booth, this volume), were clustered on the same alignment in association with Ditch 1, some with definite coffins, and some without. At least one of these people may have been adorned with beads prior to burial. Dress accessories were also a feature of rites that preceded a developing cremation tradition, probably during laying out and/ or pre-burning display, and there may well have been some ideas of status, gender or other cultural paradigms at work here, through the filter of the idealised ritual context. Some of the early Romano-British inhumations that continued this tradition focused on Ditch 1 also contained objects that apparently derived from preparatory rather than interment rites, including clothing and pottery. In one interesting case ('Cr94') an *unguentarium* was deposited in the burial, which may represent preparations of the corpse at home, while the grave was provided with post and plank shoring, perhaps to embellish it visually as a controlled, tomb-like, structure, and even allow continued access. The main emphasis appears to have been on display prior to taking up 'residence' in the cemetery, however, a pattern to which the late inhumations apparently conform. This is in stark contradiction to the trajectory of the cremation ritual represented at St Dunstan's Terrace.

Certainly, cremation rites appear to have been largely homogeneous, with generally small(?) oak pyres apparently the norm. There are a very few but notable exceptions, particularly Cr49 and Cr57, which may been derived from the same ash-dominated pyre structure. The same animal offerings (pig) were commonly made on these pyres, and other food offerings were detected in unsorted pyre material, such as nuts and possibly bread. There could easily have been more ubiquitous food offerings that were not picked out with the sorted bone deposits comprising the majority of burials. Hand selection of larger and recognisable cremated bone fragments from a cooled pyre seems to have been the norm, but some clearly collected and incorporated unsorted pyre material in burials.

And it is following cremation that diversity appears to have come to the fore, involving improvisation on established themes. The different types of dual burial, the emphasis on sorted bone deposits vs. the *Brandschuttgräber*, variation in cremated bone quantities, the diverse numbers and types of accessory vessels and other accessories, all seem to speak of a re-establishment of 'personhood', albeit within the context of the grave. This is highly suggestive of a different agency at work. Perhaps, following the specialised and perhaps socially regulated process of cremation, the family of the deceased again took possession of and responsibility for the dead. Of the many examples, reconsider, for example, the adjacent burials Cr30 and Cr31, which might be an adult and an infant derived from the same pyre but buried in separate deposits side by side. If this is the case, a few bones (8 g) of the infant at least were carefully selected, and accompanied by a miniature and naively fashioned pot. It is hard to avoid the suggestion that the small quantity of bones and crude little vessel may in some way have been directly related to or symbolic of the infant commemorated, and that these gestures were originated by somebody with a personal connection to the deceased.

There are plenty of burials at St Dunstan's, as elsewhere, marked out by particularised placed deposits, such as the burial with a small glass beaker (Cr55), the amphora burial (Cr56), the burial with a deliberately modified pipe clay figurine and beads (Cr61), another with a speculum mirror (Cr14) and a burial containing a probably gaming board with four glass counters (Cr76). While the latter and other lids and covers contribute to a continued variety in the arrangements for the dead, they also hint at a shared form of commemorative ritual, where the stabilised cremated remains remained accessible within amphora cists, under wood or tile covers, or perhaps the lids of protruding vessels?

The cremated remains especially at St Dunstan's Terrace seem eloquent of a form of ritual which took the diverse dead (albeit from a particular group among the living), *recognised* them in a laying out display, but then more generally processed them in a homogenising and perhaps symbolically 'communal' way. There may well have been a change of agency, from those in the immediate society of the deceased in life to specialists, at the cremation stage, but I would suggest that the initial participants probably took possession again of the remains of the dead following cremation. This is surely more than hinted at by the diversity of placed deposits in the burial practice, at least that which shows in the archaeological record. Continued possibilities for access built into the burials and of possible purely commemorative placed deposits and horizons, variously manifested, probably attest a new ongoing relationship with the deceased as among 'the dead'; the question as to whether such commemorative rites were more or less improvised must await further evidence. This overall pattern of meaning(s) is redolent of a funerary modification phase of liminality and *communitas*, prior to structural reinstatement: something I consider in more detail elsewhere in this volume.

With the latest inhumation burials at St Dunstan's Terrace, there is still a hint of variety in the early stages of the funeral. Vessels and potentially other objects accompanied the clothed deceased in coffins but perhaps in these cases, once the coffin was closed, the dead were already nearly gone, up the road and then up the hill to the old cemetery, by then an ancient place of the dead. Then there were last placed deposits of vessels associated with the ditch;[54] could these have commemorated the dead more generally, or even denote closure?

Notes

1. I continue to develop my method (Weekes 2008; forthcoming a), which builds on the work of Pearce (1997; 1998), Fitzpatrick (2000) and Cool (2004).
2. A full catalogue of the cemetery and its contents is to be made available online (Diack and Weekes forthcoming).
3. Gravel make up for this road, where it crossed the marshy flood plain in the forum area, has recently been identified in excavations at the Beaney Institute, dated to the early conquest period; Alison Hicks, pers. comm.
4. In light of this, some twenty-eight 'burials' (Cr4, Cr5, Cr6, Cr11, Cr12, Cr19, Cr20, Cr22, Cr28, Cr35, Cr37, Cr38, Cr51, Cr53, Cr54, Cr67, Cr70, Cr71, Cr72, Cr74, Cr75, Cr78, Cr81, Cr84, Cr96, C97, C98 and Cr99) are therefore considered here under the broader term of 'funerary feature or deposit', with further interpretation afforded here as and when relevant. Seven of the most destroyed or uncertain contexts (Cr17, Cr29, Cr42, Cr44, Cr82, Cr89, Cr95) were omitted from this study, and three other burial numbers were in fact not originally used (Cr24, Ih5 and Ih7).
5. All the burials on the site are listed with further details in Appendix 1.
6. Ih1, Ih3/4, Ih6, ?Ih9, Ih10, Ih13 and Ih14.
7. Ih20 and Ih24.
8. Cr26 and Cr41.
9. Cr81 and Cr96.
10. Cr 25, Cr30, Cr31, Cr48 and Cr91.
11. Ih11, Ih19, Ih21, Ih22 and Ih25.
12. Ih12, Ih18, Ih23, 'Cr86' and '?Cr92'.
13. Cr8, Cr9, Cr10, Cr14, Cr47, Cr50, Cr52, Cr60, Cr61, Cr68, Cr77, Cr79, Cr83, and Cr93.
14. Cr27, Cr40, Cr43, and Cr85.
15. 'Cr94', Ih16 and Ih17, dated by residual pot to c. AD 70–150.
16. Cr1, Cr3, Cr7, Cr13, Cr16, Cr18, Cr23, Cr34, Cr46, Cr59, Cr62 and Cr64.
17. Cr15, Cr32, Cr36, Cr39, Cr49, Cr57, Cr58, Cr66, Cr69, Cr80 and Cr88.
18. Cr35, Cr38, Cr42, Cr51, Cr54, Cr67, Cr70, Cr71, Cr72, Cr74, Cr75, Cr84 and Cr98, not annotated on figures.
19. Cr2, Cr21, Cr33, Cr45, Cr63, Cr73, Cr76 and Cr87.
20. Other stray deposits of cremated bone and burnt material associated with pottery located in the immediate vicinity (Cr19 and Cr20, not illustrated) could also derive from such disturbance.
21. Ih1 and Ih2.
22. Ih1, Ih6, Ih10, Ih13 and Ih14.

23. Deposits in probable non-burial contexts are excluded from this analysis for obvious reasons (Cr35 [74g], Cr54 [0.3g], Cr67 [22g], Cr70 [8g], Cr71 [32g], Cr72 [6g], Cr75 [45.7g]), as are those from destroyed or redeposited contexts (Cr4 [338g], Cr6 [30g], Cr42 [1g] and Cr74 [164g]).

24. A postulated 20 deaths per 1000 population annually is actually about the upper limit of international crude death rates in the 'developing world' today; see https://data.un.org/Data. aspx?d=PopDiv&f=variableID%3A65.

25. Cr55, Cr56, Cr64, Cr62 and Cr46.

26. This incorporated recognised concentrations of small amounts of cremated bone: Cr35, Cr54, Cr67, Cr70, Cr71, Cr72 and Cr75.

27. This in itself could be significant and should not be ruled out as a primary funerary deposit of some variety; 'pyre debris' might be an extreme form of tokenism in terms of collection and deposition of remains following cremation, see below (cf. Weekes 2008, 153).

28. This was thought during excavation to be a boxed cremation burial: the suggested identification of post and plank shoring within a partially seen north-south inhumation burial is much more in keeping with the evidence.

29. Cf. nails from cremation deposits in Cr50, Cr79, Cr85 (by c. AD 150), Cr49 (by c. AD 200) and Cr73 (by c. AD 250).

30. Cr23, Cr62, Cr49, and Cr63.

31. The skull and brain present particular difficulties for pyre cremators, who generally solve the problem with force (Weekes 2008, 150, note 51).

32. Cf. burials Cr30 and Cr30.

33. Burials incorporating cremation deposits of unsorted pyre material (Pearce 2002).

34. As with other environmental analyses, these findings form part of the assessment report: analysis that could consolidate these results has not yet been funded.

35. Cr50, Cr79, Cr85, Cr49 and Cr73.

36. Even if a 'Roman' styled street grid was continued the north side of the Stour, it need never have been wholly 'filled in'. Recent excavation at Westgate Gardens has indicated some metalled roadside developments that follow on from the known Iron Age occupation of this area (pers. observation), suggesting a similar picture to the early occupation outside the Worthgate at Wincheap (Helm and Weekes 2014).

37. Cr13, Cr34, Cr59 and Cr62.

38. In chronological sequence: Cr52, Cr77, Cr93, Cr62, and Cr73.

39. Cr5, Cr81 and Cr96.

40. Where sorted cremated bone in containers may be accompanied by loose pyre material within the burial.

41. *Brandschüttungsgraber*: Cr34 and Cr62.

42. Possible *Brandgrubengräber* Cr13, Cr49 and Cr59.

43. Cr31 and Cr30, and Cr49 and Cr57.

44. This is undoubtedly part of a wider tradition in south-east England (Weekes 2005b); by the second century a jar form, and less often a bowl, seems to have been accepted as appropriate in this specialised context.

45. Plenty of other meanings are afforded by the same piece concurrently, and this goes for all objects associated with the funerals: see Weekes this volume.

46. Cr55 before c. AD 150, Cr40 before c. AD150, Cr62 before c. AD 175 and Cr33 before c. AD 250.

47. Apart from distinctive examples in Cr91 and Cr30 (before c. AD 100), a small but apparently significant minority of burials contained them across the phases up to c. AD 200 (Cr50, Cr62, Cr16, Cr36 and Cr49).

48. Cr55 and Cr56 (by c. AD 100), Cr9, Cr40, Cr43, Cr47, Cr60 and Cr77 (by c. AD 150), Cr1, Cr18, Cr23 and Cr59 (by c. AD 175), Cr87 (by c. AD 200) and Cr65 (by c. AD 250).

49. Cr60, Cr85, Cr 46, Cr64 and Cr23.

50. Inverted in Cr41, Cr93, Cr57 and Cr73.
51. Cr50 and Cr62.
52. Or even a focus for board games with the dead, especially given the four gaming counters associated.
53. Cr30, Cr91, Cr16, Cr62 and Cr36. The hole drilled in the base jar in Cr65 would have allowed liquids to drain (Weekes 2008, 155).
54. ?Cr28 and Cr90, if these were not simple cenotaphs or disturbed burials.

Acknowledgements

I gratefully acknowledge funding from the Roman Research Trust, as well as the Friends of the Canterbury Archaeological Trust, which has allowed reassessment of this significant cemetery as a case study in *funerary archaeology*. Mick Diack led the excavation and post-excavation project, and he and the other excavators of the site showed great care and dedication in sometimes difficult circumstances. I also thank the team of specialists who worked on the material and provided so many further insights into the rituals discussed here. A fully illustrated catalogue and all specialist reports will appear as an online resource.

Bibliography

Challinor, D. (2014) St Dunstan's Terrace, Canterbury (SDT01). The Wood Charcoal from the cremation deposits. Canterbury Archaeological Trust Archive Report.

Cool, H. E. M. (2004) *The Roman Cemetery at Brougham, Cumbria. Excavations 1966-67*. Britannia Monographs Series No. 21. London: Society for the Promotion of Roman Studies.

Curruthers, W. (2014) St Dunstan's Terrace, Canterbury (SDT01). Assessment of the charred plant remains. Canterbury Archaeological Trust Archive Report.

Diack, M. (2003) St Dunstans Terrace, Canterbury, Stratigraphic Report. Unpublished client report. Canterbury: The Canterbury Archaeological Trust.

Diack, M. and Weekes, J. (forthcoming) *A Romano-British Cemetery at St Dunstan's Terrace, Canterbury*.

Duday, H. (2009) *The Archaeology of the Dead: Lectures in Archaeothanatology*, trans. A. Cipriani and J. Pearce, Oxford, Oxbow.

Fitzpatrick, A. P. (2000) Ritual, sequence, and structure in Late Iron Age mortuary practices in North-West Europe. In J. Pearce, M. Millet and M. Struck (eds) *Burial Society and Context in the Roman World*, 15–29. Oxford: Oxbow Books.

Frere, S. S., Bennett, P., Rady, J. and Stow, S. (1987) *Canterbury Excavations: Intra- and Extra-Mural Sites, 1949-55 and 1980-84*. Maidstone: Kent Archaeological Society.

Helm, R. and Weekes, J. (2014) Excavations at Nos 19 and 45–7 Wincheap, Canterbury. *Archaeologia Cantiana* 144, 235–250.

Jolicoeur, H. (2014) Varanasi Cremation 'Meditation on death' https://www.youtube.com/watch?v=n2ZUzLg_2l0. Uploaded 14 Oct. 2014 [Accessed 2015].

Lepetz, S., Van Andringa, W, Duday, H., Joly, D., Malagoli, C., Matterne, V. and Tuffreau-Libre, M. (2011). *Publius Vesonius Phileros vivos monumentum fecit*: investigations in a sector of the Porta Nocera cemetery in Roman Pompeii. In M. Carroll and J. Rempel (eds) *Living Through the Dead. Burial and Commemoration in the Classical World*, 110–133. Oxford: Oxbow Books.

Metcalf, P. and Huntington, R. (1992) *Celebrations of Death: the Anthropology of Mortuary Ritual*. Cambridge: Cambridge University Press.

McKinley, J. I. (1993) Bone fragment size and weights of bone from modern British cremations and its implications for the interpretation of archaeological cremations. *International Journal of Osteoarchaeology* 3, 283–287.

McKinley, J. I. (2008) 'St. Dunstan's, Canterbury, Kent. Human Bone Publication Report'. Canterbury Archaeological Trust Archive Report.

Pearce, J. (1997) Death and time: the structure of Late Iron Age mortuary ritual. In A. Gwilt and C. Haselgrove (eds) *Reconstructing Iron Age Societies. New approaches to the British Iron Age*, 174–180. Oxbow Monograph 71. Oxford: Oxbow Books.

Pearce, J. (1998) From Death to deposition: the sequence of ritual in cremation burials of the Roman period. In C. Forcey, J. Hawthorne and R. Witcher (eds) *Proceedings of the Seventh Annual Theoretical Roman Archaeology Conference Nottingham 1997*, 99–111. Oxford: Oxbow Books.

Pearce, J. (2000) Burial, society and context in the Roman world. In J. Pearce, M. Millett and M. Struck (eds), *Burial, Society and Context in the Roman World*, 1–12. Oxford: Oxbow Books.

Pearce, J. (2002) Ritual and interpretation in provincial Roman cemeteries. *Britannia* 33, 373–377.

Philpott, R. (1991) *Burial Practices in Roman Britain. A Study of Grave Treatment and Furnishing. A.D. 43–410*. BAR British Series 219. Oxford: Tempus Reparatum.

Rady, J. (2000) An Archaeological Evaluation at 27, St Dunstan's Terrace, Canterbury (Telephone Repeater Station). Unpublished client report. Canterbury: The Canterbury Archaeological Trust.

Ucko, P. J. (1969) Ethnography and the archaeological interpretation of funerary remains. *World Archaeology* 1, 262–280.

Weekes, J. (2005a) Reconstructing syntheses in Romano-British cremation. In J. Bruhn, B. Croxford and D. Grigoropoulos (eds) *TRAC 2004: Proceedings of the Fourteenth Annual Theoretical Roman Archaeology Conference. Durham 2004*, 16–26. Oxford: Oxbow Books.

Weekes, J. (2005b) Styles of Romano-British Cremation and Associated Deposition in South-East England. Unpublished PhD Thesis, University of Kent.

Weekes, J. (2007) A specific problem? The detection, protection and exploration of Romano-British cremation cemeteries through competitive tendering. In B. Croxford, N. White and R. Roth (eds) *Proceedings of the Sixteenth Annual Theoretical Roman Archaeology Conference. Cambridge 2006*, 183–191. Oxford: Oxbow Books.

Weekes, J. (2008) A classification of Romano-British cremation related features. *Britannia* 39, 145–160.

Weekes, J. (2014) Cemeteries and funerary practice. In M. Millett, L. Revell, and A. Moore (eds), *The Oxford Handbook of Roman Britain*. Oxford: Oxford University Press. I:10.1093/oxfordhb/9780199697713.013.025.

Weekes, J. (forthcoming) Evidence for personalisation of cult in early Romano-British cremation burials in south-east England. In R. Haeussler, A. King, G. Schörner and F. Simón (eds) *Religion in the Roman Empire: The Dynamics of Individualisation*. Oxford: Oxbow Books.

Whiting, W. (1927) A Roman cemetery at St Dunstan's, Canterbury. *Archaeologia Cantiana* 39, 46–54.

Appendix 1. Mixed rite burials at St Dunstan's Terrace

Ih/Cr number	TAQ	Age/sex[1]	Primary container	Secondary container[2]	Modification remains	Accessory vessels (n)[3]	Types of AVs[4]	Other Accessories (n)	Types of OAs[5]	Other
Ih1	50	Au	Coffin stain?	—	—	0	0	0	0	A partial body stain: skull cast with fragments of tooth enamel at the eastern end of the grave
Ih3/4	70	U	—	—	—	0	0	1	D	Originally interpreted as two burials
Ih6	60	Au	—	—	—	2	DJ	0	0	Small amount of the base of the skull had survived at the eastern end of the grave
										Pig bone: fragments of skull and tooth
Ih10	70?	Au	Coffin nails	—	—	0	0	0	0	Legs to the west, upper half of the body removed
Ih13	LIA?	Au	—	—	—	0	0	0	0	Legs had been at the eastern end
Ih14	70	Au	Coffin nails	—	—	0	0	0	0	Soil stain and some tooth fragments
Ih20	LIA?	Au	—	—	—	0	0	0	0	—
Ih24	LIA?	Au	—	—	—	0	0	0	0	—
Cr26	70	Afp	Jar	—	448g cb	0	0	0	0	—
Cr41	70	Af	Jar	—	698g cb[6]	2	BJ	0	0	Inverted jar as lid
'Cr5'	70	Cen?	—	—	<1g cb	1	C	0	0	—
'Cr96'	100	Cen	—	—	No bone	1	C	0	0	—

(Continued on next page)

Appendix 1. Mixed rite burials at St Dunstan's Terrace (Continued)

Ih/Cr number	TAQ	Age/sex[1]	Primary container	Secondary container[2]	Modification remains	Accessory vessels (n)[3]	Types of AVs[4]	Other		Other
								Accessories (n)	Types of OAs[5]	
'Cr81'	100	Cen	—	—	No bone	1	S	0	0	—
Ih11	100?	Au?	Coffin nails?	—	—	0	0	0	0	—
Ih19	100?	Au	—	—	—	0	0	0	0	Clear body stain represented bones from the skull to the proximal ends of the femurs, the legs extending beyond the section. A few, very small fragments of bone, probably tooth enamel were retrieved
Ih21	100	Au?	—	—	—	1	F	0	0	—
Ih22	100?	Au?	Coffin nails and stain	—	—	0	0	0	0	—
Ih25	100?	Au?	—	—	—	0	0	0	0	—
Cr25	100	Amp	Beaker	—	775g cb; 25.9g pig; charcoal	2?	CS?	0	0	—
Cr30	100	I	Bowl	—	8g cb	1	S	0	0	Miniature vessel in primary container; fragment of unburnt pig tooth
Cr31	100	Au+I	Bowl	—	269g cb; 3.8g pig; charcoal	1	C	0	0	—
Cr48	100	Au	Beaker	—	324g cb	0	0	0	0	—
Cr55	100	Af	Loose/bag?	—	1190g cb; charcoal	1	D	2	FG	Wood cover
Cr56	100	Af	Jar	A	844g cb; 0.9g bird (partridge?); charcoal	0	0	1	F	—

						1	S	2	D	Miniature vessel in primary container
Cr91	100	Af	Jar	—	158g cb	0	0	1	F	—
Ih12	150	Au?	Coffin nails	—	—	0	0	1	0	—
Ih16	150	C?	—	—	—	0	0	0	0	—
Ih17	150	U	—	—	—	0	0	0	0	
Ih18	150	C?	Coffin nails	—	—	0	0	1	F	—
'Cr94'	150	Au?	—	Post and plank shoring?	—	0	0	2	FS	—
Cr8	150	Af	Jar	—	258g cb; 5.2g pig	0	0	0	0	—
Cr9	150	Afp	Jar	—	84g cb; charred grain; charcoal	1	C	1	F	—
Cr10	150	U	Jar	—	2g cb	1	C	0	0	—
Cr14	150	Amp	Jar	—	606g cb; b/g;	0	0	1	M	Mirror as lid?
Cr27	150	Au	Jar	—	30g cb	0	0	0	0	—
Cr40	150	Af	Bowl	—	420g cb; b/g;	2	FD	3	FGD	—
Cr43	150	Afp	Jar	—	732g cb; ?Cu alloy frags; 4g mammal	1	F	1	F	—
Cr47	150	Au	Jar	—	332g cb	1	F	1	F	—
Cr50	150	Af	Jar	Casket	262g cb; 21.6g pig and dog; Fe nail	4	FCDS	0	0	Dish as lid
Cr52	150	Au+C	Jar	—	378g cb; 2.3g mammal	1	C	0	0	—

(Continued on next page)

Appendix 1. Mixed rite burials at St Dunstan's Terrace (Continued)

Ih/Cr number	TAQ	Age/ sex[1]	Primary container	Secondary container[2]	Modification remains	Accessory vessels (n)[3]	Types of AVs[4]	Other Accessories (n)	Types of OAs[5]	Other
Cr60	150	Yau	Jar	—	234g cb; b/g;[7] 12.5g pig; charcoal	3	FC	1	F	Lid
Cr61	150	Yamp	Jar	—	244g cb; charcoal	1	F	2	DS	—
Cr68	150	Au	Loose	—	578g cb; nail fused to femur; 6.1g Pig	2	DJ	0	0	—
Cr77	150	Afp + Amp	Two jars	—	1142g cb; Fe;	2	FB	1	F	—
Cr79	150	I	Bowl	—	102g cb; 2.3g pig; Fe nail	1	C	0	0	—
Cr83	150	Au	Bowl	—	394g cb; 2.6g pig	1	C	0	0	—
Cr85	150	Au	Jar	—	370g cb; Fe; Fe nails; charcoal	0	0	0	0	Lid
Cr93	150	Au+I	Jar	—	428g cb	1	J	0	0	Inverted jar as lid
Ih23	150+	Au	Coffin nails?	—	—	0	0	0	0	—
'Cr86'	150+	Au?	Coffin nails	—	—	1	F	0	0	—
'Cr92'	150+	Au?	—	—	—	0	0	0	F	—
Cr1	175	Afp	Jar	—	436g cb; Fe frags; 0.4g mammal	1	F	1	F	—
Cr3	175	I	Jar	—	6g cb; charcoal	1	C	0	0	—

Cr7	175	Au	Jar	—	206g cb; 2.2g pig (mandible); charcoal	3	FC	0	0	—
Cr13	175	Ou	Loose/B	—	434g cb; 4.2g mammal; possible walnut shell; charcoal	0	0	0	0	*Brandgrubengrab?*
Cr16	175	Au	Jar	—	236g cb; charcoal	2	S?	0	0	Miniature vessel in primary container
Cr18	175	Au	Jar	—	52g cb; charcoal	1	C	1	F	—
Cr23	175	Yafp	Jar	—	220g cb	2	FC	1	F	Lid
Cr34	175	Au	Jar	—	542g cb; b/g; 0.3g bird?; charcoal	0	0	0	0	*Brandschüttungsgrab*
Cr46	175	Omp	Jar	—	472g cb; b/g;	0	0	0	0	Lid
Cr59	175	Au	Loose/B	—	388g cb; b/g; 6.5g chicken and mammal; unusual mixture of charred tubers	0	0	1	F	*Brandgrubengrab?*
Cr62	175	Ofp+I	Jar/B	—	630g cb; b/g; charred grain; charcoal	4	FCDS	0	0	*Brandschüttungsgrab*; dish as lid; miniature vessel in primary container
Cr64	175	Of	Jar	—	606g cb; charcoal	0	0	0	0	Lid
Cr15	200	Au	Jar	—	246g cb; 8.3g pig (immature)	0	0	0	0	—

(Continued on next page)

Appendix 1. Mixed rite burials at St Dunstan's Terrace (Continued)

Ih/Cr number	TAQ	Age/ sex[1]	Primary container	Secondary container[2]	Modification remains	Accessory vessels (n)[3]	Other Types of AVs[4]	Other Accessories (n)	Other Types of OAs[5]	Other Other
Cr32	200	Au	Jar	—	120g cb; 4.1g pig	0	0	0	0	—
Cr36	200	Amp	Jar	—	1026g cb; 21.8g pig	2	BS	0	0	Miniature vessel in primary container
Cr39	200	Ou	Jar	—	352g cb; hazelnut shell	0	0	0	0	—
Cr49	200	Afp	Loose/B	—	446g cb; b/g; Fe; bone pin frag; footwear; large, round pulse seeds (cf. Pisum sativum); possible bread fragments; Fe nails	1	S?	0	0	*Brandgrubengrab?*
Cr57	200	C	Jar	—	262g cb; large, round pulse seeds (cf. Pisum sativum); charred seeds; charcoal	3	FCJ	0	0	Inverted jar as lid
Cr58	200	Au+Au	Jar and Flagon	—	578g cb; charred grain; charcoal	1	D	0	0	—
Cr66	200	Yau	Jar	—	66g cb	0	0	0	0	—

Cr69	200	Amp	Jar	—	366g cb	0	0	0	0	—
Cr80	200	Am	Loose	—	888g cb; 7.7g pig	0	0	0	0	—
Cr88	200	Ofp	Jar	—	232g cb	0	0	0	0	—
Cr4	200	Ou	Jar	Redeposited	338g cb; charred grain; charcoal	0	0	0	0	—
Cr2	250	Au	Flask	—	184g cb; b/g	1	D	0	0	—
Cr21	250	Au	Loose	—	200g cb; 1g mammal (immature)	0	0	0	0	—
Cr33	250	Au	Jar	—	264g cb; b/g; 17.7g pig	2	FD	0	0	—
Cr45	250	Afp	Jar	—	490g cb; charcoal	1	C	0	0	Modified tile as lid
Cr63	250	Ou	Jar	Box?	442g cb; bone pin frags	0	0	0	0	—
Cr73	250	Ou+I	Jar	—	536g cb; fe nail; charcoal	2	FJ	0	0	Jar as lid
Cr76	250	Au	Jar	—	315g cb; charcoal	1	F	1	S	Gaming board as lid?
Cr87	250	Au	Loose	—	358g cb	0	0	4	FGS	Wood cover?
Cr28	275	Au/Cen?	Beaker	—	<0.1g cb; charred grain; charcoal	2	FD	0	0	—
Cr65	275	Amp	Jar	—	288g cb	1	J	1	F	Hole drilled through base of jar
Cr90	400	Au/Cen?	Bowl	—	3g cb	0	0	0	0	—

(Continued on next page)

Appendix 1. Mixed rite burials at St Dunstan's Terrace (Continued)

Ih/Cr number	TAQ	Age/ sex[1]	Primary container	Secondary container[2]	Modification remains	Accessory vessels (n)[3]	Other			
							Types of AVs[4]	Accessories (n)	Types of OAs[5]	Other
Ih2	400	Au	—	Coffin stain	—	1	F	2	FS	Skull cast (with some tooth enamel) at the east end of the grave, human remains in the form of soil staining and tooth crowns
Ih8	400	Au	—	Coffin nails and fittings	—	0	0	1	F	The head at the northern end
Ih15	400	Au	—	Large coffin nails	—	1	C	1	S	—

[1]Af=adult female, Afp=adult, possibly female, Am=adult male, Amp=adult, possibly male, O=older adult, Ya=young adult, C=child, Ci=infant, Cen=cenotaph and U=unknown.

[2]A=amphora and W=wood (casket or box).

[3]Vessels of glass or other materials are considered 'other' than typical ceramic accessory vessels.

[4]F=flagon or pouring form, C=cup, beaker or drinking form, D=dish form, B=bowl, J=jar and S=speciality.

[5]F=footwear, G=glass vessel, D=dress accessory, M=mirror, S=other.

[6]cb=total bone weight in cremated bone deposit.

[7]b/g=blue/green staining, Fe=rust staining.

Chapter 5

Buried Batavians: Mortuary rituals of a rural frontier community

Joris Aarts and Stijn Heeren

Introduction

Some elements of mortuary practices seem to suggest that the performed rituals are meant specifically for the deceased. The grave goods are, for instance, supposed to serve them during the journey to the afterlife. However, rather than for the deceased, mortuary rituals are ultimately meant for the living. The death of a person shocks a local community, disrupting the balance of daily life. The social role and tasks of the deceased are to be redistributed among the community, as well as his or her personal belongings. Mortuary rituals usually serve two goals: one is of course to cope with the feelings of loss and bereavement of the people staying behind, the other is to re-shape the social order. This is done symbolically in two ways. Firstly, the dead person is transformed into a new social persona, living on in the afterlife and at the same time in the memory of the living. In many societies, the deceased moves to the anonymous world of ancestors, and takes on new responsibilities for the care of the community of the living. Secondly, the recreation of the social order is attained by re-stating religious and cosmological beliefs. This is why mortuary practices usually comprise a complex of ritual acts around the death of one (or several) members of a community and are much more than just the disposal of human remains. This paper aims to study the set of mortuary rituals in a rural Batavian setting in the Roman period, by analyzing its archaeological remains in both the cemetery and the settlement.

Batavians in the Dutch river area

During large-scale excavations at Tiel-Passewaaij (1995–2003) a rural settlement and a large cemetery came to light, dating to the Roman period (Fig. 5.1). The community that lived and died there in the first three centuries AD consisted probably of Batavians, as the village lay in the heart of the tribal area of this group, which was

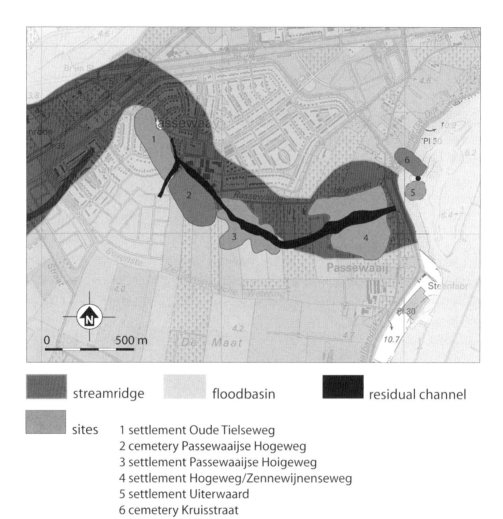

streamridge   floodbasin   residual channel

sites   1 settlement Oude Tielseweg
         2 cemetery Passewaaijse Hogeweg
         3 settlement Passewaaijse Hoigeweg
         4 settlement Hogeweg/Zennewijnenseweg
         5 settlement Uiterwaard
         6 cemetery Kruisstraat

Fig. 5.1. The microregion of Tiel-Passewaaij with four settlements and two cemeteries (1, 2 and 3 are excavated; 4, 5 and 6 known from surface finds).

formally organized into the *civitas Batavorum* with Nijmegen as its capital later in the first century AD. The settlement at Tiel-Passewaaij existed already from the early Iron Age. The nature of the settlement in its earliest phase is largely unknown to us, since it has been excavated only fragmentarily. From the period between 125 until 50 BC we know only of some dispersed graves. Houses begin to appear around 50 BC. Almost a century later, the people of Tiel-Passewaaij began to bury their dead in a central cemetery, which was excavated nearly completely. Eventually, this burial ground consisted of almost 400 graves, dating from the mid-first until the mid-third century AD. The village itself consisted of five to seven farms which were occupied at the same time. In the last few decades of the third century AD, the population of the settlement changed dramatically, which is attested by changes in the material culture. It is thought that new groups of Franks settled here, who had little or nothing in common with their Batavian predecessors. We do not know where these Frankish people buried their dead. This group continued to live there until around AD 450, after which the settlement was abandoned.

The *civitas Batavorum*, which was the administrative district in which the village lay under Roman rule, comprised the eastern Dutch river area and part of the sandy soil area which lie to the south of it (Fig. 5.2; where exactly its southern border lay is still a matter for discussion). Thus, it was part of the frontier zone of the Roman empire. In the time of the emperor Augustus, the northern half of the present-day Netherlands were claimed by the Romans but were never occupied effectively. As a result of a string of military disasters, the river Rhine became a linear barrier under Tiberius. From about AD 40, the frontier was transformed into a line of defences (the *limes*), consisting of camps, watchtowers and a road (Van Es 1981; Willems 1981; 1984). The large number of soldiers which was stationed in the camps needed supplies, which were for a small part provided by the immediate hinterland of the *limes*. This was one of the driving forces behind the gradual introduction of market exchange into small agrarian communities that had known a self-sufficient economy with only limited trade until that time (Groot *et al.* 2009; Vossen and Groot 2009).

The Batavians are relatively well known from literary and epigraphic sources (Tacitus, *Histories*, in particular books IV and V; for epigraphic sources see Derks 2009). Tacitus describes the Batavians as a *gens foederata*, i.e. as having a special treaty with the Roman authorities. In short, Tacitus tells us that the Batavians were exempt from paying taxes but provided the Roman army with soldiers. He mentions the fame of Batavian auxiliary units, who were known for their bravery and strength. The Batavians provided no less than ten auxiliary units (nine cohorts and an *ala*), as well as soldiers for the imperial bodyguard in Rome, more than any other ethnic group in the frontiers of the empire (Roymans 1996; 2004). Recent calculations estimate that each Batavian family had on average one or two able-bodied young men serving in the Roman army (Vos 2009, 221 and 352). This was a heavy burden to carry, and we might wonder if the Roman authorities did not consider the troops as a form of taxation. However, it is important to note here that the Batavian rural population became quickly imbued with Roman provincial military culture and this undoubtedly led to substantial social change.

The Batavian area is also well-known from archaeological finds and excavations. Some of the military camps along the Rhine and the central foci of Nijmegen (the Late Iron Age and early 1st century AD central place known as *Oppidum Batavorum* and the newly built city of *Ulpia Noviomagus* in the Middle Roman period) have been investigated, but the area is really unique in terms of its large numbers of surveyed and excavated rural settlements (Willems 1981; 1984; Roymans 2004; Vos 2009). In contrast to the Rhineland and Picardy, stone-built villas are very rare in the Batavian area. Small hamlets consisting of a few farms, with wattle-and-daub walls and

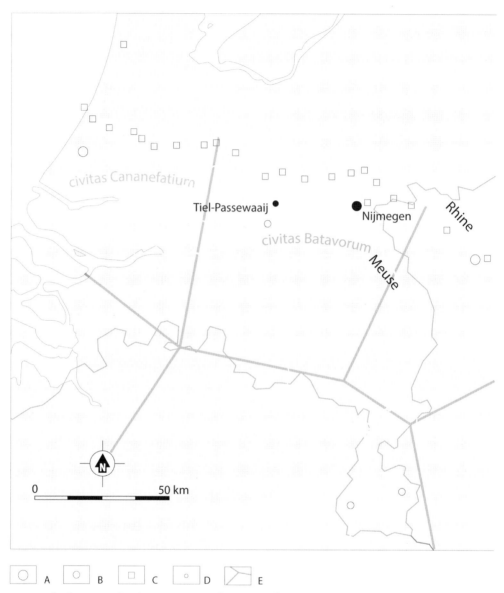

Fig. 5.2. The location of Tiel-Passewaaij in the Roman frontier zone. A Civitas capital. B Small town. C Roman fort. D Rural settlement (only Tiel is shown). E hypothetical civitas border (Thiessen polygons).

thatched roofs, are by far the most numerous settlement type (Roymans 1996; 2004). Traditionally, archaeologists associated Roman-style villas with surplus production and the small rural villages with self-sufficiency. However new evidence suggests that these small villages in the immediate hinterland of the *limes* were an important source of supply for the military forts (Groot *et al.* 2009; Heeren 2009). Many supplies were of course still transported over large distances to the *castella* along the Rhine, but some staple foods (barley, cattle) as well as horses for the cavalry were provided by the rural communities in the immediate hinterland of the forts (Vossen and Groot 2009). Because of this twofold military connection – the supply of agrarian products to military forts on the one hand and the service of many Batavian recruits in Rome's armies on the other – material culture and new ideas, customs and beliefs must have spread over the rural communities of the Batavian *civitas* very quickly (Heeren 2009; Aarts 2003; Aarts 2005), and no doubt led to change in their social and cosmological structures. In this article one of the central questions will be if we can perceive these changes in the ritual sphere, by looking closely at the development of the burial practices of the Batavian community at Tiel.

In the following sections the finds and features of the cemetery are discussed. When the nature of the local mortuary practices is discussed, we should bear in mind that close ties between the Roman military and the local communities existed – do these also appear in the burial rites?

Mortuary practices as a rite of passage

The set of rituals around the death of a member of any community can be described as a *rite of passage*. This is usually performed at moments which are regarded by members of a society as crucial experiences in their life cycle, like birth, coming of age, marriage, and death (Hertz 1960: English translation of the 1906 original; Van Gennep 1909). Their intention is to confer a new social identity on a person. All rites of passage make use of more or less the same symbolic structure: the death of the former social person, followed by rebirth into a new social identity. Hertz points out that this is given form by the twofold nature of mortuary rituals. In the first phase the deceased is detached from the world of the living and the community tries to cope with the loss, in the second phase s/he is transferred to the society of the dead and takes on a new role; in the community, a new equilibrium is reached. The situation between the detachment of the deceased and his/her 'rebirth' is often perceived as dangerous, because the deceased may become hostile towards the community and the mourners are detached from everyday life. Both the deceased and the mourners are outside normal time and space (see for instance Harris description of 'Laymi' mortuary rituals; Harris 1982, 53–54). The final phase of the rite of passage is constituted by the passing of the deceased to the afterlife: in many societies, they become part of the community of ancestors. Rituals are performed by the members of the living community throughout the process and these are particularly important at the liminal stages when the deceased individual moves from one phase to the next, or, in other words, crosses a threshold. In most societies, the first passage is marked by the denial or destruction of the individuality that the deceased possessed during his/her life, for instance by exposing the body to fire (cremation) or to air and weather, leading to decay and disintegration (excarnation) (Fig. 5.3). The second passage into the world of ancestors is often symbolised by the burial of the (cremated) remains in

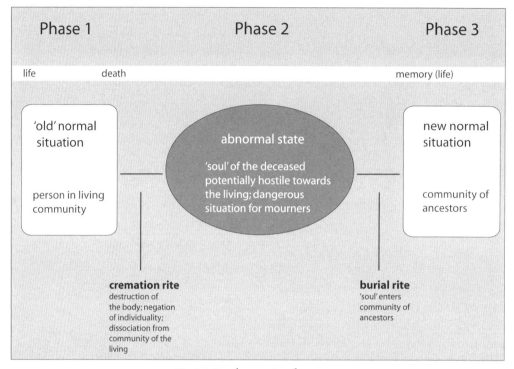

Fig. 5.3. Death as a rite of passage.

a final resting place (Hertz 1960; Bloch and Parry 1982). Sometimes the body is buried and left to decompose, to be dug up again and re-buried when the bones are clean (see for instance Hertz' study of the death rituals of the Dajak (1960), or Danforth's description of re-burial in Greek-Orthodox communities (1982).

In a more or less explicit way, the deceased makes a journey from one world to another: it is no surprise that the metaphor of travelling is a recurrent theme in the mortuary ritual of many societies. This journey of the deceased is often mirrored in the rituals performed in the course of the funeral, in which the body of the dead is moved from place to place before it reaches its final destination. When a new place is reached, a virtual or real threshold is crossed. Connected to the journey metaphor is the theme of crossing boundaries, also frequently encountered in burial rites in many societies. Frequently, when a boundary or threshold is crossed, and a new ritual place is being entered, the number of initiates present at the new location decreases. The burial of princess Juliana of Orange in the Netherlands (2004) provides us with a clear example. First, the deceased was placed on a bier in a chapel, accessible to all members of the Dutch community who wished to take leave of their former ruler. From there, a procession took place which moved the body to the place where the transformation to the community of ancestors would take place, the Oude Kerk (Old Church) at Delft where all members of the house of Orange lie embalmed in a crypt.

The taking leave and procession are public phases of the ritual, in which every member of the community may take part, but when the doorstep of the Old Church is crossed, only selected people are allowed to be present. A further decrease in the number of initiates occurs when the body is placed in the crypt, accessible to the family members only. It is remarkable that although it concerns a Christian ritual in which the deceased supposedly is united with God (or, to use a Christian metaphor: goes to heaven), the deceased Orange is symbolically included in the world of the ancestors. The transformation of the status of the deceased is underlined by a change in title. From her abdication in 1980 (another rite of passage) onwards, Juliana was officially titled princess instead of queen. After her death and placement in the family crypt (2004), she received the title of queen once again.

Another theme and metaphor which is almost inherent in the concepts of dying and death is that of rebirth and fertility. Every death poses a threat to the idea of perpetual existence of the community and its values. Every community needs to believe that its existence will not end with the death of one of its members. As Bloch and Parry (1982, 5) state: 'The rebirth which occurs at death is not only a denial of individual extinction but also a reassertion of society and a renewal of life and creative power...'. Depending on the degree of importance ascribed to the individual in a society, the notion of rebirth may also respond to the psychological need of the individual to be assured that his/her existence will not end at death (for the effect of individualism on the *forma* of burial rites, see also Chapman 1994, 44–46). According to Bloch and Parry, in many societies life is considered a 'limited good' and death is needed to create new life. As such, death is often strongly associated with fertility in its widest sense: it may refer to the fertility of agricultural land and the resources of the community, or to the fecundity of people. For instance, in their 'All Saints festival' the Bolivian Laymi do not just remember the dead, but also make sure that the ancestors watch over the harvest of the following months (Harris 1982, 67–68). But 'in most cases, what would seem to be revitalized in funerary practices is that resource which is culturally conceived to be the most essential to the reproduction of the social order' (Bloch and Parry 1982, 7). Intimately connected with this is the notion of time. If the social order is meant to be eternal, individuality and unrepeatable (i.e. linear) time are to be negated: this negation of the individual is represented by the destruction of the body in the first phase of the ritual. The second phase restores the lost energy or life force to the collectivity of the ancestors and the rite of passage as a whole refers to the cyclical nature of time, thereby negating the unpredictability of death.

With the above we have tried to show that:

1. We are dealing with a complex set of ritual actions and cosmological concepts which are hard to understand for outsiders, especially if they are far removed from the object of study in time.
2. It is clear that the archaeological record may leave huge gaps when we aim to reconstruct the complete mortuary rites of a community.

3. By using a model based on anthropological evidence, we can overcome some of these gaps, or at least try to place the snippets of evidence we have in their correct context: the rite of passage as a whole.

What we must do next is examine what the archaeological record has to offer to understand Batavian mortuary ritual. How do the archaeological features of the cemetery of the Roman period relate to the mortuary rituals that were performed? Is it possible to link them to the anthropological model described above?

Archaeological remains of mortuary rituals: the case of Tiel-Passewaaij

Archaeologists have historically tended to focus on a single and archaeologically the most visible part of the mortuary ritual, its second phase, namely the final burial of the remains of the body. Often a classification of graves is made based on their visual properties and grave goods, while the relevance of these properties for the ritual as a whole is not a point of discussion. The analysis of the mortuary ritual is mostly left at that, although in some cases comments are made about the nature of grave goods in relation to the gender or age of the buried person, if such information is available. The lack of a theoretical framework, however, frequently hampers the archaeologist in contextualizing the different material remains in terms of the complete ritual (although see Pearce 1997; Weekes 2008). Also, the number of (almost) completely excavated and published cemeteries remains very scarce indeed. However, there are a few exceptions in which new approaches have been taken in the last decade. In his publication of two Roman-period cemeteries in the south of the Netherlands, Hiddink proves that an anthropological-historical approach to the archaeological remains offers useful insights into the mortuary rituals of local communities (Hiddink 2003). His analysis of the relation between the visual properties of graves and their location within the cemetery, as well as the location of the cemetery within the landscape, reveals interesting patterns concerning ancestor worship, group identity and territorial claims. The excellent handbook by Parker-Pearson (2003) on the archaeology of death also features many possible methods and insights for tackling the subject, including anthropological theory.

   We will try to start not from the archaeological remains, but from the model provided by anthropological research, hoping to create a new interpretative framework for the burial practices of rural Batavian communities. As we remarked earlier, the most recognizable part of the ritual is the burial of the cremated remains of the body, which in our model must be located in the second phase of the rite of passage. Concerning the rituals before (placing on a bier, ritual procession, cremation) or after the burial, far fewer clues are found. This can of course be explained by the fact that the burials are meant to be recognized and remembered: the monuments consisting of a small mound and surrounding ditch were visible elements of the landscape and respected for a considerable time afterwards. Most other actions of

	Phase 1	Phase 2	Phase 3	Phase 4
	pre-cremation rites	cremation	burial	post-cremation rites
act	excarnation? processions? meal/feast?	burning of the body	meal/feast? erection of monument	visit (grave robbery) (destruction of the grave)
archeological form	excarnation platforms?	pyre location burned gravegoods pyre debris pits	grave pit cremated remains complete grave goods mound demarcation (ditches)	complete grave goods

Fig. 5.4. The various acts around a burial and their archaeological visibility.

the mortuary ritual were passing stages of the ritual and not intended to leave traces. We shall see that in some cases the cemetery of Tiel-Passewaaij offers insights in the more ephemeral aspects of the ritual. The different phases described below are schematically summarized in Figure 5.4.

The first phases of the mortuary ritual

Generally speaking, archaeology cannot trace the rituals performed immediately after the death of an individual. The mourning, procession or other rituals do not leave any archaeological features. Probably the rituals started in or near the house where the deceased used to live: this means that some are likely to have been performed within the settlement rather than the cemetery. Realizing this, it may be possible to look for possible remains of these rituals when excavating a settlement, although they are unlikely to have left many permanent traces. We may also infer from this that some sort of procession will have taken place, because, whatever form the treatment of the body took, its destruction will certainly have taken place outside the settlement boundaries and probably also outside those of the cemetery

More can be said about the actual destruction of the body. This could be done in several ways: either the cremation of the body, or excarnation (de-fleshing) by exposure in open air or temporary burial. Excarnation and cremation can be seen as two different ways of destruction of the body leading to the same result. Excarnation and cremation are in general mutually exclusive, but modern practice shows that excarnation (in the form of temporary burial) takes place while a cremation rite and accompanying festival is prepared over weeks, months or even years (see for instance McKinley 2006, 82; Niblett 2000, 99).

There is some possible evidence for excarnation rituals at Tiel-Passewaaij. During the excavation of the settlement of the 2nd century AD, a cluster of burials dating to the Late Iron Age (late second or first century BC) was recovered. Thirteen cremation burials, two inhumations of small children (4–6 months old) and one pit containing some cremated remains and an unburnt spine of an adult individual belong to this small burial site. The spine consists of 11 vertebrae in anatomical connection. Although burned bone and charcoal were present in the same pit as well, the spine itself showed no signs of burning. Examination by a physical anthropologist proved the absence of any cut- saw- or gnawing marks (Baetsen 2006). The only way the vertebrae connected by tissues could have been separated from the other parts of the body is by exposure in the open air or temporary burial and subsequent retrieval and reburial. The cremated remains of other body parts in same pit as the spine may suggest that excarnation was followed by cremation. This does seem strange, however: one would expect the defleshing and cremation of the body to be more ideologically separated processes or stages in the ritual. Also, excarnation followed by cremation would mean a double destruction of the body. The earlier mentioned excarnation is better seen as a postponement of cremation; this would mean a temporary exposure of the body rather than an active process of excarnation. Although clear evidence of exposure of the body before the cremation is extremely scarce, more may be found in the wider region around Tiel. Unburnt human bones are found more often in Late Iron Age settlement contexts of the Low Countries, and these are possibly connected to exposure as well (Roymans 1990; Hessing 1993). The cemetery of Lamadelaine, belonging to the Late Iron Age and Roman period oppidum of the Titelberg in modern Luxemburg, provided more evidence, in the form of disjointed body parts, four-post structures at the edges of the cemetery (interpreted as platforms for excarnation), and weather-worn iron brooches (Metzler *et al.* 1999). In the Tiel-Passewaaij cemetery, some four-post structures were also found. They were situated between the graves, or in the middle of circular ditches. Two of the four structures were associated with pyre debris, indicating that they were used to support the cremation pyre. Although this does not exclude the possibility that they were used as excarnation platforms before the cremation, application of Ockham's razor would postulate an interpretation as relating to pyre construction.

We cannot be sure that the pit with the unburnt spine represents a specific set of circumstances rather than a standard practice. Until we find more examples of such a burial, the evidence of Tiel is too thin to assume excarnation as being part of the rite of passage.

Cremation

There are some literary sources that inform us about mortuary practices in the Roman period, but these concern mostly the civic elite in Italy (e.g. Pliny, *Natural History*). It is extremely doubtful whether the information from such sources can be applied to the rural communities of distant provinces. There are some general comments on Gallic funerals made by Caesar (*Gallic Wars* VI, 19):

> Their [i.e. the Gauls'] funerals, considering the state of civilization among the Gauls, are magnificent and costly; and they cast into the fire all things, including living creatures, which they suppose to have been dear to them when alive; and, a little before this period, slaves and dependents, who were ascertained to have been beloved by them, were, after the regular funeral rites were completed, burnt

together with them. (translation W. A. McDevitte and W. S. Bohn, http://classics.mit.edu/
Caesar/gallic.6.6.html)

This passage seems to relate to the burial of special persons rather than to the
average person we are dealing with in the Batavian community of Tiel, but it is
clear cremation was part of the ritual, and that at least in some cases, objects were
cremated together with the corpse. This of course was already known to us, since
burnt or molten objects are a regular find among the cremated remains which were
buried later. Another passage which is more relevant to our area of research we find
in the *Germania* of Tacitus (*Germania 27*):

> *In their funerals there is no pomp; they simply observe the custom of burning the bodies of illustrious
> men with certain kinds of wood. They do not heap garments or spices on the funeral pile. The arms
> of the dead man and in some cases his horse are consigned to the fire. A turf mound forms the tomb.
> Monuments with their lofty elaborate splendour they reject as oppressive to the dead. Tears and
> lamentations they soon dismiss; grief and sorrow but slowly. It is thought becoming for women to
> bewail, for men to remember, the dead.* (translation A. J. Church and W. J. Brodribb, London 1877)

This observation by Tacitus gives us little more information, but we must realize
that Tacitus was no participating observer in these rituals, and that the *Germania* is
riddled with the *topos* of the noble savage (the German people). His observations about
the absence of pomp at Germanic funerals and the public display of grief probably
represent more a critique of his own society than a detached view of things. The use
of low burial mounds is again in accordance with what we know from archaeological
data. In short, the ancient sources tell us not much more than we already know.

Also, archaeologically speaking, little is known about the actual cremation, since
the material remains are consumed by fire and most of these remains were not
deposited later. The cremation took place outside the boundaries of the cemetery
and settlement and thus have in only a few cases been touched by excavation. But
even if the immediate surroundings of cemeteries are excavated, cremation sites are
not easily identified, since the pyre debris will have remained at the surface and is
not likely to have been preserved in the archaeological record (see McKinley, this
volume). Additionally, it is possible that cremation did not always take place in the
same location, in which case it becomes even harder to find any traces of it. The only
information we usually have about the actual cremation concerns the remains (bone
and some fragmented pyre gifts) that were collected afterwards and later buried.

At Tiel-Passewaaij, however, a few locations have been identified at which
cremation may have taken place. The first possible cremation site is a more or less
open space east of the centre of the cemetery, where no 'normal' graves are situated
(Fig. 5.5 location A). There were a number of pits here, but none lay within a circular
or rectangular ditch, unlike the 'normal' graves. The pits did contain charcoal, some
cremated bone and fragmented ceramics, but the weight of the cremated bone was
very low (always less than 100 grams). The pits most probably represent deposits of
pyre debris that were the result of cleaning up after the descendants of the deceased
selected the larger pieces of bone to be transferred to a grave monument. Most

Joris Aarts and Stijn Heeren

Fig. 5.5. The central cemetery of Tiel-Passewaaij.

finds in these pyre debris depositions date to the earliest period of the cemetery, c. AD 50–90. To the east and south of this cluster of pits, later graves (with circular or rectangular ditches) were situated. This implies that this cremation site was on the eastern periphery of the cemetery in its earliest phase. After some time, the space used for cremations was needed to locate more graves and the cremations were executed somewhere else.

Another cremation site lay on the western edge of the cemetery (Fig. 5.5 location B). There too the graves were not as close together as in other parts of the cemetery and left an open space, filled with pyre debris deposit pits rather than 'real' graves. The sherds that were collected from some of these pits could be matched to the same vessel. This implies that the remains were once scattered over the surface and became interred eventually. It is also in this location that several post-built structures were found, which possibly supported the pyres. The features of this cremation site and the surrounding graves date to the period c. AD 120–190.

Burial

By far the most archaeologically visible stage in the mortuary ritual is the burial. The remains of the deceased are moved from the cremation site to a final resting place, made visible by a small monument. This stage in the funeral marks the passage of the deceased into the world of the ancestors. Although the term 'burial' may sound like one action, this phase too comprises a sequence of several steps.

After the pyres were extinguished, some of the human remains were collected and moved to the burial site. The remains varied in weight from several grams to one kilogram, much less than the 'normal' weight of a cremated skeleton (2,700 g for males; 1,800 g for females; Holck 1996). The selection and removal of some larger pieces of bone was apparently enough for a 'proper' ritual. The cremated remains were mostly white, without any sign of charcoal, indicating that they were burnt at a high temperature and washed after selection.

Next, a grave pit was dug, in which the washed selection of bones was deposited. We assume that the bones were deposited in a cloth, since they were often recovered as ball-shaped concentrations. Sometimes a closed and unburnt brooch lay on top of the cremation deposit. This brooch most likely closed the textile container of the cremated bone and is not to be seen as part of the dress of the deceased (see below). Complete unburnt grave gifts, mostly pottery, were placed close to the cremated remains. The pottery consisted almost exclusively of tableware (plates, bowls, jugs, beakers); cooking pots or storage jars are rare exceptions. Sometimes animal bone was found on top of plates, indicating that the tableware actually contained food (meat on the bone in this case). Grave-gifts other than tableware occurred (like shoes with hob-nails, coins, knives) in smaller numbers. Items like finger-rings, brooches, toilet-instruments like strigils, spatulas and *unguentaria* were only sporadically included.

Burying a selection of washed bone ('clean' cremation) together with grave gifts was the most common way to treat the cremated body (type A) but not the only one. There were also graves that contained both a clean cremation deposit and pyre debris (type B), the latter consisting of much charcoal, fragmented burnt ceramics and also some scattered cremated bone. A third type (C) consisted of pyre debris only, in which case the cremated bone was not separated from the rest of the pyre debris and not washed. It is not exactly clear how we should interpret these variations in burials. They may have been meaningful, in which case we cannot reconstruct why they occurred. There seems to be a slight shift from type A burials to type C over the period of time during which the cemetery was used: less time was invested to separate the cremated bone from the rest of the pyre waste and clean it. This burial type (C) is related to age, since a relatively large proportion were children's graves (0–6 years). However, the fact that these children were cremated, and that sometimes a substantial number of grave objects was included, suggests that they were considered to be members of the community, and were not excluded from the ritual which also occurs frequently (see for instance Hertz 1960, 84: 'Since society has not given anything of itself to the (new-born) child, it is not affected by its disappearance and remains indifferent').

Sometimes the quantity of cremated bone was very small; this may be accounted for by the disturbance of the grave through later agricultural activities but this was not always the case. In such cases, we are faced with the question of whether a pit containing 2–10 g of cremated bone must be counted as a burial (concerning the ritual through which the essence or soul of the deceased is transformed into an ancestor) or as a way of disposing of the pyre waste (which may also be a ritual act, but of another order). We chose to count as burials only those pits which were enclosed by a ditch.

Another phenomenon, the opposite of the one just discussed, was the presence of ditches (grave monuments) that seemed to contain no burial. This occurred especially in the latest phase of the cemetery. This may be explained in two ways. First, it is possible that a profound change occurred in the mortuary ritual, in which the actual burial of the cremated remains was no longer necessary. This seems the most far-fetched explanation, which in any case must be checked against data from other Batavian cemeteries (regrettably there are none at this moment which are suitable for such a comparison). The second interpretation is that we did not find any trace of grave pits because they were not dug into the surface level, but into the top of the mound, thus being lost through later levelling of the burial mounds. A variation of the second explanation would be that the cremated remains and grave goods were set on top of the burial mound. Recent – yet unpublished – excavations of a Roman cemetery at IJsselstein, in which the burial mounds were still intact, have shown that in some cases the grave was indeed dug into the top of the mound.

As well as the variation in the treatment of the remains of the body, there was also variation in the objects which were included in the burial. Some graves contained no grave goods at all, some included unburnt objects, and others only contained burnt objects or a mixture of both. We may assume that the gift objects which were burnt

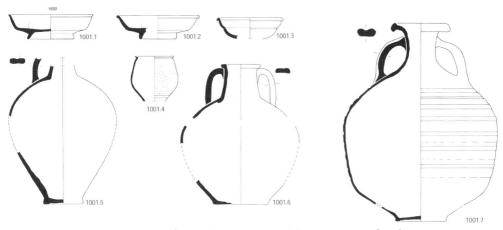

Fig. 5.6. A typical grave inventory, containing pottery vessels only.

together with the body had a different significance to those which were later placed in the grave, because they belong to a different phase in the rite of passage. The repertoire of gifts offered at the cremation and the time of burial did not however seem to differ: in both cases the largest part consisted of table ware, perhaps containing food and drink for the dead (Fig. 5.6 for examples).

Perhaps the only important and standard practice was to inter some remains of the cremated body as *pars pro toto*. The way that this was done, or whether grave goods were included or not, were matters of personal choice. Although funerals were primarily a public ritual, this does not mean that there was no room for personal ritual as well: perhaps the latter is the most important source for the variation in the details which show up in the archaeological record, though as noted above the association with pyre debris graves (type C) suggests that young children apparently received a less elaborate funeral.

Finally, the grave was filled with clean soil again and a low mound was erected above it. To enclose the mound a shallow ditch was dug, mostly of a circular or rectangular form. The excavated soil from the ditch was probably used for the mound. Figure 5.7 shows a reconstruction of what the graves would have looked like. Frequently a space was left open in the ditch, giving access to the mound. In most cases, this entrance was located on the west or northwest side of the enclosure. Of the ditches surrounding the graves, 122 were round, 149 were rectangular and nine had an irregular form. We do not know what determined the form of the ditches. The results of the physical-anthropological analysis of the bones provided no clues: there was no clear connection between the form of the ditch and the sex or age of the individual that was buried inside it. There was also no clear development over time, i.e. no difference between the ditches surrounding 1st century or 3rd century graves. Clearly factors that cannot be studied by archaeology determined the form of the ditches (for instance the season in which the grave was dug). Sometimes the ditches of a number of monuments seem to

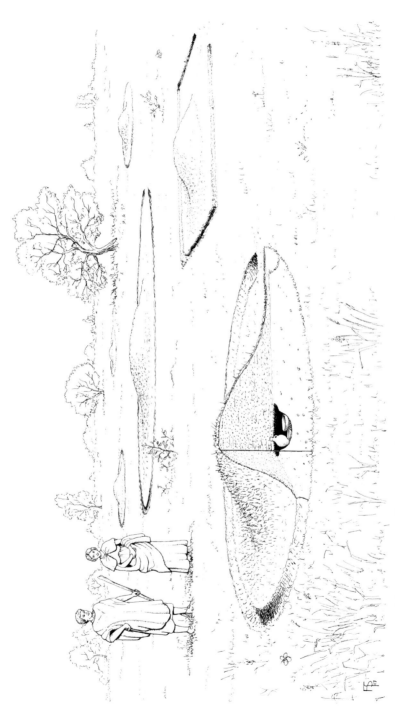

Fig. 5.7. An impression of the graves as small monuments (drawing F. Spangenberg, Illu-Atelier Konstanz). Courtesy Rheinisches Landesmuseum Bonn.

have been incorporated into a larger cluster. It is tempting to interpret these clusters as family groups, but this regrettably, cannot be proved.

Memory and ancestral relations

In most societies, the interaction with the dead does not end with the funeral; ancestors are supposed to take an active role in protecting and caring for the community of the living (Davies 1997, 99; Chapman 1994, 46). This may take the form of consulting the dead, or winning their attention or approval so they may help the living with their existence. In the latter case, interaction with the ancestors takes place on a regular basis, as, for instance, on All Saints day with the Bolivian Laymi: 'They pour *chicha* discreetly into the tomb so that the dead may drink, and they make a show of doing him honour, placing cooked meals and roasts upon the grave for him to eat' (Harris 1982, 46, citing a 17th-century priest). Roman society itself had two yearly festivals for the dead: the *Parentalia* and the *Lemuria*. Although in the official calendar, 'they were legally classified as private rituals, the business of clubs and families' (Davies 1999, 146). This duality in the relationship between the living and the ancestors is a recurrent phenomenon in many societies. Undoubtedly, also in the ancient world some of the rituals performed during such festivals for the dead referred to the ancestral community as a whole, while families were responsible for their own forebears and performed private rituals at their monument. The degree to which attention is paid to the collective dead or to the individual may differ; in rural Basque society for example, the individual graves are neglected and all attention is focused on the *sepulturie*, a symbolic burial plot for the collective dead of a household in the local church (Bloch and Parry 1982, 33).

In the cemetery at Tiel-Passewaaij, there are some indications of a continuing relationship between the Batavians and their ancestors. The most evident ones are of course the burial monuments themselves, and the larger spatial grouping of monuments into a cemetery. While the monument constitutes a visual memory of individual ancestral spirits (although in some cases, more than one individual was buried in one grave), the central cemetery itself is symbol of the community of ancestors. Deposits in the ditches surrounding the grave are also clear indicators of post-funerary rituals. They consisted mostly of complete pottery vessels. Sometimes a coin was also thrown into the ditch. There is an important difference between the pottery assemblages of the grave-pits and these deposits: the grave pits usually contained tableware (beakers, jugs, plates, fine bowls), whereas the ditch deposits consisted for the most part of cooking ware and storage vessels (large coarse ware bowls and pots). This combination, together with the contrast with the pottery in the grave pits, seems significant. The cooking vessels may have been used for preparing food for the dead, symbolizing the gift in the reciprocal relationship with the ancestor. The storage vessels, however, may be representative of the obligations of the ancestor(s), who for example might deliver a good harvest, and can be interpreted as symbols of fertility.

The living community: the cemetery in context of the settlement(s)

In the case of Tiel-Passewaaij, we are in the enviable position of being able to compare the data of the cemetery with those of the settlements, so we may learn more about the way the inhabitants treated their dead by looking at the living community. First, we will consider the demographic data of the cemetery and compare them with those of the settlement; this may reveal if the Batavians of Tiel-Passewaaij were the only ones who buried their dead in the cemetery. Discrepancies may tell us more about the nature of mortuary ritual: were all the deceased members of the community actually buried, or only a selection? Or were there more people buried in the cemetery than were living in the village? Secondly, by comparing the material culture of the settlements with that of the cemetery we may learn more about the nature of funerary rituals, for instance, by looking at processes of selection and the contextual meaning of objects in the settlement. For the purposes of this article, three categories of metal artefacts were selected for closer study, brooches, military objects and coins.

Demography: the size of the local population

The size of the local population can be studied in two ways. The size of the burial community can be estimated using the well-known formula of Acsádi and Nemeskéri (1970): population = *number of graves × correction × life expectancy/time*. Since most of the Tiel-Passewaaij graves contained pottery they could be dated reasonably well, and therefore this formula could be executed for four groups of graves, dating to the later first century (AD 50–90), the early second century (AD 90–150), the later second century (AD 150–210) and the third century (AD 210–270) (table 5.1). The community of the settlement(s) can be estimated by taking the number of simultaneously existing houses and multiplying this by the average size of a pre-modern family (five to eight persons). Table 5.2 gives a low as well as a high estimate of the inhabitants of both settlements bordering the cemetery. Both tables include the excavated features as well as an estimate of the features in parts of the site that could not be excavated. Figure 5.8 combines the two tables into a curve indicating the size of the local community over time.

A first observation is that the trend of growth and decline is the same in the settlement and the cemeteries. The trend as a whole seems to be reliable. A second important observation is that the size of the population in the cemetery differs from that of the two settlements: it is substantially higher, even the highest estimate of the settlement population not coming close to the numbers of burials in the cemetery. What explanations can be found for this discrepancy? One approach could be to question the assumptions underlying the settlement estimates (table 5.2), for instance the period of use of the wooden farmhouses, or the size of the families inhabiting the farms. This solution offers little help, since even unrealistic adjustments like an average family size of 12 persons would not completely close the gap between settlement and cemetery populations. Another approach would be to question the basic assumption that only the inhabitants of the two excavated settlements buried their dead in the

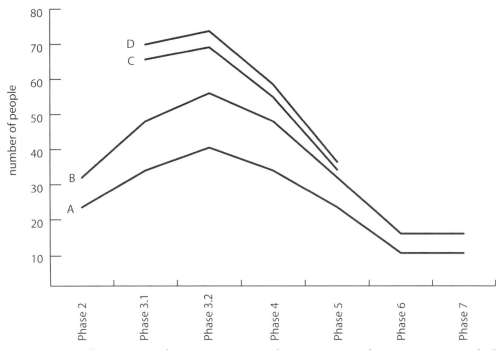

Fig. 5.8. Demographic curve. A: settlements communities – low estimate. B: settlements communities – high estimate. C: burial community – low estimate. D: burial community – high estimate.

Table 5.1. The size of the Tiel-Passewaaij burial community

	A	B	C	D	E	F	G	H	I	J	K
AD 50–90	84	12	72	5	77	10	87	1.10	25	40	59.8
AD 90–150	125	12	113	7	120	23	143	1.10	25	60	65.5
AD 150–210	106	19	87	1	88	20	108	1.10	25	60	49.5
AD 210–260	65	14	51	2	53	2	55	1.10	25	50	30.3
Total	*380*	*57*	*323*	*15*	*338*	*55*	*393*				

A gross number of funerary features (B and C)
B number of pyre debris pits
C number of graves
D double interments
E excavated individuals
F correction for not-excavated terrain
G d = estimated total number of burials
H k = correction for underrepresented groups (children)
I e = life expectancy
J t = time, number of years
K p = k × d × e/t (Acsadi/Nemeskeri 1970)

Table 5.2. *The size of the Tiel-Passewaaij settlements-communities*

	low estimate		high estimate	
	farmhouses A	*× 5 persons*	*farmhouses B*	*× 8 persons*
50 BC–AD 40	2	10	4	32
AD 40–90	4	20	6	48
AD 90–150	5	25	7	56
AD 150–210	4	20	6	48
AD 210–240	3	15	4	32
AD 240–270	1	5	2	16
AD 270350	1	5	2	16

cemetery. It is possible that the cemetery was also used by the inhabitants of other settlements: other excavations of rural settlements and cemeteries in the Batavian *civitas* suggest the same thing. In the region around Oss-Ussen it was established that several settlements surround a large cemetery (Wesselingh 2000). In a article about five Roman sites around modern Zaltbommel Blom, Veldman and Zuidhoff (2007) showed that the inhabitants of at least four small settlements used the central cemetery together. If we may extrapolate from these extensively excavated examples in the Dutch river area, it is not unlikely that these collective cemeteries were a normal occurrence in the Roman period.

Material culture: brooches, military equipment and coins

During the Tiel-Passewaaij excavations, thousands of metal items were found, including brooches, coins, finger-rings, bracelets, military equipment (including horse gear), toilet instruments, iron tools, seal boxes and others. The recovery of large numbers of metal items can be explained by two factors, firstly the excellent conservation properties of riverine clay soils (for bronze in any case, somewhat less so for iron objects), second, the intensive use of metal detectors: not only were the archaeological levels checked for metal finds but also the topsoil as it was removed layer by layer by machine.

Brooches

Let us consider the brooches, and primarily their spatial distribution, in more detail. No less than 1007 brooches were recovered from all excavated sites of Tiel-Passewaaij. Only nine can be dated to the Iron Age with certainty; 16, or perhaps 18, pieces probably belong to the Late Roman period. If we leave out most of the Early Roman brooches as well (since the cemetery started to be used from around AD 50), 588 brooches remain for the two settlements and the cemetery in the period c. AD 50–270. The vast majority of these were hooked wire brooches, rounded wire brooches and hammered wire brooches. Enamelled plate brooches were also present in some numbers, while German knee-brooches and British trumpet brooches were very scarce.

While the brooches appear in the settlements in large numbers, relatively few were found in the cemetery (Fig. 5.9). Of the 55 recovered there; 12 pre- or post-dated the cemetery (earlier than AD 50 or later than AD 270), leaving us with 43 pieces from funerary contexts. Three brooches were found complete, unburned and in closed position on top of the cremated remains: these were most probably used to close a textile containing the cremated bone. Sixteen pieces were partially melted and therefore most likely belonged to the attire of the deceased on the pyre, although it seems strange that they would have survived the high temperatures at all. A further three were found among the complete and unburnt grave goods.

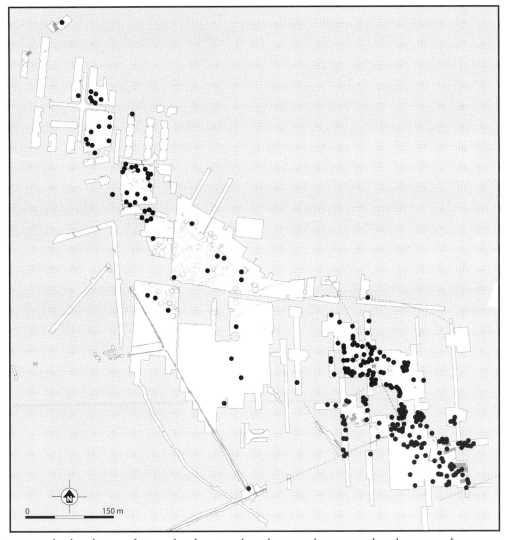

Fig. 5.9. The distribution of 1st- and 2nd-century brooches over the excavated settlements and cemetery.

The other 21 brooches were found close to the graves but their connection with the grave ritual is unclear and some may have been lost during later visits to the cemetery. Only 19 of the brooches can therefore be seen as grave gifts for the deceased with certainty and a further 21 are possibly connected to the grave ritual. Of all the 588 brooches dating to the period of the cemetery, this is only a small fraction (3.5 or 6.8%).

Military equipment

Military equipment includes (pieces of) weapons and armour as well as horse gear and both functional and decorative pieces. In the past finds of military equipment were generally taken as evidence for the presence of a fort or at least a watchtower. In recent decades so many pieces have been found on Batavian rural sites that this explanation has become untenable. A recent study by J. Nicolay (2007) made an inventory of all military finds in non-military contexts and provided new models for their interpretation. Nicolay distinguishes between 'military' use of these objects – in the context of military service – and 'social' use, i.e. what the soldiers did with their gear after their honourable discharge. The soldiers were the owners of their equipment and had two options after their discharge, to re-sell it to the army or to take it with them when they returned to the countryside. Since much military equipment is found in rural settlements and in sanctuaries, two patterns of 'social' use can be discerned. Soldiers deposited elements of their equipment to deities in sanctuaries, perhaps as a ritual gift during yet another rite of passage, namely when their identity as soldier was shed and life as a civilian was resumed (see also Roymans and Aarts 2005), perhaps to fulfil an earlier promise to a god for ensuring the soldiers' safe return after 25 years of military service. Other elements of military gear were taken to their new place of residence to serve as souvenirs of their long life of service. Following Nicolay, finds of military equipment in rural settlements like Tiel-Passewaaij are interpreted as evidence for the presence of veterans. The fact that so many finds of military gear are found in the Batavian rural area is another illustration of the large-scale recruitment of Batavian young men into Roman auxiliary units (see above).

In Tiel-Passewaaij a total of 145 pieces of military equipment was found. One sword hilt dates to the Late Iron Age, 12 parts of weapons and armour and 43 pieces of horse gear to the 1st century or early 2nd century (AD 1–120), three parts of weapons and armour and 56 pieces of horse gear to the 2nd and 3rd centuries (AD 120–270) and one piece to the Late Roman period. Fourteen parts of weaponry and fifteen pieces of horse gear could not be dated. Of all these 145 items, only five were found in the cemetery and all the remaining 140 pieces in the settlements, primarily located around the farmhouses. The five pieces found in the cemetery date to the 2nd and 3rd centuries. All 1st-century military objects (AD 1–120) were found in the settlements (Fig. 5.10). To understand the large numbers of military items found in Tiel-Passewaaij, we have to reflect a little further on the meaning of these pieces in a rural environment. The farmers' sons who entered military service most likely kept

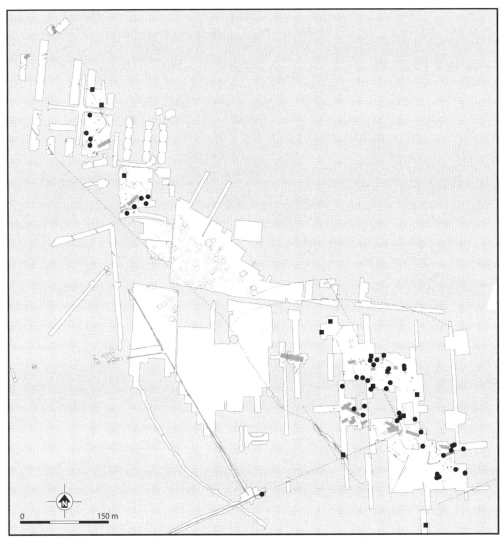

Fig. 5.10. The distribution of military equipment and horse gear of the period AD 1–120 over the excavated settlements and cemetery.

contact with their families for the period of service. Some of the Vindolanda tablets may bear evidence of contacts between soldiers and family far away (for instance *Tab. Vindol.* II, 310 and 346; Bowman and Thomas 1994), and since that fort was occupied by the 9th cohort of Batavians, this could very well apply to young men from Tiel-Passewaaij or similar settlements. This contact between soldiers and their families is one explanation why veterans returned to their places of origin after 25 years of service. During their military service, the Batavian young men became familiar with Roman military culture with respect to all aspects of daily life, including eating and drinking, religious practices, bathing and body care, reading and writing. They potentially also travelled extensively within and beyond the Roman world with their unit. When veterans returned to their families, possibly bearing Roman citizenship, they had many stories to tell and experiences to share. It is very likely that pieces of military equipment not only served as souvenirs of life in forts for veterans themselves but that they also were symbols of their knowledge of distant places and experiences of Roman lifestyles. In this perspective, the presence of Roman military gear in rural settlements can be seen as an expression of soldier (or veteran) identity. Its virtual absence among the material culture of the cemetery is significant.

Mortuary ritual and personal identity

It was argued above and in the introduction to this paper that a military identity was very important to the Batavian population. The Batavians supplied more troops to the auxiliary troops than any other tribe. Service in Rome's armies was one of the Batavian ways of being Roman. Yet we can conclude that this military identity was not expressed in the grave ritual. In the early phase of the cemetery (AD 50–120), not a single piece of military equipment was deposited in graves, while at the same time large numbers of military items circulated in the settlements.

A similar situation occurs with the brooches. Several authors have argued the importance of *fibulae* for the expression of the bearer's identity, including the number and type of brooches worn and the manner of fastening them. Apart from the Aucissa brooch, which was often worn by soldiers, we do not really know which brooch-types were worn by whom, but it is assumed that each age/gender/ethnicity/profession-group will have had their own fibula-types or ways of bearing brooches. (Jundi and Hill 1998; Eckardt 2005).

Like military equipment, brooches are hardly found in the cemetery. Of the 19 pieces found, three served to close the cloth which held the cremated bone and the other sixteen were worn during cremation and were destroyed in the fire along with the other symbols of personal identity or individuality. If our idea is correct that the cremation ritual represented the negation of individual identity, it is not to be expected that items referring to this individuality will show up in the burial. Other items possibly connected to the expression of personal identity like toilet instruments, finger-rings, valuables, are also very scarce in the cemetery.

To conclude, it seems that personal identity was not expressed in the Batavian burial ritual. References to age, sex, social standing or profession are scarce, if not completely absent from the grave goods. Standardized sets of tableware refer to a meal in general, not to the deceased as a person. The function of the grave ritual seems therefore to be aimed at the transformation of the deceased into an (anonymous) ancestor.

Coins and boundaries

A total of 347 coins from the Late Iron Age and Roman period were found in the settlements and cemetery of Tiel-Passewaaij. The Late Iron Age coins were probably part of the coin pool of the early Roman period and are not representative of a pre-Roman circulation. Two hoards were found, one consisting of over 44 third-century *denarii* buried in one of the houses, another more than 60 barbarous radiates and some official *antoniniani* dating to the late third or early fourth century. The latter was found in the area of the cemetery, to which we shall return later. Twenty-two coins were either found in grave pits or in the ditches surrounding the graves. No more than one coin was ever found in a burial. All coins in the cemetery context were bronze. We also encounter the same predilection for low-value coins in other ritual settings, like votive deposits in sanctuaries or ritual deposits at river crossings and bridges.

Although some of the coins dated to the 1st century AD, the act of depositing coins during mortuary rituals did not start before the 2nd century. It seems to have been a matter of personal choice for some families in the 2nd century to offer coins in a funerary context and not a standard ritual. As for their meaning, we must differentiate between the coins deposited in the grave pits and in funerary ditches. A possible third group leaves no trace, i.e. those coins which accompanied the body when it was cremated. If we may project back in time the placing of coins in the hands or on the eyes of the dead which we know from later inhumation graves, these coins – if present – will have melted during the cremation and would turn up if at all in the form of occasional bronze droplets amongst the cremated bone in the grave pit. It is of course not easy to comment on the significance of coins in these funerary contexts. However, research shows that money in funerary contexts often refers to the journey the deceased must make (for instance the find of a purse hoard in the Sutton Hoo ship burial; the *viaticum* in Christian burials). This travelling money could possibly have been used by the dead for paying tolls when crossing important boundaries, like the well-known Greek myth of the obol for Charon which was due in order to cross the underworld river Styx (see Stevens 1991). If this is true, coins as grave objects are part of the traveling metaphor often seen in mortuary rituals.

Another possibility is that the coins in grave pits were symbols of fertility. As we have seen before, this is another recurrent theme in mortuary rites of many societies. Bloch and Parry (1989) have argued that one of the qualities commonly attributed to money is its power to replicate itself, as such being an apt symbol of fertility. The Celtic god Cernunnos, sometimes depicted vomiting a stream of coins, may also be an expression of the fertility ascribed to coins or money (for the concept of fertility in funerary rites, see Bloch and Parry 1982; Cernunnos, see for instance Stevens 1991, 228).

The coins in the ditches are from a different ritual activity. They belong to the 'memory-phase' of the Batavian mortuary ritual and were thrown in the ditches during later visits to the grave, mostly near the opening in the ditch which allowed access to the burial mound. This is a practice we also encounter on sanctuaries in the Roman period, where coins were regularly deposited at the entrance of the cella. The same liminal symbolism can be seen too in the coins offered at river crossings and bridges. Apparently, coins in a ritual context are often used when crossing virtual or real boundaries.

Finally, something must be said about the hoard found within the territory of the dead. Its burial date probably lies somewhere in the last decade of the 3rd or possibly in the early 4th century AD. This means that it was buried there long after the people ceased to bury their dead in the central cemetery. The most recent grave can be dated to the mid-third century AD at the latest. Furthermore, the hoard was not buried by the Batavian community which lived in the settlement in the first three centuries AD but by a group of newcomers, possibly Franks from north of the Rhine, who re-settled at Tiel-Passewaaij when the Batavians had left. The location they chose for the hoard is not without significance, since savings hoards are usually concealed in or close to the house in which the owner lived, not some hundred metres from it, as in this case. The hoard was also buried in the ditch of one of the graves of their Batavian predecessors, i.e. it lies in the realm of the ancestors of the people who lived there before them. It is not unlikely that the hoard represents an expiatory offering of the new group of Franks to these ancestors, to ask for their permission to use their land or 'pay' for their stay. This association between cemeteries and territoriality is a recurring phenomenon in the burial practices of many people in many places through time (Chapman 1994, 53; Parker-Pearson 2003, 132–141; Hiddink 2003, 62).

Changes over time

So far, the mortuary rituals have been analysed without much discussion of chronology. Nonetheless, changes over time can be addressed since not only the cemetery of the period AD 50–270 was excavated, but also small clusters of graves dating to the Iron Age as well as some late Roman inhumation burials.

The Late Iron Age and first decades of the Roman period

Two small clusters of graves dating to the Iron Age have been excavated at the location of the later settlements from the Roman period. One cluster consisted of four 'clean' cremation deposits, without any grave goods or visual markers (ditches). The second cluster consisted of thirteen pits containing scattered cremated remains, two inhumations of children several months old, and the previously discussed spine of an adult individual. Only one of the cremation graves contained a grave gift and none of the graves was marked by ditches. The first cluster was discovered in the low-lying (marshy) parts of the Tiel-Passewaaij area when a new road was constructed, the second during excavation of the Roman settlement when the Iron Age graves appeared between Roman period postholes. Since the graves were not at all conspicuous they were very hard to find and both clusters were discovered by chance. It is therefore very likely that more clusters of graves were present and remain undiscovered.

The small numbers of graves are an indication that the locations for burial were in use by one family group for only a limited period of time. The fact that the graves were not marked by ditches and were probably not meant to be seen for a long period of time may mean that the actual burial site did not mean much in remembering the dead. Perhaps this was done in rituals which leave no archaeological trace, or perhaps in the house where a little shrine was kept. We simply do not know. Cremation seems to have been the normal procedure, in some cases possibly after a period of excarnation. Small children received an alternative ritual, involving burial without cremation. Since grave goods are almost absent, they either consisted of organic materials (foodstuffs, wooden plates or bowls) which have disappeared over time or were not part of the mortuary ritual. In the latter case, since it is assumed that the same descent groups continued to live there in the Roman period, one wonders why this change in the mortuary practice occurred. It is important to realize that in the Iron Age dwelling places and houses moved from time to time within a certain territory. There was little continuity of place, unlike the Roman period: it probably did not survive the time span of a house. Perhaps this was the main reason that not much attention was paid to the burials themselves. This all changed in the mid 1st century AD.

The collective cemetery (AD 50–270)

As a result of the Roman occupation of the Dutch river area from 19/16 BC onwards, fundamental changes took place in the Batavian area. Rural populations experienced military rule at first and civic government later, as well as new economic developments, with social and cultural changes following suit. One of the changes occurred in the social organization of rural communities, perhaps as a result of the different way land-ownership was defined under Roman rule. This new social organization (or at least social self-definition) of rural communities is expressed by the emergence of place continuity in the settlements. Instead of short-lived farmsteads moving through the landscape, settlements continued to exist at the same place for generations. This increasing place continuity was also reflected in the appearance of large collective cemeteries. Instead of small burial sites used by one family for a short period of time, several groups of co-resident families cremated and buried their dead together. The graves in the collective cemetery were marked with ditches and small mounds, made to last and to be seen. The large numbers of grave monuments grouped together were meant to express a long history of occupation, implying that this community rightfully claimed the lands their ancestors had lived in for a long time. In this manner, ancestral relations probably shifted from the house of farmstead to the burial site, which by its monumentality was an extension of the houses of the living. More attention was devoted to the burial site: in more and more graves, gifts accompanied the dead. The cemetery itself probably became a place of ancestor worship, collective memory and collective claims on land. Burial sites but also the permanent settlements became imbued with place value (Chapman 1994, 53: '...place value ... the nexus of stored meaning of past activities and traditional usage associated with a significant place').

The cemetery of Tiel-Passewaaij was in use for more than two centuries. The most profound changes occurred in the early Roman period we have just discussed. Another possibly significant change was that young children were treated differently. We have seen that they were excluded from the cremation ritual in the Iron Age burials, but in the communal Roman period cemetery we regularly find cremated remains of children, sometimes buried together with the bones of their mothers, perhaps indicating death in childbirth. Apparently, children became full members of the community in the course of the 1st century AD, thus deserving the same treatment as adults. Apart from these major changes associated with the emergence of the communal cemetery, little change is noticeable in the mortuary ritual. The main structure of the rite of passage remained the same: the destruction of individuality by the cremation rite, followed by the insertion into the community of ancestors by burial of the cremated remains. We have already discussed the gradual shift from burials in which the bones were washed and separated from the rest of the pyre debris to graves which only contained unsorted pyre waste in the later second and third centuries AD. Some changes occurred in the repertoire of grave goods, but this was more a result of changes in the material culture within the settlement than a product of changing funerary rituals.

The changing way of depositing the cremated remains and the grave goods is possibly connected to the decreasing care for the cremated remains. In the early periods of the cemetery, the grave pit – containing the cremated remains and the grave goods – was dug into the surface and a small mound was erected above the pit. In the graves from the later second and third centuries AD, the pit is often absent while the ditches normally surrounding the pit are well preserved. Earlier in this text we presented two possible explanations, the first (in our view the most unlikely) being that burying the cremated remains was no longer deemed necessary. The implication that the second phase of the rite of passage was thoroughly restructured takes the argument too far. The second indicates a minor change in the burial rite, i.e. that the grave was no longer dug into the surface but in the top of the mound, many later being lost through agricultural activities. Possibly this was connected with changes in the physical landscape, such as increasing wetness. A third explanation, that grave pits are lacking because there were no bodies, would mean that a large part of the community died far from the settlement but the percentage of cenotaphs would be implausibly high.

The Late Roman period (AD 270–450)

In the material culture of the Late Roman period, two phases can be distinguished, from AD 270 to AD 350 and from AD 350 to 450. In the first half of the Late Roman period a small settlement of two farmhouses, several outbuildings, ditches and wells was present at Tiel-Passewaaij. However, not a single grave connected to the inhabitants of this settlement was found. The cemetery was not in use in this period, at least not for burial. The coin hoard of late third or early fourth century date found between the much older graves indicated that the 'place value' of the cemetery remained high: it was probably still recognised by the newly settled Franks as the domain of the ancestors of the departed Batavian group (see above).

Two isolated inhumation graves dating to the period AD 350–450 were found; one complete skeleton was buried on the fringe of the former cemetery, another was recovered more than one kilometre to the south. No settlement dating to this period is known in the immediate vicinity. This situation in Tiel-Passewaaij, where late third/early fourth century graves are lacking and late fourth/early fifth century inhumation graves are present only in small numbers, seems to be representative of the wider region around Tiel. Although the relationship between settlement and burial customs cannot be studied for this period, it is clear that the drastic changes in the political/military, economic and social sphere of the late Roman period had their impact on the local population and their burial customs. Recent work sheds a new light on the dating of the Late Roman remains of Tiel-Passewaaij (Heeren 2017).

Conclusion

With this article, we have tried to explore new ways to study the archaeological remains of mortuary rituals. We started out from a model based on anthropological research. Doing so, it becomes clear that all mortuary ritual can be regarded as a rite of passage, which transforms the deceased from the person s/he was during life into a new social persona, living on in the memory of the living community but also playing an active role in society. The theoretical framework offered new perspectives on cremation and burial, placing the different rituals in a larger context. The model used made it possible to link the archaeological remains to different stages in the rite of passage: archaeological features which would be previously ignored or conceived of as isolated phenomena could be placed in this context.

The cemetery of Tiel-Passewaaij was used as case study because it is one of the few sites at which a cemetery and settlement have been almost completely excavated. Although in Tiel (like elsewhere), large parts of the rite of passage probably left few if any archaeological traces and the best-represented phase is that of the burial, evidence for the first phase of the cremation ritual was present in the burials. Some scattered remains of potential cremation sites were identified. Although there are some indications that the body was exposed temporarily, excarnation (as an active process of destruction of the body) was deemed unlikely. The objects which were found in the ditches which surrounded the burial mound were interpreted in the context of the relationship of the living community with their ancestors.

By comparing the features and finds of the cemetery with those of the settlement, it was possible to obtain even more information about the funerary rituals. Discrepancies between the demographic data from the settlement and from the cemetery showed that the latter was probably used by more communities than that living at Tiel-Passewaaij. It was also possible to develop new ways of looking at material culture in funerary contexts. By comparing find categories from settlement and cemetery, more could be learned about their use in the communities of the living and of the dead. Three types of metal objects were selected for a closer study. One of the conclusions was that objects used for display of personal identity in the context of the settlement (brooches and military objects), were absent in the cemetery or used in a different way. This supported the idea of the cremation ritual as a destruction of the individual in order that s/he become part of the more anonymous community of ancestors through burial of the cremated remains. Coins in funerary contexts were interpreted as objects which referred to the metaphors of travelling and the crossing of boundaries (liminality) which are very common in the mortuary rituals of many societies.

Finally, we sketched the changes to the mortuary rituals of the community at Tiel over time. The most substantial occurred at the transition from the Late Iron Age to the Roman period. Of these the most important was the appearance of a communal central place for burial in the mid-first century AD, instead of isolated clusters of graves which were bound to the house and which moved through the landscape together with the settlements. Through this continuity of place, attention shifted more to the burial site itself. Burials became more elaborate and more objects accompanied the dead. The space of the cemetery itself became imbued with meaning, or place value, illustrated by the coin hoard which was ritually deposited by Frankish settlers at Tiel in the second half of the third century, probably to appease the ancestral community of their Batavian predecessors. The position of children in society also seems to have undergone change, because they were previously excluded from the cremation ritual. The cremating of young children meant that they were now seen as full members of society. During the period in which the communal cemetery was used, however, only limited change was noted in funerary rituals. These included a growing preference for burying the remains of the cremation unsorted and an absence of grave pits in the latest monuments, perhaps because burials were made into or on top of the mounds rather than beneath them, though it remains unclear what inspired this change in burial practice.

Apart from the fact that the cemetery and settlements of Tiel were almost completely excavated, the archaeological evidence of mortuary ritual is not so unique. However, we hope to have shown that using a new perspective on the archaeological remains leads to a more complete understanding of the set of rituals which surrounded death in Batavian society.

Bibliography

Aarts, J. G. (2003) Monetization and army recruitment in the Dutch river area in the early 1st century AD. In R. Wiegels and S. Seibel (eds) *Kontinuität und Diskontinuität. Germania Inferior am Beginn und am Ende der römischen Herrschaft*, 145–161. Berlin: De Gruyter.

Aarts, J. G. (2005) Coins, money and exchange in the Roman world. A cultural-economic perspective. *Archaeological Dialogues*, 12.1, 1–27.

Acsádi, G. and Nemeskéri, J. (1970) *History of human life span and mortality*. Budapest: Akadémiai Kiadó.

Baetsen, S. (2006) Fysisch-antropologisch onderzoek. In S. Heeren (ed.) *Opgravingen bij Tiel-Passewaaij 1. De nederzetting aan de Passewaaijse Hogeweg*, 172–180. Zuidnederlandse Archeologische Rapporten 29, Amsterdam: Archeologisch Centrum Vrije Universiteit.

Bloch, M. and Parry, J. (1982) Introduction. In M. Bloch and J. Parry (eds) *Death and the regeneration of life*, 1–44. Cambridge: Cambridge University Press.

Bloch, M. and Parry, J. (1989): Introduction. Money and the morality of exchange. In M. Bloch and J. Parry (eds) *Money and the morality of exchange*, 1–32. Cambridge: Cambridge University Press.

Blom, E., Veldman, A. and Zuidhoff, F. (2007) Inheems-Romeinse bewoning te Zaltbommel. *Westerheem* 56.6, 425–438.

Bowman, A. K. and Thomas, J.D. (1994) *The Vindolanda Writing Tablets (Tabulae Vindolandenses II)*. London: British Museum Press.

Chapman, J. (1994) The living, the dead and the ancestors. Time, life cycles and the mortuary domain in later European prehistory. In J. Davies (ed.) *Ritual and remembrance. Responses to death in human societies*, 40–85. Sheffield: Sheffield Academic Press.

Danforth, L. M. (1982) *The Death Rituals of Rural Greece*. Princeton: Princeton University Press.

Davies, D. J. (1997) *Death, ritual and belief: the rhetoric of funerary rites*. London: Cassell.

Davies, D. J. (1999) *Death, burial and rebirth in the religions of Antiquity*. London: Routledge.

Derks, T. (2009) Ethnic identity in the Roman frontier: the epigraphy of Batavi and other Lower Rhine tribes. In T. Derks and N. Roymans (eds) *Ethnic constructs in antiquity. The role of power and tradition*, 239–282. Amsterdam Archaeological Studies 13, Amsterdam: Amsterdam University Press.

Eckardt, H. (2005) The social distribution of Roman artefacts: The case of nail-cleaners and brooches in Britain. *Journal of Roman Archaeology* 18, 139–160.

Es, W. A. van (1981) *De Romeinen in Nederland*. Bussum: Unieboek b.v.

Gennep, A. van (1909) *The Rites of Passage*, London: Routledge and Kegan Paul (Reprint 1960).

Groot, M., Heeren, S., Kooistra, L. I. and Vos, W. K. (2009) Surplus production in rural settlements in the Dutch River Area in the Roman period: integrating the evidence from settlement archaeology, botanical archaeology and zooarchaeology. *Journal of Roman Archaeology* 22, 231–252.

Harris, O. (1982) The dead and the devils among the Bolivian Laymi. In M. Bloch and J. Parry (eds) *Death and the regeneration of life*, 45–73. Cambridge: Cambridge University Press.

Heeren, S. (2009) *Romanisering in de Bataafse civitas. De casus Tiel-Passewaaij*. PhD thesis Vrije Universiteit Nederlandse Archeologische Rapporten 36, Amersfoort: Rijksdienst voor het Cultureel Erfgoed.

Heeren, S. (2017) From Germania Inferior to Germania Secunda and beyond. A case study of migration, transformation and decline. In N. Roymans, S. Heeren and W. De Clercq (eds), *Social dynamics in the Northwest Frontiers of the Late Roman empire. Beyond decline or transformation*, 149–178, Amsterdam.

Hertz, R. (1960) The collective representation of death. In R. Needham and C. Needham (eds) *Death and the right hand*, 27–86. Aberdeen: Cohen & West.

Hessing, W. A. M. (1993) Ondeugende Bataven en verdwaalde Friezinnen? Enkele gedachten over de onverbrande menselijke resten uit de IJzertijd en Romeinse tijd in West- en Noord-Nederland. In E. Drenth, W. A. M. Hessing and E. Knol (eds) *'Het tweede leven van onze doden': voordrachten gehouden tijdens het symposium over het grafritueel in de pre- en protohistorie van Nederland gehouden op 16 mei 1992 te Amersfoort*. 17–40. Nederlandse Archeologische Rapporten 15, Amersfoort: Rijksdienst Oudheidkundig Bodemonderzoek.

Hiddink, H. A. (2003) *Het grafritueel in de Late IJzertijd en Romeinse tijd in het Maas-Demer-Scheldegebied, in het bijzonder van twee grafvelden bij Weert*. Zuidnederlandse Archeologische Rapporten 11, Amsterdam: Archeologisch Centrum Vrije Universiteit.

Holck, P. (1996) *Cremated bones*. Antropologiske Skrifter 16. Oslo: Anatomical Institute, University of Oslo.

Jundi, S. and Hill, J. D. (1998) Brooches and identity in first century AD Britain: more than meets the eye? In C. Forcey, J. Hawthorne and R. Witcher (eds) *TRAC 1997. Proceedings of the seventh annual theoretical Roman Archaeology Conference Nottingham 1997*, 125–137. Oxford.

Metzler, J., Metzler-Zens, N., Méniel, P., Bis, R., Gaeng, C. and Villemeur, I. (1999) *Lamadelaine. Une nécropole de l'oppidum du Titelberg*. Dossiers d'Archéologie du Musée National d'Histoire et d'Art 6, Luxembourg: MNHA.

McKinley, J. I. (2006) Cremation ... the cheap option? In R. Gowland and C. Knüsel (eds) *Social Archaeology of Funerary Remains*, 81–88. Oxford: Oxbow.

Niblett, R. (2000) Funerary rites in Verulamium during the Roman period. In J. Pearce, M. Milett and M. Struck (eds) *Burial, society and context in the Roman world*, 97–104. Oxford: Oxbow.

Nicolay, J. A. W. (2007) *Armed Batavians. Use and significance of weaponry and horse gear from non-military contexts in the Rhine delta (50 BC to AD 450)*. Amsterdam Archaeological Studies 11, Amsterdam: Amsterdam University Press.

Parker-Pearson, M. (2003) *The archaeology of death and burial*. Stroud: Sutton.

Pearce, J. (1997) Death, time and Iron Age mortuary ritual. In A. Gwilt and C. Haselgrove (eds.) *Reconstructing Iron Age Societies*, 174–180. Oxford: Oxbow.

Roymans, N. (1990) *Tribal societies in northern Gaul. An anthropological perspective*. Cingula 12, Amsterdam: Albert Egges van Giffen Instituut voor Prae-en Protohistorie, Universiteit van Amsterdam.

Roymans, N. (1996) The sword or the plough. Regional dynamics in the romanisation of Belgic Gaul and the Rhineland area. In N. Roymans (ed.) *From the sword to the plough*, 9–126. Amsterdam Archaeological Studies 1, Amsterdam: Amsterdam University Press.

Roymans, N. (2004) *Ethnic identity and imperial power: the Batavians in the early Roman empire*. Amsterdam Archaeological Studies 10, Amsterdam: Amsterdam University Press.

Roymans, N. and Aarts, J. G. (2005) Coins, soldiers and the Batavian Hercules cult. Coin deposition at the sanctuary of Empel in the Lower Rhine region. In C. Haselgrove and D. Wigg-Wolf *Iron Age Coinage and Ritual Practices*, 337–359. Studien zur Fundmünzen der Antike 20, Mainz: von Zabern.

Stevens, S. T. (1991) Charon's obol and other coins in ancient funerary practice. *Phoenix* 45.3, 215–229.

Vos, W. K. (2009) *Bataafs platteland: het Romeinse nederzettingslandschap in het Nederlandse Kromme-Rijngebied*. Nederlandse archeologische rapporten 35, Amsterdam: Vrije Universiteit.

Vossen, I. and Groot, M. (2009) Barley and horses: surplus and demand in the civitas Batavorum. In M. Driessen, S. Heeren, J. Hendriks, F. Kemmers, and R. Visser (eds) *TRAC 2008. Proceedings of the Eighteenth Annual Theoretical Roman Archaeology Conference Amsterdam 2008*, 89–104. Oxford: Oxbow.

Weekes, J. (2008) Classification and analysis of archaeological contexts for the reconstruction of early Romano-British cremation funerals. *Britannia* 39, 145–160.

Wesselingh, D. A. (2000) *Native neighbours. Local settlement system and social structure in the Roman period at Oss (The Netherlands)*. Analecta Prehistoria Leidensia 32, Leiden: Faculty of Archaeology, University of Leiden.

Willems, W. J. H. (1981) Romans and Batavians, a regional study in the Dutch eastern river area I. *Berichten van de Rijksdienst voor het Oudheidkundig Bodemonderzoek* 31, 7–217.

Willems, W. J. H. (1984) Romans and Batavians: a regional study in the Dutch eastern river area II. *Berichten van de Rijksdienst voor het Oudheidkundig Bodemonderzoek* 34, 39–331.

Chapter 6

They fought and died – but were covered with earth only years later: 'Mass graves' on the ancient battlefield of Kalkriese

Achim Rost and Susanne Wilbers-Rost

In and around the district of Kalkriese, situated at the northern edge of the Wiehengebirge 20 km north of Osnabrück, countless Roman finds indicate the site of a large ancient battlefield which we can today associate with the Varian disaster, the battle of the Teutoburg Forest of AD 9. In this battle, Germans and their leader Arminius caused the total defeat of Publius Quinctilius Varus and his three legions who were returning from a summer camp in northern Germany, perhaps on the river Weser, to the left bank of the Rhine. Ancient authors only described this disaster vaguely (for example Cassius Dio, *Historia Romana* 56, 18–24, 1; Florus 2, 30, 29–39; Tacitus, *Ann.* 1, 55–72,1; Velleius Paterculus, *Historia Romana* 2, 117–120), causing the wildest speculation over the location of the battlefield until archaeological research began. Since the discovery of the Kalkriese site in 1987, intensive archaeological research – field survey and excavations – have brought to light many Roman coins and pieces of military equipment across an area of more than 30 km², indicating that the battle probably took place along a defile (Fig. 6.1). In the centre of this combat area the Oberesch field seems to be one of the principal sites of activity, if not the main place of annihilation (Wilbers-Rost 2009; Wilbers-Rost *et al.* 2007). A sand and turf rampart 400 m long (Fig. 6.2) was used by Germanic warriors in order to ambush Roman troops who had to negotiate the narrow passage between hill and bog. The evidence suggests that the Roman soldiers had already been attacked at different places before they reached this ambush at the Oberesch where they were encircled by the rampart to the south, the wet area to the north and two creeks limiting the field to the east and the west. Under these circumstances, they must have been unable to withstand the German ambush effectively since they could not fight in their normal formation on this small field, while the Germans could start assaults and retreat behind the rampart when the Romans tried to return

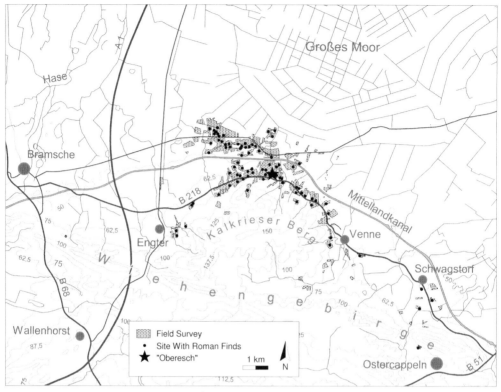

Fig. 6.1. Area under investigation at the Kalkriese Hill.

the attacks. More than 5,000 fragments of military equipment, most of which were found on the Oberesch, indicate the total defeat of the Roman army. The range of artefacts shows that legionaries, auxiliaries, horsemen and the baggage train participated in the action.

In addition, the bones of victims were discovered, of men, mules, and horses. Most had lain on the surface for many years, being exposed to sun, rain and wild animals; some were covered by material from the rampart, some sections of which had collapsed quite soon after the battle. In this respect the mule skeletons are very important: one was nearly complete, while another also yielded many metal harness fittings. Such features show that the rampart must have collapsed before looters could collect the metal or wild animals could tear away parts of the carcasses. Unfortunately, we have not yet found comparable human bone assemblages: we have merely discovered, for example, a row of human teeth and a few fragments of a skull near the rampart which indicated the remains of a body – probably a Roman soldier. In one of the drainage ditches behind the wall two rows of human teeth lay in their original position, but nothing from the jaw was left except the dental enamel. These observations indicate that many bodies must have lain on the surface after the battle,

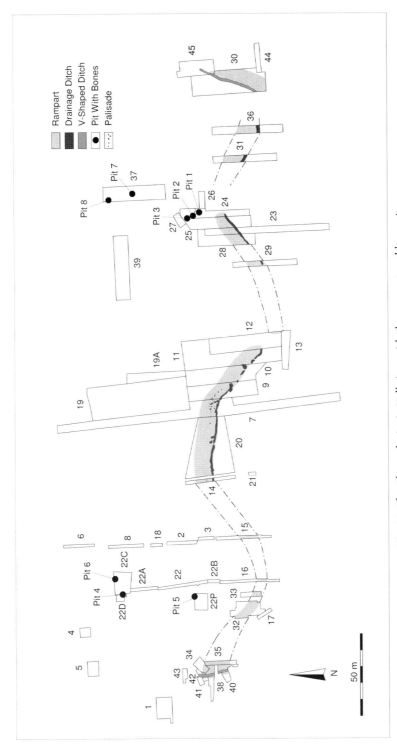

Fig. 6.2. The Oberesch site in Kalkriese with the rampart and bone pits.

but because of the poor preservation conditions only very few survived until today. This is one of the reasons why we cannot decide how many soldiers died in the battle at Kalkriese by archaeological and osteological analyses.

In addition to the mule skeletons and the remains of exposed bones, there is a very unusual feature at the Oberesch which does not belong to the immediate context of this event. So far eight pits have been found which contained a mixture of human and animal bones (Figs. 6.3–6.7). The skeletons are never complete and most of the bones are only small fragments in very bad condition, suggesting that they must have lain on the surface for several years before their deposition – probably between two and ten years according to osteological studies (Großkopf 2007, 176; Uerpmann and Uerpmann 2007, 112). Some bones show sword cuts (Fig. 6.8) and all human bones except one fragment are from men who were between 20 and 40 years old and well nourished. A few Roman artefacts were also found in the pits, namely nails, fragments of silver leaf and a spear butt. Anthropologists and zoologists suggest that these remains must be the bones of Roman soldiers, rather than of Germans (Großkopf 2007, 173), and of animals – mostly mules – used for baggage transportation (Uerpmann and Uerpmann 2007, 136).

How might we explain the existence of these pits? Who would have bothered about burying the remains of the victims of the Varian disaster? Considering the long exposure, we may interpret the pits as part of the activities of the Roman commander,

Fig. 6.3. Bone pit 1 during excavation.

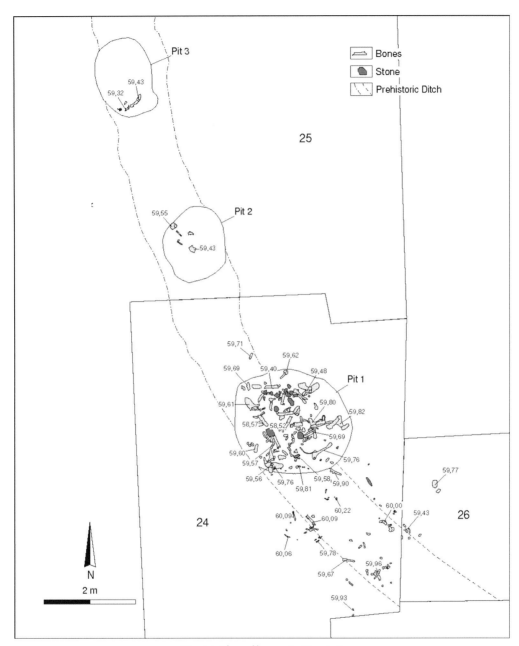

Fig. 6.4. Plan of bone pits Nos. 1–3.

Germanicus, who is said to have visited the place of the battle and buried the dead soldiers in AD 15 (Tacitus *Ann.* 1, 60–62). We may thus consider these pits as a form of mass grave for Varus' legionaries, though not mass graves in the common sense, since only individual bones were buried and not the complete corpses.

Fig. 6.5. Bone pit 5 during excavation.

The size of the pits and the quantity of bones deposited vary significantly. The diameter of the pits varies from one to four metres; some are almost completely filled with bones, while others contain only a few (Wilbers-Rost 2007, 84–95). But how can we explain the fact that all pits contain a mixture of human and animal bones?

Osteological analyses indicate that there are more human bones than animal bones in the pits, while on the surface the relation is inversely proportional. Therefore, it seems likely that those who wanted to bury the remains of the dead first of all tried to collect human bones and that they often could not decide whether a bone was from a soldier or a mule. Most of the bones are very fragmentary, but there is one pit (No. 5; cf. Figs. 6.5 and 6.6) which contained bones of mules and of only two men. Limestone in the soil must have preserved the skeletons much better than in other parts of the site and more fragments of these two individuals therefore survived to be found and buried (Großkopf 2007, 168–173). This explanation seems more likely than the idea that this pit indicates a preferential treatment of these two men, perhaps because of their rank. Many small bones like those from fingers and toes, however, were missing and one can imagine that these small pieces were not collected when Romans tried to bury the remains of the dead: the army of Germanicus was after all on campaign in enemy territory.

Bone-pits were only discovered at the Oberesch, and in combination with the rampart and the large quantity of finds they might show that this was one of the

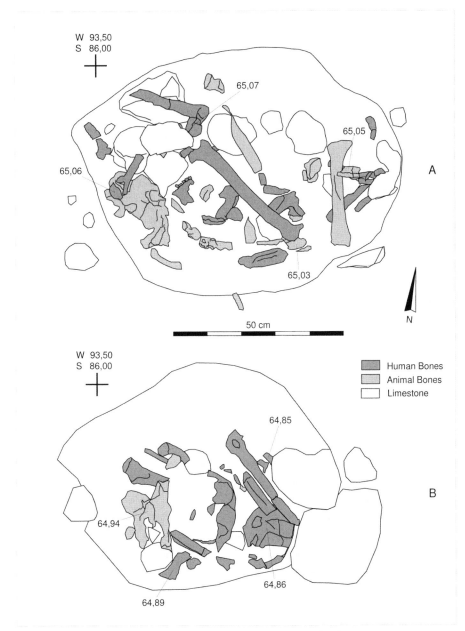

Fig. 6.6. Bone pit No. 5, upper and lower levels.

main sites of the Varian disaster. The pits do not have an equal distribution across the field. Five were arranged almost in a row in the eastern part and three in the western, none in the centre. This may be explained by differences in vegetation at the Oberesch at the time of the battle and during the time of burial. It is a plausible

Fig. 6.7. Bone pit 7 during excavation.

scenario that where the vegetation consisted of open grassland during the battle, bushes and small trees must have grown considerably during the following years, i.e. a dense vegetation of this type might have prevented Germanicus' troops from recognising and collecting the bones that remained here. However, in the wooded areas with rather open undergrowth more bones would have been visible on the surface, even several years after the battle. This may explain the presence of five pits in the eastern part of the field where, because of the wet terrain, the presence of trees can be expected at the time of the battle and during the following years. In other words, the opportunity to find and collect the remains of Roman soldiers some years after the battle depended largely on differences in vegetation rather than on the distribution of bodies. The distribution of bone pits does not therefore reflect the distribution of bodies immediately after the battle but the situation years later when the remains of the dead were to be collected.

From this general overview of the Oberesch bone-pits we move on to consider the interesting arrangement of bones in one of these pits, which does not in fact fit our earlier general observation that the pits only contain single-bones without any

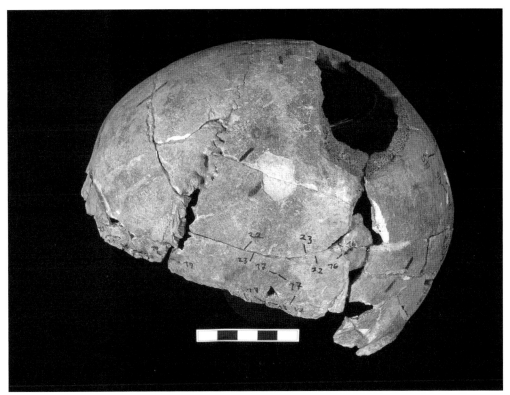

Fig. 6.8. Human skull with a sword cut in the rear region of the cranium. The cut measures 6.7 × 4.0 cm on the interior of the cranium and 8.8 × 5.0 cm on its exterior (other marks relate to restoration of the cranium fragments).

skeletal context. In pit 1 (Figs. 6.3 and 6.4) the osteologists found a few concentrations of bones from fingers and hands, which allowed the reconstruction of three nearly complete human hands (Fig. 6.9); there was also one forearm (Fig. 6.10) and a human skull with the mandible in its original position, together with two vertebrae from the upper neck, and a part of the shoulder with fragments of the chest (Großkopf 2007, 166f.). Since there are no natural causes for these more complete skeletal elements, we have to look for possible anthropogenic explanations (Großkopf 2007, 167, 175), for example, the use of special military equipment like gloves, but the Romans did not use such objects. Besides, not only hands but different parts of the body are represented by the bone combinations, like the head with neck and upper extremities. These are zones with the highest risk of being wounded for a fighting Roman legionary. Considering this aspect, we have to take into account other factors which might be responsible for these unusual bone finds.

The most likely reason seems to be medical treatment. The battle of Kalkriese took place in a defile. Marching from east to west through the narrow passage between the bog and the Kalkriese hill, the Roman force was attacked at several locations along

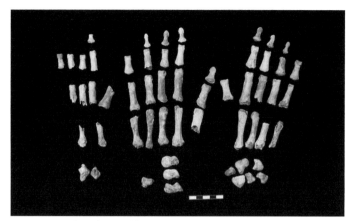

Fig. 6.9. Phalanges, metacarpal and carpal bones from bone pit 1 which allowed reconstruction of three human hands (reconstruction B. Großkopf, photo S. Hourticolon, Göttingen).

Fig. 6.10. Human forearm from bone pit 1 (photo Th. Finke, Göttingen).

a route of more than 10 km. Medical instruments found in Kalkriese (Harnecker and Tolksdorf-Lienemann 2004, 58 No. 792, Pl. 3; Harnecker 2008, 21 No. 328, 329, Pl. 22) show that this army was attended by a medical field service. We may assume that at least at the beginning of combat – as long as Roman military logistics functioned – wounded legionaries must have received medical treatment.

The explanation for these unusual bone finds from pit 1 might be that Roman soldiers, having been wounded and bandaged some kilometres further east, were transported with the intact parts of the Roman army until they reached the Oberesch where the army with all its military services and transport systems collapsed. Even those soldiers who had been wounded and bandaged in earlier encounters and had reached this place were killed in this critical phase of the battle. Like all the other dead and wounded soldiers, they were looted and left on the surface, unburied. Some years later when any skeletal context was gone the last remaining bones were buried in pits. As we shall see, remains of bandages may have survived until that time. When the bones were collected to be buried in pits, the bandages may have contained bones of those parts of the bodies which had been injured some years before. Germanicus' soldiers may have collected, for instance, the bones of a hand together with the bandage which functioned like a bag in which they could be easily carried.

To prove this pattern, we need more detailed information about the bandages which were common in the Roman period. Furthermore, we have to investigate whether bandages lying on the ground could still be preserved after six years. Illustrations of Roman medical field service are rare. There is one well known example from Trajan's column in Rome which shows medical attention for injured soldiers, but it gives limited information. We do, however, know astonishing details about Roman medicine from written sources. During the first half of the first century AD Celsus, a Roman encyclopaedist, wrote eight books about medicine (*De Medicina*). They contain quite precise information about the knowledge of Roman surgeons and their methods of treating wounds. Celsus mentions bandages not only for wounded arms and legs but also for heads and shoulders (VIII 4, VIII 8.1). For the preparation of bandages, he wrote:

> *"The bandage too for binding up a wound is best made of linen, and it should be so wide as to cover in a single turn, not the wound alone but somewhat of its edges on either side.... Moreover, the wound is to be bandaged so that it is held together, yet not constricted.... In winter there should be more turns of the bandage, in summer just those necessary; finally, the end of the bandage is to be stitched by means of a needle to the deeper turns; for a knot hurts the wound, unless, indeed, it is at a distance from it."* (Celsus V 26.24).

In the case of a broken forearm he describes an additional sling comparable with those still in use today:

> *"...and after applying the bandage to the forearm it is most comfortably placed in a sling, the broader part of which encloses the forearm, whilst its tapering ends are knotted around the neck. And thus the forearm is comfortably slung from the neck..."* (Celsus VIII 10.3).

These descriptions do not only reflect a very substantial medical knowledge, they also show that such bandages must have been quite resistant: they consisted of several turns and were also stitched. In addition, the linen used by the Romans consisted of much more stable fibres than modern cotton. Especially in combination with slings the bandages could have been preserved - even after some years - in a condition which allowed them to be carried together with the bones they still contained and be deposited as small 'bags' in the pit. During the last 2,000 years, the bandages have perished but may be indirectly identified by the unusual preservation of these bone ensembles.

Studying Celsus reveals further information that supports our hypothesis. He describes several ointments and plasters which were used as applications for bleeding wounds to suppress inflammation. Some of the most popular consisted of ingredients like wax, verdigris, resin, pitch and bitumen (Celsus V 19), which were not only absorbed by the wound but also by the bandages. One can imagine that a piece of linen soaked in pitch or bitumen can resist the process of decomposition much longer.

Assuming that bandages were responsible for the exceptional preservation of skeletal fragments in one of the Oberesch bone-pits, this would imply that the place where a soldier died may not be identical with the location where he fought and was injured. It seems unlikely that the medical treatment of the wounded soldiers took place in the vicinity of the pit since it is located only 20 m north of the rampart (cf. Fig. 6.2), i.e. in an area of intensive combat which could hardly be used for medical care. Thus, even the corpses and bones of dead soldiers do not inform us about the intensity of the action at a particular site. These considerations have an impact on the interpretation of battlefield remains in general.

The transport of wounded soldiers by their comrades also implies the transport of equipment since the men were not recovered without their clothes. In particular equipment attached to the body should therefore have been removed from sites of first skirmishes. This means that the military equipment of injured soldiers might be transported by the intact parts of the army from the place where they were wounded to the site where large parts of the army collapsed. We have to realise that the transport of wounded soldiers may have resulted in a reduction of military equipment in those areas of the defile where the military organisation was still intact. On the other hand, equipment belonging to those who had been transported as invalids must have been left at the place of the final military debacle. Therefore, we have to conclude that the number of archaeological finds which we usually interpret as indicators for the intensity of fights in different parts of the battlefield may be affected by such processes to a considerable degree. In Kalkriese nearly 90% of the finds are concentrated on the Oberesch, a phenomenon which presumably is not only the result of intensive combat in this place but was also produced by the transport of wounded soldiers and their equipment from the site of earlier fighting. Roman finds west of the Oberesch may in contrast reflect the areas of final skirmishes and flight (Rost 2007, 56; 2009, 70 f.).

However, the main reason for the great quantity of fragments of Roman equipment which were left at a central place like the Oberesch to survive until today seems to be that the Germans as victors were able to strip the bodies systematically and brutally when they looted this site; this seems to have produced many of the small fragments of soldiers' equipment fixed to the body which were sometimes overlooked by the looters, such as buckles and plates from armour, hooks from ring mail shirts, scabbard fittings, belt buckles and apron fittings (Fig. 6.11).

Taking into account all the indicators that suggest that Roman troops were completely annihilated in the area of Kalkriese, we can deduce that this must be the site of the Varus battle. Later campaigns under Germanicus and Caecina would not have resulted in any comparable archaeological remains, since these commanders did not suffer a total defeat and were therefore able to care for and retrieve their wounded and dead soldiers (Tacitus *Ann.* 1, 64).

As we have seen, a number of methodological problems have to be solved in order to analyse an ancient pitched battle by modern archaeological and multidisciplinary methods for the first time. The burials in the bone-pits, which may also give indirect hints of logistic processes in the course of action in the defile, allow us to understand

Fig. 6.11. Roman legionary with equipment; shading indicates those items that were found in Kalkriese (following Horn, H. G. (ed., 1987) Die Römer in Nordrhein-Westfalen, Figure 1. Stuttgart, Theiss Verlag).

a series of events beyond the action which took place some years after the end of the fighting. These 'graves' are an exceptional kind of archaeological resource, and we want to examine them in further detail.

Without doubt the corpses had lain on the surface and had been left to decompose for several years before fragmented remains of the bones of men, mules and horses were deposited in the pits. Human bones, however, predominate over animal bones in the pits, and we notice, for example, that some human skulls were deposited carefully. Crucial for further classification of these features is the answer to the question of who carried out the funerals. At the current stage of research there are many arguments in favour of burial by Roman soldiers. It is less plausible to interpret the burial as an act of the Germans. If the Germans had wanted to clean the battlefield to use the land for farming purposes they could have buried the dead in mass graves, and they would have done that a short time after the battle to avoid epidemics and the penetrating smell of rotting bodies. However, we do not have any indication of an immediate use of the field for farming, neither from archaeological sources (Wilbers-Rost 2007, 83; Tolksdorf-Lienemann 2007, 180) nor from the description of Tacitus of the visit of Germanicus who still noticed fragments of bones and weapons lying on the surface (Tacitus *Ann*. 1, 61).

If the Germans had created sacrosanct areas, like the cult and ritual activities Tacitus describes (*Ann*. 1, 61; von Carnap-Bornheim 1999, 504; Rost 2009, 73ff.), it would also be elusive why they should have made huge efforts after some years to inter these tenuous near-rotten remains of skeletons: for the destruction of the fragments simple ploughing would have sufficed. At best, we might reconstruct a pattern of rites which caused the battlefield at first to be a zone of taboo, suspended years later in combination with disposal of the last surviving bones in pits. Such transformations of cult sites where dead soldiers from a battlefield were deposited are known from Celtic contexts – for example Ribemont-sur-Ancre (Brunaux 2008, 339 ff.) – but this procedure is not known in a Germanic setting.

On the contrary, Tacitus (*Ann*. 1, 60–62) reports that Germanicus visited the location in AD 15, i.e. six years after the Roman defeat, to bury the remnants of the dead soldiers. The Romans noted fragments of bones and weapons scattered across the field, and they buried bones without being able to decide between members of the family or strangers:

> "And so the Roman army now on the spot, six years after the disaster, in grief and anger, began to bury the bones of the three legions, not a soldier knowing whether he was interring the relics of a relative or a stranger, but looking on all as kinsfolk and of their own blood, while their wrath rose higher than ever against the foe. In raising the barrow Caesar laid the first sod, rendering thus a most welcome honour to the dead, and sharing also in the sorrow of those present." (Tacitus *Ann*. 1, 62, 1).

With the bone pits on the Oberesch we may have indications of this event, but there are some discrepancies between the written sources and the archaeological features at Kalkriese. Germanicus is said to have dug the first grass sod to build a grave mound (*tumulus*), but we have not yet found any traces of a *tumulus*. This does not undermine our interpretation, because the chances of discovering the remains of such a mound are not very good, especially if we take into account the fact that according to Tacitus (*Ann.* 2, 7, 3) this *tumulus* was destroyed by the Germans soon after and was not restored by Germanicus during his campaign in the following year. Thus, we may expect traces of a ditch at best, if such a structure was dug at all.

The distribution of the bone pits on the Oberesch site – there is a distance of about 200 m between the pits in the east and in the west (cf. Fig. 6.2) – indicates that they may not be interpreted as a central burial below a collective grave mound. But we should critically review Tacitus' text and ask to what extent we can actually assume a central *tumulus* as the result of these burials. The erection of such a monument on a battlefield is, in any case, exceptional in the Roman period (Hope 2003, 90f.). Among the three examples of war memorials Valerie Hope mentions only one – at Adamklissi – that was in fact realised (ibid. 91 f.). The second – and earliest – example Hope describes was never built; it was merely an idea of a monument developed by Cicero (*Phil.* XIV, according to Hope 2003, 90f.). With this in mind we have to ask, in relation to the third example, i.e. the tumulus erected by Germanicus (Hope 2003, 91), how a rhetorical element of this type might have affected Tacitus' report too. Therefore, one should not exclude the possibility that the description of Tacitus was exaggerated as virtually a *topos* for an adequate burial. That the mound, in contrast to the Drusus Monument, was not re-erected after its alleged destruction (Tacitus *Ann.* 2, 7, 3) may hint at its marginal importance (Hope 2003, 91).

Another aspect concerns the typical Roman burial rite of cremation: the bones in the pits from the Oberesch have not been burnt, though we should expect this according to Roman burial custom at the beginning of the first century AD. Some have taken this as a hint that it may not have been Romans who buried the bones in Kalkriese (Zelle 2008). Primarily however we have to ask if burial rules from civil or military settings under peaceful circumstances may be taken for granted when dead soldiers were to be buried in the context of a battle. Valerie Hope has demonstrated that the burials of war dead could differ significantly from ideal conceptions due to the need to make compromises (Hope 2003, 85 f., 87 f., 93). In fact, cremation of bodies was sought for (ibid. 87), but the rule could be neglected when the situation required it. Cremation could be abandoned, but it was necessary to cover the dead with earth to let them find their peace (ibid. 87 f.). Under special circumstances one could even avoid this (ibid. 88).

The Varian disaster was obviously a comparable exception; the dead had to be left on the battlefield to whiten. The attempt of Varus' followers to burn his body on the battle site after he had committed suicide (Velleius 2, 119) cannot be taken as the common way of behaving towards the soldiers killed in this action, as the corpse of a Roman senator and governor will have undergone a special treatment and care to prevent it from being desecrated by the Germans – which actually happened when they took his skull and sent it to Maroboduus (Velleius *ibid.*). The situation which apparently resulted in the cremation of dead soldiers after attacks by Batavians against the Roman camp of Gelduba (Krefeld-Gellep) in AD 69 (Reichmann 2006, 502) was quite different from that at Kalkriese: in Gelduba the Roman troops prevailed and had the opportunity to burn the bodies as they remained in the camp for some time afterwards.

After a battle, it seems to have been customary that the dead were buried on the battlefield (Hope 2003, 87 f.; Carroll 2006, 160). These burials did usually not pay attention to the individual, as they were frequently mass graves (Hope 2003, 87), except when officers were to be buried on the battlefield; they may have received special treatment and a separate burial, remaining near to their men (Hope 2003, 90). In contrast to the war dead, soldiers who had died during their service in peacetime (Hope 2003, 85, 87, 90, 93) were buried in regular cemeteries reflecting common Roman burial practices. From the Augustan period, cemeteries with graves of Roman soldiers have been found in Germany east of the Rhine, for example at the fortress of Haltern (Zelle 2008, 15 f.; Berke 1991). In a regular cemetery, however, cremation and burial of the dead soldier, the choice of grave goods and erection of a tombstone, were a private matter that involved relatives or fellows of the deceased, mostly paid by a special burial fund (Hope 2003, 86). In other words, we cannot expect burials, created in an exceptional situation on a battlefield, to follow standard Roman customs. The bone pits in Kalkriese need not contradict Roman practice but reflect the particular conditions of a battlefield.

The conditions of Germanicus' visit to the battlefield and his attempt to bury the remains of the Roman victims were in general rather unusual, even when considering its military context; this made an ideal burial with cremation virtually impossible. Being on campaign in enemy country, Germanicus could be attacked by the enemy at any time. Furthermore, he would have faced particular problems if he had attempted to cremate the remains of the soldiers killed in the Varus battle. At the time of his visit the corpses must have been completely skeletonized and they had no flesh or fat on them. These substances are, however, necessary for the cremation of bodies since the fire needs to be nourished. A significant amount of wood would have been necessary to burn these bones to ash (Großkopf 2009). Besides, seasoned firewood would not have been available; thus, the Romans would have had to expend a very great effort on burning, and the production of smoke would have alerted the Germans.

Therefore, the evidence from the bone pits on the Oberesch in Kalkriese does not contradict their interpretation as the remains of the funerals conducted by the troops of Germanicus which Tacitus records. When Germanicus avoided the cremation of the bones of the dead, which is quite likely according to the written sources (Tacitus *Ann.* 1, 62) and also common practices in case of a war, this did not mean a dereliction of duty. Six years after the disaster Germanicus put right what Romans had not been able to do for their comrades before: he had the human remains covered with earth and thus met the minimal entitlements of the dead as victims of a battle.

An alternative explanation, which sees these bone pits as burials of dead soldiers from the Germanicus campaigns, seems rather unlikely. Based on the evidence from the written sources that describe these later campaigns, we can largely exclude the need for interment only some years after the action. The armies of Germanicus and Caecina, as we said before, never collapsed as completely as the army of Varus and must have continued to be able to take care of their casualties (Tacitus *Ann.* 1, 64). Only the annihilation of an army as in the Varian disaster can create the conditions that prevented the Roman army from immediately caring for their dead and wounded. Only in such a situation do we find large-scale looting and stripping of the bodies and the Roman incapability even to bury their dead provisionally.

Among the historical and archaeological sources which provide an insight in the treatment of the dead soldiers from the Varus battle, there is a monument bearing an inscription which refers to this event: the tombstone of Caelius who died in the *bello variano* (CIL XIII 8648 = ILS 2244 = CSIR D III.1, 1). The stone was put up by his brother, and also commemorates the two freedmen of Caelius. It is uncertain whether this tombstone marked a grave of the *centurio* and – perhaps - his freedmen (Wolters 2008, 100 f.) or a cenotaph, which is more probable (Carroll 2006, 160 f.; Hope 2003, 89; Komp 2009, 42). A cenotaph would fit the interpretation of dead soldiers and even officers left behind on a battlefield who were, in the special case of the Varus battle, interred later by Germanicus, while their relatives occasionally erected tombstones as private memorials.

The investigations in Kalkriese cannot add much to the question of whether Caelius or his freedmen were really buried under the stone, but according to the historical sources (Tacitus *Ann.* 1, 62) and the archaeological and anthropological record, at the time when Germanicus visited the battlefield, the condition of the bones did not allow recognition of individuals. It would therefore have been impossible to bring home the bones of particular individuals and bury them in individual graves.

Summary

In the current state of research, the identification of the Kalkriese battlefield as the site of the historically recorded Varus Battle is not 100% certain but is highly probable. It also seems conclusive that the bone pits from the Oberesch must represent Germanicus' attempts to bury the dead of the Varian disaster as recorded by Tacitus, though lack of comparable 'burials' from other ancient sites of conflict means that we are entering uncharted waters. In any case, we get outstanding insight into the sequence of events of the battle, including the treatment of dead and wounded, starting with the initial care for the wounded, their transport with the intact parts of the army, the abandoning of wounded and dead soldiers when the army collapsed, and the looting by the victors after the fighting had ended, as well as the decomposition of the bodies until only a few remains were left to be buried in pits. It is astonishing to what extent the archaeological evidence from Kalkriese can help us to reveal the complexity of the human activities on this site during and after the battle, and we hope to be able to continue our multidisciplinary analyses in order to clarify not only the course of the fighting, but also the post-battle processes.

Acknowledgements and bibliographic note

We must thank the textile biologist Dr. Inhülsen from the Landeskriminalamt Lower Saxony for the discussion of bandages and plasters, Dr. Ralph Häussler (University of Wales Trinity Saint David) for the improvement of our manuscript and Julia Wonne (University of Münster) for the bibliographic research. This article was finished in October 2009. Since then, some articles and two books have been published which deal with finds and features from the Kalkriese battlefield and with aspects of battlefield archaeology, including the interpretation of the finds distribution, namely: Großkopf *et al.* 2012; Harnecker 2011; Rost and Wilbers-Rost 2010; Rost and Wilbers-Rost 2012; Rost and Wilbers-Rost 2013.

Bibliography

Berke, S. (1991) Das Gräberfeld von Haltern. In B. Trier (ed.) *Die römische Okkupation nördlich der Alpen zur Zeit des Augustus*, 149–157. Münster: Aschendorff.

Brunaux, J. L. (2008) Das Tropaion und Denkmal von Ribemont-sur-Ancre – Von der keltischen Schlacht bis in die Kaiserzeit. In A. Abegg-Wigg and A. Rau (eds) *Aktuelle Forschungen zu Kriegsbeuteopfern und Fürstengräbern im Barbaricum,* 331–344. Neumünster: Wachholtz Verlag.

Carroll, M. (2006) *Spirits of the Dead. Roman Funerary Commemoration in Western Europe*. Oxford: Oxford University Press.

Von Carnap-Bornheim, C. (1999) Archäologisch-historische Überlegungen zum Fundplatz Kalkrieser-Niewedder Senke in den Jahren zwischen 9 n. Chr. und 15 n. Chr. In W. Schlüter and R. Wiegels (eds.) *Rom, Germanien und die Ausgrabungen von Kalkriese*, 495–508. Osnabrück: Universitätsverlag Rasch.

Celsus, *De Medicina* (3 vols). English translation by W. G. Spencer (1960–1979). Cambridge, Massachusetts: Harvard University Press and London: William Heinemann Ltd.

Cornelius Tacitus, *The Annals*. English translation by Alfred Church and William Brodribb. New York: Random House, Inc. reprinted 1942 (www.perseus.tufts.edu).

Großkopf, B. (2007) Die menschlichen Überreste vom Oberesch in Kalkriese. In Wilbers-Rost *et al.* 2007, 157–178.

Großkopf, B. (2009) Kalkriese – Schlachtfeld ohne Massengräber? In H. Meller (ed.) *Schlachtfeldarchäologie. Battlefield Archaeology. 1. Mitteldeutscher Archäologentag vom 09. bis 11. Oktober 2008 in Halle (Saale)*, 81–88. Halle: Landesamt für Denkmalpflege und Archäologie Sachsen-Anhalt.

Grosskopf, B., Rost, A. and Wilbers-Rost, S. (2012) The ancient battlefield at Kalkriese. In M. Harbeck, K. v. Heyking and H. Schwarzberg (eds) *Sickness, Hunger, War, and Religion. Multidisciplinary Perspectives. RCC Perspectives 2012/3*, 91–111. Munich: Rachel Carson Center.

Harnecker, J. (2008) *Kalkriese 4. Katalog der römischen Funde vom Oberesch. Die Schnitte 1-22.* Römisch-Germanische Forschungen 66, Mainz: Philipp von Zabern.

Harnecker, J. (2011) *Kalkriese 5 Katalog der römischen Funde vom Oberesch. Die Schnitte 23-39.* Römisch-Germanische Forschungen 69, Darmstadt/Mainz: Philipp von Zabern.

Harnecker J. and Tolksdorf-Lienemann, E. (2004) *Kalkriese 2. Sondierungen in der Kalkrieser-Niewedder Senke. Archäologie und Bodenkunde.* Römisch-Germanische Forschungen 62, Mainz: Philipp von Zabern.

Hope, V. (2003) Trophies and tombstones: commemorating the Roman soldier. *World Archaeology* 35.1, 79–97.

Komp, J. (2009) Leere Gräber: Das Kenotaph. In H.-J. Schalles and S. Willer (eds) *Marcus Caelius. Tod in der Varusschlacht*, 38–43. Darmstadt: Primus Verlag.

Reichmann, Chr. (2006) Kriegsgräber. In R. Pirling and M. Siepen (eds) *Die Funde aus den römischen Gräbern von Krefeld-Gellep*, 497–512. Stuttgart: Steiner.

Rost, A. (2007) Characteristics of ancient battlefields: Battle of Varus (9 AD). In D. Scott, L. Babits and Ch. Haecker (eds) *Fields of Conflict. Battlefield Archaeology from the Roman Empire to the Korean War. Volume 1. Searching for War in the Ancient and Early Modern World*, 50–57. Westport: Praeger Security.

Rost, A. (2009) Das Schlachtfeld von Kalkriese: Eine archäologische Quelle für die Konfliktforschung. In Varusschlacht im Osnabrücker Land (ed.) *2000 Jahre Varusschlacht: Konflikt*, 68–76. Stuttgart: Konrad Theiss Verlag.

Rost, A. and Wilbers-Rost, S. (2010) Weapons at the battlefield of Kalkriese. *Gladius* 30, 117–136.

Rost, A. and Wilbers-Rost, S. (2012) *Kalkriese 6. Die Verteilung der Kleinfunde auf dem Oberesch in Kalkriese. Kartierung und Interpretation der römischen Militaria unter Einbeziehung der Befunde.* Römisch-Germanische Forschungen 70, Darmstadt/Mainz: Philipp von Zabern.

Rost, A. and Wilbers-Rost, S. (2013) Bestattungen auf dem Schlachtfeld von Kalkriese. In M. Sanader, A. Rendić-Miočević, D. Tončinić and I. Radman-Livaja (eds) *Proceedings of the XVIIth Roman Military Equipment Conference, Weapons and Military Equipment in a Funerary Context*. 37–48. Zagreb: Faculty of Humanities and Social Sciences/Archaeological Museum Zagreb.

Tolksdorf-Lienemann, E. (2007) Ergebnisse der Bodenkunde und der Untersuchungen an bodenlagernden Knochen vom Oberesch. In Wilbers-Rost *et al.* 2007, 179–188.

Uerpmann, H. P. and Uerpmann, M. (2007) Knochenfunde aus den Grabungen bis 2002 auf dem Oberesch in Kalkriese. In Wilbers-Rost *et al.* 2007, 108–156.

Wilbers-Rost, S. (2007) Die archäologischen Befunde. In Wilbers-Rost *et al.* 2007, 1–107.

Wilbers-Rost, S., Uerpmann, H. P., Uerpmann, M., Großkopf, B. and Tolksdorf-Lienemann, E. (2007) *Kalkriese 3. Interdisziplinäre Untersuchungen auf dem Oberesch in Kalkriese. Archäologische Befunde und naturwissenschaftliche Begleituntersuchungen.* Römisch-Germanische Forschungen 65, Mainz: Philipp von Zabern.

Wilbers-Rost, S. (2009) The site of the Varus Battle at Kalkriese. Recent results from archaeological research. In A. Morillo, N. Hanel and E. Martín (eds) *Limes XX, XXth International Congress of Roman Frontier Studies, Anejos de Gladius 13,* 1347–1352. Madrid: Consejo Superior de Investigaciones Cientificas.

Wolters, R. (2008) *Die Schlacht im Teutoburger Wald. Arminius, Varus und das römische Germanien.* Munich: Beck.

Zelle, M. (2008) Überlegungen zum Grabtumulus für die Varusschlacht. *Lippische Mitteilungen aus Geschichte und Landeskunde* 77, 13–21.

Chapter 7

Some recent work on Romano-British cemeteries

Paul Booth

Introduction

Recent developments in Roman cemetery studies in Britain, in terms of theoretical approaches, synthesis and publication of major sites, have provided important material for the subject as a whole. It remains the case, however, that the number of even modestly-sized and well-published cemetery excavations is relatively small. Many new excavations of Romano-British cemeteries and even individual burials therefore have the potential to reveal something unexpected and/or important, on the basis that the existing dataset still gives a far from comprehensive picture, particularly of regional variation in burial practice, and that, consequently, new evidence may fill gaps in the existing picture or cause us to reconsider what we thought we knew. Realising this potential is dependent not only on high quality data relating to graves and their material contents, but the highly variable preservation of skeletal remains means that these may also benefit from close *in situ* observation (albeit not as a substitute for detailed examination in laboratory conditions). The importance that is now rightly attached to the close integration of osteological study with that of all other aspects of burial underlines the advantages gained in having one or more trained osteologists on site during cemetery excavations, rather than leaving all consideration of the skeletal material to the post-excavation phase of the project.

Despite advances in the field and the laboratory, however, the reality of most recent Romano-British cemetery excavations is that they provide relatively little evidence that relates directly to death as a process. This is principally for the simple reason that in most such excavations, and certainly in the cases referred to here, the ground surface contemporary with the burials in question does not survive, so that evidence for a whole range of potential actions carried out in the vicinity of individual burials has been lost – for the most part irretrievably. There can be methodological reasons for this, particularly related to the ways in which overburden is removed from excavation sites (Weekes 2007), but in most cases the relevant horizons have long since been truncated by agricultural and other activity and are not available for examination, even with the most sensitive excavation techniques. This problem is particularly applicable in the case of inhumation burials – a variety of activities associated with the processes of cremation may leave traces in subsurface features (see McKinley; Lepetz; Ortalli; Pearce; this volume), but with inhumation it is only very rarely possible to identify stages in the rituals which resulted in the placement of a body within the grave. What is usually encountered is therefore the phase of longest duration in the 'process' – the final resting place, but it is generally isolated from what may have been a complex sequence of stages leading up to it. As has been observed before (e.g. Esmonde Cleary 1992, 29) the actual process of interment may have been, for many participants in the rites, one of the less important stages.

Nevertheless, evidence may be recoverable for broader patterns of activity which, when considered on the basis of a sufficiently large sample of sites, and sufficiently closely, may provide a more nuanced understanding of the range of burial practices which can in turn help to illuminate more detailed aspects of funerary ritual. The present discussion is therefore more concerned with these broader patterns of practice than with fine detail and will consider some aspects of regional variation in Romano-British burial practice and of the spatial context of selected individual cemeteries.

In view of the compelling evidence for regionality in Britain with regard to settlement patterns, their character and chronology (e.g. Taylor 2007), and the new insights revealed by a range of artefact studies (e.g. Crummy and Eckardt 2003), it seems reasonable that we should expect regional variation in burial practice throughout the period. While such variation has long been recognised in the Iron Age (although on the basis of apparently very localised traditions), consideration of its implication for the Roman period has been less explicit, with a few obvious exceptions. Recent work is beginning to emphasise diversity of tradition in this respect in the late pre-Roman Iron Age of south-eastern England, for example (Hill 2007, 28–30; Hamilton 2007, 90), but although indications of the degree of variation in Romano-British practice are to be found in Philpott's (1991) major survey, and elsewhere (see particularly Esmonde Cleary 1992), this theme can now be developed much further. Equally, consideration of the location of burials in relation to associated settlement sites can be informative. The question has been discussed in part by John Pearce (e.g. 1999; 2013, 79–110) and Simon Esmonde Cleary (2000) but can also be thought of in different ways, perhaps particularly in relation to movement through local landscapes between settlement and cemetery (see further below).

The purpose of this contribution is therefore to consider aspects of the evidence from a selection of relatively recently-excavated sites principally located in three regions, the south-east (mainly Kent), the Upper Thames Valley and Gloucestershire, and the west Midlands, and set them against the generally-agreed understanding of broad trends in development of funerary practice in Roman Britain. As is well known, the 'standard' early Roman burial rite is that of cremation, gradually replaced by inhumation burial, perhaps from as early as the mid-late second century onwards, with the change largely completed in the second half of the third century (e.g. Crummy

1993, 264). While this statement parodies the known position, in which a number of local exceptions to these trends are already apparent, it can be argued that the picture is even more complex and that understanding of Romano-British burial practice will be enhanced only if this is more widely recognised. It may be useful to pay some attention to the interaction of successive 'imported' traditions with a number of pre-existing regional ones and also to consider the importance of context – in the broad sense of associated settlement type, as a significant factor affecting variation in the choice of burial rite. These are not new ideas, but some of the data are new and will help to improve present understanding.

From Kent to the West Midlands

Kent is an area with very significant potential for observing interaction between existing and new/imported practices. Here the established pre-conquest burial rite has generally been considered to be that of cremation in the Aylesford-Swarling tradition. The tradition seems to have been introduced from the near continent in the late Iron Age, perhaps no earlier than the beginning of the first century BC (e.g. Hill 2007, 28). Cemeteries are characteristically small and may be entirely pre-conquest in date or extend into the early Roman period, spanning the conquest period. Work on the Channel Tunnel Rail Link Section 1 (now known as High Speed 1 – HS1) in Kent has produced a range of evidence from sites encountered in excavations and watching briefs along the route of this major development (Booth 2011). A typical cemetery from Saltwood Tunnel (Fig. 7.1), contained ten cremation burials, all but one urned, dating from the late Iron Age to the end of the second century AD. Other burials encountered in the HS1 work, occurring individually or in small groups and for the most part loosely associated with rural settlements, are almost invariably of this type, although the proportion of urned burials varies (Table 7.1). The number of burials involved raises questions, however, because, with the possible exception of Saltwood Tunnel, none of these groups can represent even short term family burial plots, and at Saltwood the chronological range of the burials might preclude such an interpretation.

Small groups of cremation burials therefore do not tell the whole story of burial in a rural settlement context. Elsewhere in Kent, for example in Thanet (Andrews *et al.* 2015) and at Mill Hill, Deal (Parfitt 1995), it is clear that there was an important middle to late Iron Age inhumation tradition, of greater antiquity than the cremation tradition. Like the latter, the inhumation tradition also survived into the early Roman period. Given its apparent time-depth it is curious that the evidence for this tradition seems to be quite patchy and, as can be seen from Table 1, it appears to be absent from late Iron Age/early Roman rural settlements on the HS1 route, for example, with the exception of occasional neonatal burials which in any case probably fall outside this framework. This raises many interesting questions about the significance of the spatial patterning of the different burial traditions (do they reflect the distribution

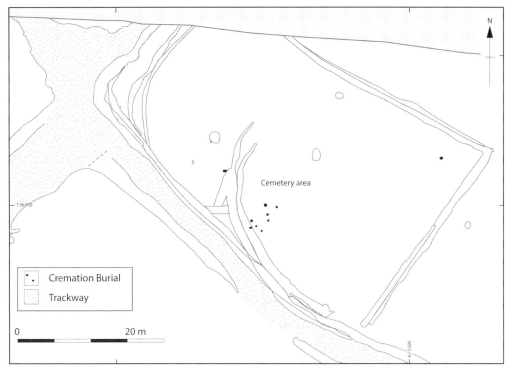

Fig. 7.1. Plan of Saltwood cemetery, Kent .

of distinct social groups, or is the difference status-related, for example?) that cannot be considered here. At the north-west end of the HS1 route at Pepper Hill, Springhead, however, a very different situation was encountered (Biddulph 2009). Here, a few hundred metres south of the small town/religious complex based around the line of Watling Street, was a major and almost completely excavated cemetery in which inhumation and cremation traditions were inextricably intermingled (Fig. 7.2, Table 7.2).

A single inhumation burial at this site was assigned to the middle Iron Age on the basis of a radiocarbon date. Otherwise, although an origin in the late Iron Age is possible, the ceramic evidence suggests that the cemetery was entirely post-Conquest in origin (i.e. after AD 43), in which case it was clearly established very early in the Roman period. Analysis of the cemetery has produced many interesting and important insights, but the simple point to be made here is that the inhumation rite was firmly and probably better established than cremation at the very start of the Roman period. There is no reason to suppose that either of the major (inhumation and cremation) traditions at Pepper Hill represented anything other than a continuation of pre-Roman practice, even if in many cases this now incorporated the use of an expanded repertoire of (for example) pottery vessel forms as grave goods. The prevalence of early Roman inhumation at Pepper Hill forms a marked contrast with the next

Table 7.1. Late Iron Age (LIA, c. 100 BC–c. AD 50) and Roman burials from High Speed 1 (HS1) sites other than Pepper Hill. Sites are listed in geographical order from north-west to south-east (Early Roman c. AD 50–125/150, Mid Roman c. AD 125/150–250)

Site	Date	Inhumation burials	Cremation burials	Disarticulated & ex situ bone	Comment
Northumberland Bottom	LIA/ERB	2 neonates	1 unurned, 1 urned	2 adults	
White Horse Stone	Roman			1 fragment	probably redeposited IA
Thurnham	c. AD 120	1 neonate			probable villa foundation deposit
	late 3rd century	4–8-month infant			in coffin in stone lined cist
Snarkhurst Wood	LIA/ERB		1 unurned, ?1 urned		pedestal urn in unexcavated feature
Chapel Mill	LIA/ERB		2 unurned		
Leda Cottages?	LIA/ERB			1 redeposited cremation	
Tutt Hill	LIA/ERB			cremated fragments	
Beechbrook Wood	ERB		5 urned		in southern part of site
	ERB		?2 unurned		in northern part of site; possibly redeposited pyre debris
	?Late Roman		1 unurned		?auxiliary vessel 120–220, C14 date 220–420
Boys Hall	LIA/ERB		3 unurned, 2 urned		2 unurned cremations have associated pottery vessels
Bower Road	MRB		1 urned		
	4th century			in 2 contexts	
Little Stock Farm	Roman uncertain		?1 unurned		
Saltwood Tunnel	LIA/ERB		1 unurned, 9 urned		'western group'
	LIA/ERB		4 unurned		'eastern group'
	4th century	1 adult			

largest individual cemetery known from Roman Kent, that at Ospringe, with some 387 burials (Whiting *et al.* 1931). Here again inhumation and cremation burials occurred side by side, although the latter were in the majority (ibid., 4, 6), but on the basis of the pottery associations none of the inhumations is dated before the later second century (M. Lyne pers. comm.) and cremation was the dominant (and potentially

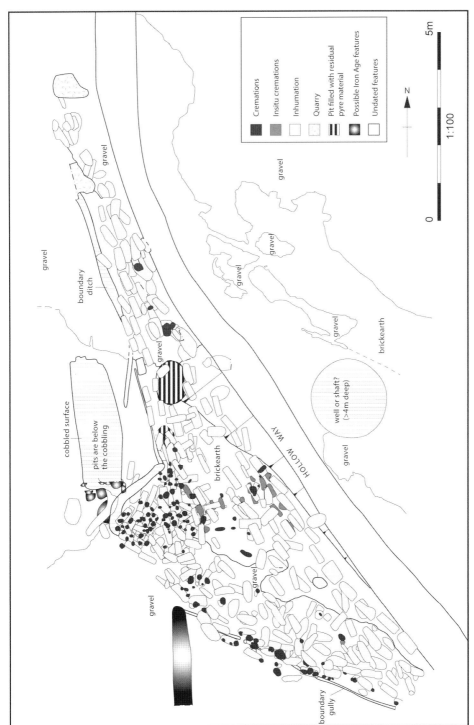

Fig. 7.2. Plan of Pepper Hill cemetery, Springhead, Kent.

Table 7.2. Pepper Hill; quantification of funerary feature type by period ('Other' = other funerary related feature)

Phase	Inhumation burials	Cremation burials	Cenotaph/ disturbed crem. burials	Busta	Pyre sites	Other	TOTAL
Middle Iron Age	1						1
Late Iron Age-early/ middle Roman	193	92	17	6	13	7	328
Middle Roman	43	34				2	79
Middle/late-late Roman	17	7	1				25
Roman uncertain	95	12	8	1	3	8	127
Total	*349*	*145*	*26*	*7*	*16*	*17*	*560*

exclusive) early burial rite. Here, therefore, at a significant roadside settlement, is an example of what might be regarded as a 'typical' Romano-British sequence, whereas at Springhead, some 40 km further west on Watling Street, a very different pattern is evident. By contrast again, elements of both patterns may be observed in recent work in Thanet (e.g. Egging Dinwiddy and Schuster 2009, 93–113; Andrews *et al.* 2105). The explanation for this contrast is not clear, but it is possible that the late Iron Age settlement and religious background of Springhead was the principal influencing factor, even though the Pepper Hill cemetery itself was not established until a little later. Notwithstanding the very significant component constituted by burials in the local tradition, the cemetery also received burials of radically different and novel character, of which the *bustum* burials (see Table 7.2) are the most obvious example.

Remarkably, more recent work on the A2 barely two km east-south-east of Pepper Hill, at a location almost certainly related to the CTRL site of Northumberland Bottom, has revealed early Roman high status cremation burials (Allen *et al.* 2012, 322–386) of a different character to the Pepper Hill graves, possibly forming a similar contrast to that seen later between the Pepper Hill burials and the well-known antiquarian finds from the walled cemetery only 300 m distant from that site (Davies 2001). The principal and perhaps earliest of the A2 burials was located in an isolated pit 2 m^2 (Fig. 7.3). The cremated human remains lay on the base of the pit. Sixteen pottery vessels, including a terra sigillata cup and cups and dishes of terra nigra and terra rubra, were set on and around an item of furniture at one side of the pit, and a large bronze cauldron and a patera and jug were also placed close by. A brooch, part of a pig (including the head) and a gaming board and glass counters were amongst the other objects placed in the grave. The other important burials lay in a north-south row some distance away. Two cremation graves were in pits roughly 1 m square. The more northerly produced another patera and jug set, a brooch and a gaming board (but apparently no pieces to go with it), as well as a box containing a cosmetic mixing palette and toilet implements and 14 pottery vessels, including a decorated

Fig. 7.3. A cremation burial dated c. AD 50–65, A2 site, near Northumberland Bottom, Kent.

terra sigillata bowl – an unusual item in a Romano-British grave – while the adjacent burial to the south had only five vessels, most of which, with the cremated remains, may have been placed in a wooden box. Also within the grave was another smaller box containing a bronze mirror and an unguent bottle. A further cremation burial immediately to the south and a north-south aligned inhumation to the north were both less well-equipped and later in date than these two, but clearly related to their location. The three richest burials all probably date to the period AD 50–65.

This group of burials juxtaposes what is a chronologically 'normal' late Roman inhumation burial with cremation burials some two centuries older, the location of which was clearly still understood in the later third century. As for the earlier, pre-Roman south-eastern inhumation tradition, its geographical extent is still unclear and it may have been much more pervasive than has been thought. Individual examples of late Iron Age or very early Roman inhumation burials are known in the London area and inhumation burials are assigned to the broadly dated Period 1 in the East London cemetery (Barber and Bowsher 2000, 300). Unfortunately, the date range of Period 1 activity in that cemetery (AD 40–197) potentially encompasses burials both in the early native tradition and instances from the mid-late second century onwards. These could represent either a survival of that tradition or the 'reintroduction' of inhumation from the continent, and it is not clear how many of each category is present. A few of the east London inhumation burials clearly predate the late 1st century, however (e.g. B435, dated AD 40–80; ibid., 193–195), although they are presumably a minority of the c. 68 inhumations notionally assigned to Period 1 (ibid., 12, table 4). Isolated early burials are known for example from the Tower of London (Parnell 1985, 5, 7) and Southwark (Dean and Hammerson 1980) and two early or mid-second century crouched inhumation burials at the Stratford Market Depot site, West Ham (Hiller and Wilkinson 2005, 17–20) may represent the survival of Iron Age practices. The Kent evidence thus appears to support the conclusion that the early burials in the eastern London cemetery possibly 'reflect[s] a pre-Roman inhumation tradition in the London region' (Barber and Bowsher 2000, 300). It is not even certain if this tradition should be seen as restricted to the region or as part of a more widespread but as yet poorly-understood pattern of late Iron Age inhumation burial.

Further west, the Upper Thames Valley is an area almost totally bereft of any significant evidence for burial in the late Iron Age or early Roman periods. In terms of indications of pre-Roman burial traditions, a group of up to 35 inhumation burials at Yarnton, mostly crouched and firmly dated to the middle Iron Age (Hey *et al.* 1999), is quite exceptional in the region, particularly in view of the number of Iron Age settlements that have been excavated in the Upper Thames Valley with no significant evidence for burial (ibid., 560). The occurrence of disarticulated human bone, a widely recognised feature, as for example at Gravelly Guy, Stanton Harcourt, appears much more characteristic, although the latter site also included a few more complete crouched inhumation burials (Lambrick and Allen 2004, 223–236). At present the Yarnton burials do not appear representative in a regional context and their significance is therefore uncertain. Identification of these burials underlines the importance of improved chronological definition, but understanding of the development of funerary ritual will be impossible if the chronology of distinct traditions is not firmly established. The dating of the Yarnton cemetery was entirely dependent upon radiocarbon. Further significant contributions from radiocarbon dating are mentioned below.

Late Iron Age to early Roman cremation burials are extremely rare in the Upper Thames. Their presence may be anticipated, particularly in association with the larger nucleated settlements such as Cirencester, Alchester and Dorchester on Thames from the early Roman period onwards; it has in fact been demonstrated very recently in a rural settlement context at Gill Mill, South Leigh (Oxfordshire) where a group of about a dozen such burials is now known. The group is dated on the basis of two characteristically first-century vessels used as urns. It is otherwise remarkable for its isolated location, apparently several hundred metres from contemporary settlement features and with no clear definition in terms of an associated enclosure (Booth and Simmonds forthcoming). Another notable early Roman rite occasionally encountered in the region is that of cremation burial within a small square ditched enclosure. Interestingly, the few known examples of this practice occur in rural contexts, at Duntisbourne Abbots, Glos (Lawrence and Mudd 1999), Roughground Farm, Lechlade (Allen *et al.* 1993, 52–53) and Appleford (Booth and Simmonds 2009, 38–40). A related practice may be seen in the late Roman period (see further below), but another unusual cremation burial from this area, not well-dated but perhaps early Roman on the basis of the rite, is one contained in a lead lined stone cist, probably associated with a villa site at Harnhill (Wright 2008), just east of Cirencester. In contrast, recent work, again at Gill Mill, has revealed the first example of an early Roman barrow burial to be excavated in the region, located at the margin of a substantial agglomerated rural settlement. The ring ditch, with an internal diameter of 16 m, surrounded a central wood-lined chamber 3 m × 1.8 m aligned north-east to south-west and containing the extended inhumation of an adult male. A single pottery vessel within the chamber (the only other grave good was a chicken) and material from the fill of the surrounding ditch indicate a second century date for the burial. The existence of a mound above the chamber is suggested by the configuration of later Roman features – ditches of a field system to the south-west and a small stone-built structure (probably domestic in function) to the north-east. These features seem to have respected the location of a mound at a time after its associated ditch had silted up (Burnham *et al.* 2008, 308–309). It seems that potentially high status burials in early Roman rural contexts in this region were marked by distinctive rites without obvious local antecedents. The Tar Barrows at Cirencester may be particularly examples of this, but there is no direct evidence for their date at present. Meanwhile the bulk of the contemporary rural population apparently remained invisible, a situation which did not change significantly before the later third or fourth century.

A little further west again, in the lower Severn Valley, significant evidence for early Roman rural cemeteries has recently been recovered for the first time. Inhumation burials are now known at some rural settlement sites in the vale (Holbrook 2006, 121) and may have links to a local pre-Roman tradition of which, however, the best-known examples are mostly crouched rather than extended inhumations, although occasional examples of the latter practice do occur (Staelens 1982, 28–29). The potential complexity of pre-conquest practice in this region has been discussed recently by Moore (2006, 68–80), but the difficulty with what is still a very small, if diverse, data set is to try and understand what may have been typical, and indeed to establish if any of the visible rites represent anything other than site- or individual-specific practice. It is this distinction that makes the new early Roman material important. Although the groups of burials are small, the numbers may represent communities rather than individuals and thus potentially provide an indication of more widespread practice. The rural settlement associations of these burials are also significant, so much of the earlier material having lacked any meaningful settlement context.

The initial impact of the Roman conquest on burial practice here as elsewhere will have been seen most clearly at the early military and urban centres. At Gloucester, parts of the main northern cemetery have been examined recently at London Road (by Foundations Archaeology at No. 122, where 19 cremation and 39 inhumation burials were recovered (see now Ellis and King 2014), and by Oxford Archaeology (OA) at Nos 118-120; Simmonds *et al.* 2008). The setting here is of course urban, alongside

Ermin Street, although at some distance both from the early fortress at Kingsholm to the west and from the later fortress and *colonia* to the south-west. Superficially the traditions are, unsurprisingly, those of mainstream urban practice, including in the OA site the erection of two tombstones, one to a legionary veteran and one to a slave (Tomlin and Hassall 2005, 474–477). However, the great majority of the burials in this site were inhumations; 66 graves (four without skeletal remains) as opposed to only eight cremation burials. The chronology of the burials is of considerable interest. On artefactual evidence, all eight cremation burials and four of the inhumations were assigned securely to a phase dated c. AD 60–120, with a number of burials definitely of pre-Flavian date, but the finds suggested that there was then a hiatus in burial activity (with one significant exception to be discussed below) until the late Roman period. Four of the 'undated' inhumations were radiocarbon dated in order to see how late in the Roman period they really were, and in all cases the dates lie within a range from the later first to early 3rd century, but firmly centred around the mid 2nd century. This underlines the importance of the inhumation rite at this time and also plugs a large part of the 'gap' in the sequence suggested by other evidence. Most of the burials were extended and supine, but at least one was crouched, and other examples of crouched inhumations from Gloucester have previously been suggested as representing a survival of native tradition (Heighway 1980, 57; Holbrook 2006, 121). On the basis that the early inhumation burials, whether crouched or extended, most likely do reflect established regional practice, these burials therefore suggest not only the presence of a local element in the population of Gloucester (in itself unremarkable), but an element which retained its burial traditions alongside those of incomers who might perhaps have been expected to dominate practice in the legionary fortress and colonia context. Although the dating is uncertain, recently-reported burials from the south gate cemetery of Gloucester might also have included early Roman inhumations (Holbrook and Bateman 2008, 94).

The most remarkable feature of the London Road site, however, was a mass grave located in the southern part of the site (Fig. 7.4). This feature was roughly 3.5 m² and 0.8 m deep and contained a tangled mass of human remains, not all properly articulated. The minimum body count is 77 individuals – 66 adults, five subadults and six unaged. The overall age range is from c. three years to 50 plus, but the majority are young adults. Both males and females are represented; there is no evidence of peri-mortem trauma or of active lesions. Limited artefact evidence suggests a date in the second half of the second century for this feature and a single radiocarbon date is entirely consistent with this. The feature therefore appears to be a mass burial probably carried out over a very short period of time (perhaps a few days at most), and may most likely represent the result of an epidemic of some kind (though see Esmonde Cleary 2009). For comparative purposes, it may be noted that the mass grave at Towton was c. 3.25 × 2.0 m and 0.65 m deep and contained a minimum of 38 individuals (Fiorato *et al.* 2000): pit volumes are c. 4.225 m³ for Towton and 9.8 m³ for the Gloucester feature. The context of the Towton grave (the aftermath of the

Fig. 7.4. Mass grave under excavation at London Road, Gloucester.

battle of 1461) is, however, quite unambiguous, unlike that of the Gloucester grave, though violent death can certainly be ruled out for the latter.

The West Midlands is an area where the pre-Roman burial tradition is even less visible than in the Upper Thames Valley and the Vale of Gloucester, and evidence even for late Roman burials is generally scarce. Work in the course of another major infrastructure project, the M6 Toll road north and east of Birmingham, has shed some light on the development of burial practice in the post-conquest period. Close to the Roman small town of Wall in Staffordshire, at the junction of Watling Street with the north-south road Ryknield Street, Oxford-Wessex Archaeology Joint Venture examined part of a cemetery in a classic roadside location adjacent to Ryknield Street (Fig. 7.5). The excavated sample comprised 42 cremation burials, at least 15 (perhaps as many as 21) inhumation burials and a variety of other features, including pyre sites, pyre dump deposits and three square ditched enclosures, only one of which, however, contained a burial (McKinley 2008; see also McKinley this volume). The excavated burials may represent only a small proportion of the total number from this cemetery but they suggest its establishment in the later first century AD, with cremation the exclusive burial rite at this stage. Inhumation burial followed later, but unfortunately the chronology of the transformation/overlap of burial rites is not very clear (although it may fit the 'standard' model of introduction of inhumation in the later second century). This part of the cemetery probably did not continue in use after the late third century, and the majority of burials can be assigned to the late first and second centuries.

Although extremely important in regional terms, in many respects the Wall cemetery is unexceptional, and it has several points of similarity with sites such as Little Chester, Derby (Wheeler 1985). Of particular interest here, however, is its

Fig. 7.5. Plan of the roadside cemetery at Wall, Staffordshire.

significance as a benchmark against which to judge burial practice in the contemporary nearby rural settlements. In fact, a single (probably early second century) cremation burial from a potentially high status site (including a large aisled building) close to Watling Street 3 km west of Wall was the only burial associated with any of the five rural settlements examined in this project. The presence of probable hobnails indicates that this burial was of broadly comparable character to many of those at Wall, that is to say that the tradition was essentially an alien one, at least in origin. It is possible that the failure to identify other rural burials in the course of this project is an accident of the location of the excavated sample, exacerbated by soil conditions inimical to the preservation of bone. It seems more likely however that, as in the Upper Thames Valley, an archaeologically largely invisible burial tradition was maintained by rural communities at least through the early Roman period and perhaps later as well.

The difference between these two regions is most marked in the late Roman period. In the West Midlands, that is to say north and west of the valley of the Warwickshire Avon, there is still no meaningful evidence for burial in a rural context at this time: the associations of late Roman burials, such as they are, are with the larger nucleated settlements. The situation in central and southern Warwickshire is clearly a little different, with known late Roman inhumation cemeteries at rural sites such as Wasperton in the Avon valley (much the most important rural cemetery in the region, and recently published; Carver *et al.* 2009) and Stretton-on-Fosse (Ford 2002, 9–21), as well as occurring in association with the larger nucleated settlements such as Alcester and Tiddington. Nevertheless, such sites are not as yet common. In the Upper Thames Valley, in contrast, burial became clearly visible in the archaeological record for this period, as the region finally caught up with 'mainstream' tradition, which by now was inhumation. Inhumation cemeteries and more dispersed small groups of burials are found widely both in rural and in nucleated settlement contexts (for Oxfordshire see Booth 2001). There are occasional and localised instances of cremation burial as well, most particularly at Radley II, near Abingdon, where 12 cremation burials, several of them urned and/or with auxiliary pottery vessels, were found alongside 57 inhumations (Chambers and Macadam 2007, 13) (Fig. 7.6). It may, however, be a mistake to see these cremation burials as a survival of early Roman practice, since such practice was at best only very poorly established in the region, particularly in rural contexts (see above). It is of interest that a majority of the cremation burials at Radley II were associated with a small square-ditched enclosure reminiscent of the rare regional tradition of early Roman date noted above. Such enclosures are also encountered in the region associated with inhumation burials, but these are generally single or at most double burials; the character of this association is more reminiscent of the well-known examples at sites further afield such as Poundbury, Lankhills, the spectacular example from Boscombe Down, Amesbury (with multiple secondary inhumations and later cremation burials in the surrounding ditch; Wessex Archaeology 2008; McKinley, this volume) and the post-Roman examples discussed by Webster and

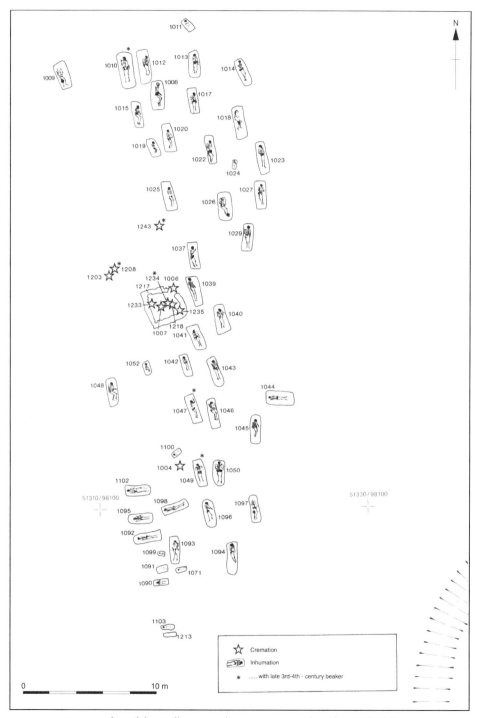

Fig. 7.6. Plan of the Radley II rural cemetery near Abingdon, Oxfordshire.

Brunning (2004). However, a remarkable site at Kempston, Bedfordshire, with both square and oval enclosure ditches associated with at least 11 of the graves in the latest phase of an enclosed late Roman cemetery (Dawson 2004, 225–227), is at odds with the generally western British evidence for this method of defining graves. This may necessitate some reconsideration of this tradition, if indeed it can be regarded in such coherent terms. Upper Thames Valley examples of small burial enclosures include one from Gill Mill, South Leigh (situated some 275 m from the round barrow burial discussed above; Booth and Simmonds forthcoming) and one from Queenford Farm, Dorchester, containing two burials (Chambers 1987, 45). Two conjoined enclosures containing burials at Claydon Pike, Gloucestershire, are of broadly similar character but one had an opening at one corner while the second was completely open on the north-east side. They contained one and two burials respectively, while another five lay close by and a further two some 20 m distant (Miles *et al.* 2007, 184).

A recent and important addition to this evidence is a group of three small square enclosures from Tubney Wood just west of Oxford (Fig. 7.7). These lay just to the west of eight unenclosed inhumation burials (five north-south, three west-east) and certainly of late Roman date. One produced a ?4th century coin and another was furnished with a composite bone comb of late 4th century date. Such objects are found occasionally at a number of late cemeteries in the region (Booth 2001, 26–27), most notably in the primary burial within the only ditched mortuary enclosure recorded in the large cemetery at Queenford Farm, Dorchester on Thames (Chambers 1987; 58; see above). The square enclosures together contained four inhumation burials, with three further burials, two prone and aligned north-south, just to the west. The principal burials within the two best-defined square enclosures, which were linked to a more extensive system of ditches, were also prone adult inhumation burials aligned approximately north-south; no skeletal remains survived in the two graves in the third enclosure. Dating evidence for the ditches was poor, but radiocarbon determinations from two of the burials gave almost identical date ranges of AD 420–540 and AD 425–545 at 95% confidence (Simmonds *et al.* 2011, 172). A 'post-Roman' date is therefore certain.

The relationship between the late and early Roman use of square ditched burial enclosures in the Upper Thames region is therefore unclear. Was the square enclosure ditch tradition a continuous one? The few early examples all seem to be of 2nd century date and are isolated features with either roadside or potentially high status settlement associations, whereas most of the later Roman examples occur in cemetery contexts of 4th century or later date, the principal exception being the isolated (and not certainly late Roman) example from Gill Mill. The examples from Tubney constitute evidence for developing regional practice in the fifth century, if not later. The occurrence of prone burials within these particular enclosures appears problematic, as the prone rite can be considered to represent marginalisation of the individuals concerned (Philpott 1991, 74–75), while the provision of an enclosure appears to be a consistent marker of distinctive and usually (in relative terms) high status. Nevertheless, there are cases where the marginal status of the individuals

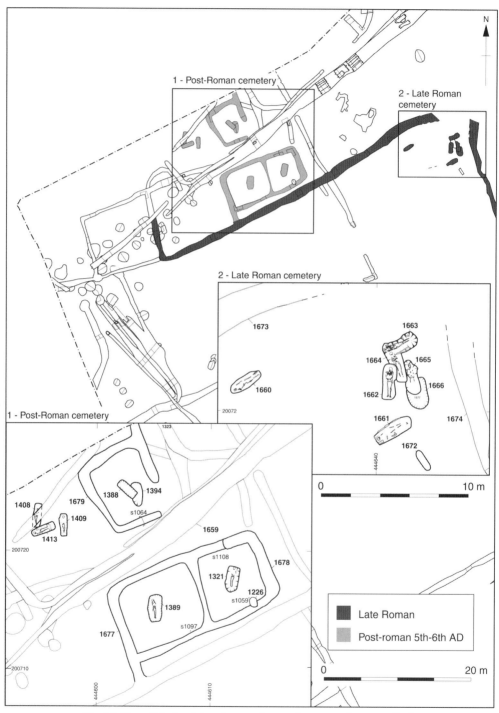

Fig. 7.7. Plan of the burial enclosures from Tubney Wood, near Oxford.

subject to prone burial is less than clear (Taylor 2008, 107–110) and these may be further examples. The provision of enclosure ditches at Tubney makes these burials particularly prominent in the archaeological record, but we may perhaps question whether it was their most important/defining characteristic in the eyes of observers at the time.

The late Roman inhumation cemeteries in the region exhibit most of the well-known characteristics of such cemeteries, including a distinct emphasis on rites such as decapitation, discussed years ago by Harman *et al.* (1981), an emphasis which continues to be well-represented in more recently excavated rural cemeteries. A particularly good example of the latter is Horcott, near Lechlade in Gloucestershire, examined in 2006. The completely excavated cemetery was in two parts, both adjacent to a NNE-SSW aligned boundary ditch, but otherwise unenclosed. The main (southern) part of the cemetery had 57 graves aligned parallel to the ditch, with remains of 61 individuals, nine of which were subadult. Thirteen burials were prone (one decapitated) and six supine burials were also decapitated. Some distance away to the north was a smaller group comprising 17 graves perpendicular to the ditch and one parallel to it, together containing 14 individuals, one prone. Eight of these are subadults and all four graves with no preserved human remains were also child-sized. There was evidence for wooden coffins or stone lining in 10 of the 18 grave cuts in the northern part of the cemetery, so these are about twice as common as in the larger group to the south. One of these coffins was crudely made of flat lead sheets nailed together. A single, isolated SSW-NNE prone burial lay between the two groups (Fig. 7.8).

The contrast in the alignments of the two groups is of note, as is the concentration of sub-adults in the northern group. It is now clear that there were chronological as well as age differences between the groups. All the (relatively limited) artefactual evidence suggests a fourth century date, but a modelled series of calibrated radiocarbon dates indicates an overall range of c. AD 250–350 for the main cemetery, the solitary SSW-NNE burial has a date of cal AD 320–430, while two burials from the northern group are dated cal AD 380–540 and cal AD 420–570 (Hayden *et al.* forthcoming). The latter dates are very reminiscent of those from Tubney (above). It is likely that the cemetery was originally associated with a modest farmstead site located c. 200 m to the west, but on the basis of very recent excavation it seems that the stone building at the centre of this complex was not in use in the fourth century, and the settlement association at that time was presumably with another nearby focus, perhaps in the unexcavated (and now lost) area to the north. It is also notable that an early Anglo-Saxon settlement, currently much the largest known of its kind in Gloucestershire (represented by over 30 sunken featured buildings and several posthole structures), lay immediately west of the north-south ditch; the northerly group of burials lay only a few metres from the two nearest sunken featured buildings. As already mentioned, therefore, assumptions about chronology need to be tested and radiocarbon dating has clarified a sequence of cemetery development that otherwise might only have

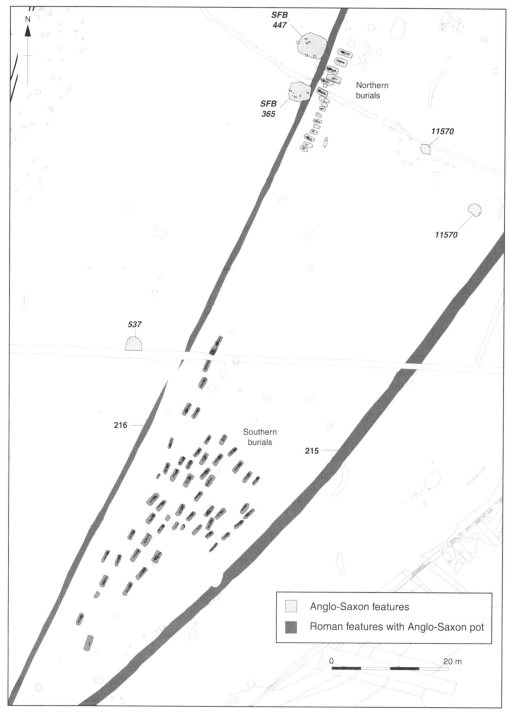

Fig. 7.8. Plan/aerial photograph of later and post-Roman burials at Horcott, Gloucestershire.

been guessed at. Overall, perhaps the most notable feature of this cemetery is the proportion of prone burials, which at 19.7% is significantly higher than values previously recorded for the region (Booth 2001, 24), although a high proportion of prone burials was also a notable feature at Tubney, discussed above (and, incidentally, also at Kempston (Boylston *et al.* 2000, 247)).

Finally, to Lankhills, Winchester. (Fig. 7.9) A second major phase of excavation at this very important site was carried out by OA in three stages from 2000 to 2005 (Booth *et al.* 2010). This added a further 304 excavated inhumation graves, plus 25 cremation burials, to the 444 inhumation graves and seven cremation burials excavated by Giles Clarke from 1967–1972 (Clarke 1979). It should also be noted that excavation by Wessex Archaeology in 2007 has revealed 56 more burials in areas immediately south and north-east of Clarke's excavation which also form part of this cemetery (Paul McCulloch pers. comm.). The general character of the cemetery remains essentially as revealed by Clarke, but it is not entirely uniform. The OA work identified the well-defined northern boundary of the cemetery. Only one further grave surrounded by a gully was encountered, but a complex area of intercutting pits was found in the northern part of the site. These may have provided a focus for a range of broadly funerary-related activities; certainly the cremation burials concentrate in this area. The latter included seven examples of in situ or *bustum*-type cremation burial, all dated to the fourth century and with at least one assignable to the very end of the century. These are important additions to a small but growing corpus of evidence for the practice of this rite in late Roman Britain, other southern examples of which are known at Bray, Berks (Stanley 1972), from work at Denham, Bucks (Coleman *et al.* 2004), and elsewhere.

In the context of discussion of 'death as a process' analysis of the new evidence from Lankhills covers fundamental topics such as the skeletal remains in relation to the orientation and character of their grave cuts, evidence for coffins and other grave structures, and grave goods, including the interpretation of the differences between objects which were worn and those placed elsewhere in the grave, this last an aspect of considerable interest for the earlier excavators (e.g. Clarke 1979, 377). A particularly good example of this distinction can be seen in the coffined burial of an unfortunately rather poorly preserved individual (presumably male) with a fine set of personal accoutrements. His crossbow brooch was found in the vicinity of the right shoulder and was probably worn in a standard approved style. His belt and riding boots, however, had been treated differently, the former laid extended between his legs and the latter, indicated by the attached spurs (the only examples of this type of object from a grave in Roman Britain), placed alongside his right leg. It seems that the distinction between the brooch and the other equipment may have been important, though the reasons for this can only be a matter of speculation at present (Fig. 7.10).

It is perhaps inevitable in a site like this that, initially at least, attention focuses upon the most striking individual burials, but these are important not only in their own right but also in helping refine understanding of what appears normative. The

Fig. 7.9. Plan of the late Roman cemetery at Lankhills, Winchester.

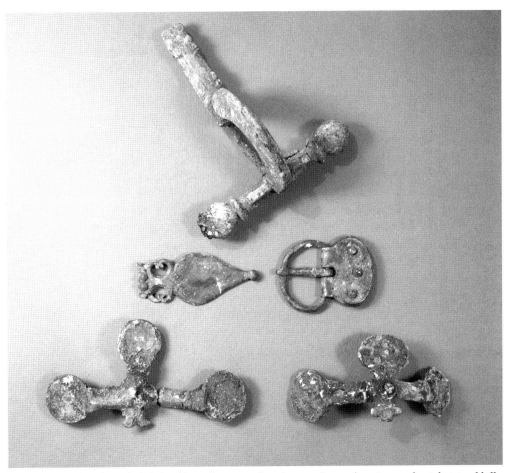

Fig. 7.10. Crossbow brooch, belt buckle and strap end and spurs from a late Roman burial at Lankhills, Winchester.

perennial fascination with minority rites such as prone and decapitated burials is a long way from being exhausted (e.g. Boylston *et al.* 2000; Taylor 2008), though here they are not particularly significant in numerical terms. The recent work identified four additional prone burials at Lankhills and four decapitated ones. Other issues for consideration include non-standard grave alignments; in OA Lankhills at least eight graves, half of which are amongst the very latest graves identified (on both artefactual and stratigraphic criteria), were aligned north–south/south–north. The only two south-north burials in this group were both prone. This association seems to echo one seen at Horcott, but is this meaningful or coincidental?

In addition to these well-established lines of investigation, however, the current analysis includes integration of isotopic evidence with the osteological data. The carbon and nitrogen isotopes give a useful indication of a population with a varied diet, but with hints of differentiation, including an apparent correlation between the more enhanced isotope values and the provision of coffins in the cemetery. This may be the first time that it has been possible to demonstrate from empirical evidence the obvious point that the relatively poor, indicated by the lower carbon and nitrogen isotope values that reflect a less protein-rich diet, were less likely to be buried in coffins than other members of the community (cf. Cummings 2009, 80); in itself this is hardly a radical conclusion, but one that helps extend our characterisation of that community in a new way. Meanwhile, understanding of the origins of members of the community has been enhanced by analysis of strontium and oxygen isotopes, significantly supplementing earlier work (Evans *et al.* 2006) which was particularly directed at examining the origins of an 'intrusive' population element identified by Clarke in the earlier excavation and attributed by him to the middle Danube. Analysis of the more recently excavated group shows that of 40 individuals sampled, as few as 17 may have grown up in the Winchester area (i.e. roughly within a 30 km radius of the town), while a further 12 are likely to have come from other parts of Britain. The remaining 11 are certainly not local to Winchester and have $d^{18}O_{dw}$ values suggestive of origin outside Britain, though some of them fall fairly close to the margins of the British range of such values. At least three individuals were brought up in relatively distant lands. One of these was quite possibly from the Danube region identified by Clarke, but all the other probable non-British individuals in this study are likely to be from warmer regions distinct from this one. Overall both the present study and that of Evans *et al.* (2006) have shown a poor correlation between isotopic signatures of origin and the archaeological criteria identified by Clarke as defining origins in particular parts of the Roman empire. In broad terms, the analysed individuals with striking grave good assemblages emerge with isotopic signatures which suggest that they were of local origin, whereas the isotopically 'exotic' individuals are for the most part not distinctive in terms of their burial rite, and therefore typically 'local' in relation to the characterisation of the 'intrusive' graves offered by Clarke (1979, 377). What the isotopes do demonstrate, therefore, is the diverse nature of the Lankhills community, and careful examination of these results alongside the archaeological material associated with sampled individuals underlines recent work which stresses the complexity of interrelationships between the factors influencing the construction of identity.

Discussion

None of these examples sheds much light on a sequence of distinct stages between the death and burial of the individuals concerned. Indeed, for most of these sites the imagination is given free rein in terms of reconstructing the details of rituals of mourning and commemoration. Nevertheless, the sites draw attention to three closely related areas which may contribute in broad terms towards improved definition of these processes. These are burial tradition, sample size and settlement context. Underpinning all of these is a theme of considerable regional variability.

First, burial tradition. Although many regions, including parts of south-east England, have very poor evidence for Iron Age burial practice, the body of data is increasing and we can additionally draw inferences about such practice, and perhaps in particular the lack of it, from early Roman evidence in a number of key regions. The interactions of established (even if in places archaeologically invisible) practices with those introduced in the aftermath of the Roman conquest are potentially complex; moreover, it is important to emphasise that 'tradition' is just as likely to be dynamic and fluid as timeless and immutable (see Weekes this volume, ch. 11). The processes and chronology of absorption or replacement of one practice by other(s), or modification in response to new influences, are variable from region to region and hugely dependent upon settlement/social/cultural context. So, for example, at Pepper Hill it is arguable that two principal, parallel pre-Conquest rites were carried into the early Roman period while the 'Welwyn style' variant of the cremation tradition was seen close by on the A2 site. The near continental background to the latter has been emphasised recently by Crummy (2007, 451–455). The *bustum* burials are most likely to be an early post-Conquest continental introduction, although whether or not directly indicative of the presence of immigrants is not known. Struck (1993, 84, 92) is rightly cautious about a possible native tradition, and a recently-reported partly-burnt burial from Latton Lands, Wiltshire (Powell *et al.* 2009, 52–53, 97) discussed in relation to this practice is clearly anomalous. Their presence raises an interesting but at present unresolved question about the extent to which other cremation burials in this cemetery derived from established 'native' Aylesford-Swarling practice or reflect a reintroduced Roman provincial tradition. The two traditions may have been of very similar character, at least in terms of the final manifestation of remains in the archaeological record. In regions without a pre-Conquest cremation tradition it is easier to see the appearance of this rite as a straightforward 'Roman' introduction – but what sort of 'Roman' do we mean? Even in a military context there is room for variation. At London Road, Gloucester the milieu is potentially military and the presence of a tombstone of a north Italian legionary backs this up, although unfortunately the stone cannot be associated directly with any of the excavated cremation burials. At Wall the earliest burials are certainly contemporary with military activity, here probably involving an auxiliary unit of unknown origin. It is not possible to say whether these burials are of military personnel, or their dependants, or of other inhabitants of a military-associated community, or how the balance of these elements may have changed during the course of the second century. Either way, the mode of burial is completely new in regional terms and indicates, at least initially, a non-local group. Whether or not local people were subsequently attracted to the nucleated centre at Wall and modified their burial practices in conformity with what they found there unfortunately cannot be known, but the evidence from the adjacent countryside suggests that these new-fangled rites had minimal impact there.

Inhumation burial was variously an established and an introduced rite. Clearly in late pre-Roman Kent it was used by some communities but not others. There is increasing evidence to suggest that traditions of inhumation burial including both crouched and extended individuals may have been widespread across southern England in the later Iron Age. At present, however, these appear to be very patchy in extent and, even allowing for a tendency in some areas for undated burials to be assigned to the Roman period on the grounds of probability, it does not seem likely that such rites were ever widely practised. In parts of Gloucestershire, for example, a similar situation to that seen in Kent may have prevailed, but without the clearly identified alternative of cremation burial. Late Iron Age and early Roman inhumations from this area may thus have represented only a minority of the rural population. Overall, therefore, it is likely that the spread of inhumation as a dominant 'Roman' rite, perhaps from as early as the late second century, may have resulted in its gradual adoption by communities living alongside others which had already practised the rite for a couple of centuries. In the Upper Thames Valley, however, the widespread uptake of inhumation may have been delayed until the fourth century, while in parts of the West Midlands it apparently never happened at all. These differences seem likely to have been conditioned by deep-rooted tradition and therefore by important differences in rural society (in particular) between these regions.

A quite different aspect of understanding that emerges from recent work is the importance of sample completeness (see also Esmonde Cleary 1992, 28). This is what makes sites such as Pepper Hill and Horcott so valuable, because the detailed osteological analysis of the buried population can be used as a basis for consideration of social aspects of the whole community represented by that population. It not only provides demographic data but also an indication of which members of society were being excluded from these cemeteries and perhaps why. The evidence for complex spatial variation within cemeteries, exemplified by sites such as Poundbury (Farwell and Molleson 1993), underlines the importance of securing 'complete' samples. This is of course more likely to be possible in a rural context, as at Horcott, Radley II, Kempston and the Boscombe Down sites. In the case of Horcott it is very interesting to see that the contrasting elements in that cemetery really were distinguished chronologically, as well as in other ways. Recent work has also underlined the importance of recovery not just of complete individual cemeteries but of multiple cemeteries within limited areas. The existence of such cemeteries in urban contexts is well understood, but they occur in rural contexts too. The two cemeteries at Radley are a case in point (Radley II, discussed above, and a closely adjacent cemetery of 35 burials excavated by Atkinson (1952/3)). It is worth considering, for example, the differences in interpretation that might have arisen if only one or the other but not both main parts of the Horcott cemetery, with their different grave orientations, age structures and body positions, had been recovered. The most striking example of closely spaced multiple burial groups, however, is the cluster of cemeteries excavated at Boscombe Down, Wiltshire, where five spatially distinct groups of burials have been recovered over a distance of some 700 m. The significance of this grouping remains to be established, though several of the cemeteries lie adjacent to (and all are south of) a major linear boundary of Late Bronze Age origin, and it is possible that all may have related to the minor nucleated settlement of Butterfield Down situated a little distance to the north of the boundary (Wessex Archaeology 2008, 8–9; McKinley this volume).

Settlement context has already been discussed. Even the small sample of sites considered here indicates that there is much more to this than just a distinction between military, urban and rural cemeteries. In broad terms, it might be expected that early Roman cemeteries will show the most clear-cut distinctions between rural and urban practice, the latter reflecting the influence of communities with a substantial component of incomers. Such elements need not have been immigrants from outside Britain or even from outside the region. It has been suggested that early Roman inhumation burials, as at London Road Gloucester, and perhaps in the east London cemeteries and (for example) Winchester (Struck 1995, 144), might reflect the presence of people drawn from local rural communities. On the other hand, isotope evidence has revealed considerable diversity of origins for the late Roman population of Lankhills (see above), and even the rural community of Wasperton does not seem to have been entirely local in composition (Montgomery *et al.* 2009, 48). The larger towns, in particular, are always likely to have been net consumers of population, although the marked imbalance in the representation of the sexes noted at some well-known sites (e.g. Cirencester (Wells 1982, 135) and Trentholme Drive (Warwick 1968, 147), see also Davison 2000) may perhaps only be partly explained in these terms and the basis of some of the records of 'unbalanced' populations may be questioned. Re-examination of the skeletal material from the 1967–1972 excavations at Lankhills, for example (Gowland 2002), suggested a much more evenly balanced population than originally thought, with a roughly 1:1 ratio closely matched by that recorded from the more recent excavations.

It would of course be simplistic to assume that there were straightforward correlations between inhumation or cremation burial rites and the presence of local or incoming population groups in individual cemetery populations. Pepper Hill may demonstrate some of the potential complexity, even leaving aside the possibility of identification of exotic individuals (Biddulph 2009). Its size reflects the association with the adjacent small town of Springhead, but were the people buried there all inhabitants of that site, or was it also used by the local rural population, in a way that has been suggested, for example, for the Poundbury cemetery at a slightly later date (Woodward 1993, 239)? An important spatial characteristic of many of the cemeteries considered here is also introduced. Pepper Hill lay beyond, rather than at, the margin of the associated settlement. The same is clearly the case with the potentially villa-related cemeteries of Barrow Hills and Horcott. Particular factors perhaps constrained the location of these cemeteries in relation to their associated settlements, but it is unlikely that they all needed to be placed several hundred metres away. While we cannot at present identify the specific reasons for the selection of these locations, a significant implication is that the 'process' involved a funeral procession, as is likely to have been the case in the urban settings of Gloucester and Winchester, and even the small town at Wall, and was indeed unavoidable in an urban context where the living and the dead were segregated.

Esmonde Cleary (2005) has drawn attention to the potential importance of regular cycles of religious activity, including processions, in conditioning aspects of the plans of a number of major towns in Britain. At a more individual level, the funeral procession was an integral part of Roman practice (Toynbee 1971, 46–48). For late Roman Gaul there is, for example, the story of St Martin of Tours encountering a pagan funeral procession; interestingly this takes place in a rural context (Sulpicius Severus, *Vita Sancti Martini* 12). While documentary evidence for such activity is lacking from Britain, the spatial evidence makes it certain that some of the same ideas prevailed here. What is less clear is the extent to which these actions occasioned significant public display, and how far variation in this display was occasioned by wealth or other aspects of status. One aspect of this issue is illuminated by the evidence for funeral biers from sites such as Brougham (Greep 2004). While distinctive inlays of the type found at Brougham were lacking, evidence from the nails, tacks and studs suggests the use of biers at Pepper Hill and Wall. Widespread use elsewhere is likely, but the phenomenon is not necessarily a straightforward one. Why was the use of a bier preferable to that of a coffin? The deceased could as easily have been carried to the pyre site in a coffin as on an open bier, and it is of course possible that this was done by many communities. The use of a bier of this kind would have made it easier for observers to see the body, however it may have been shrouded or clothed. This aspect of display was perhaps particularly important in this context and was subsequently enhanced by the conspicuous consumption (in every sense) of the decorated bier on the pyre. Conversely, in the context of inhumation burial, the deceased could have been carried on a bier to the graveside, where a coffin awaited. This would have allowed the aspect of display already referred to, and would probably have been physically easier for those doing the carrying, given the very substantial weight likely for at least some Romano-British coffins; a large empty coffin made from oak boards 45 mm thick could have weighed 200 kg, two to three times the weight of the body inside (Crummy and Crossan 1993, 34–35). There is, however, no direct evidence for this practice – although if it did occur it is possible that the bier was regarded as an item like a hearse, to be reused – and it is perhaps unlikely. If so, this implies a potentially significant distinction in the approach to the funeral procession between those communities cremating their dead on decorated biers and those inhuming their dead in coffins – albeit that the coffin may well have been carried to the grave without a lid (see further below).

Another important aspect of the cemetery at Brougham was its hilltop location, emphasised by Cool (2004, 25). This can be seen both as a landmark and a destination point for processions. That such ideas were not confined to potentially 'intrusive' communities (for the suggestion that the community buried at Brougham comprised military personnel and their families originating in or close to Pannonia see for example ibid., 463–467) is demonstrated at Pepper Hill (Biddulph 2009). Approached from the north from Springhead, down the minor road alongside which the cemetery lay, the site was on a slight eminence and would have been clearly visible to the members of the funeral procession in the same way as Brougham. It may therefore be worthwhile to consider more systematically the topographical aspects of cemeteries, in the way that has been done for some temple sites (e.g. Fulford and Rippon 1994, 198–200; Bird 2004, 83–88).

Selection of a prominent location for cemeteries was clearly not a universal characteristic, but it certainly seems to have been a consideration of some importance at a number of sites. Such a setting is an obvious factor in the siting of mausolea such as Bancroft (Williams and Zeepvat 1994, 89). The cemetery at Radley II, for example, would have been clearly visible from the associated villa site, but it is interesting to note that the one at Horcott would probably not have been seen from the adjacent farmstead because the building and the burial ground lay on opposite sides of a slight rise. A ditched trackway led into the defined area within which the cemetery lay, but while aligned upon the farmstead the ditches did not extend all the way to it. A large tree hole was located close to the trackway near the high point of the rise and it is tempting to speculate that this may have been the location of a significant feature in the local landscape serving, inter alia, to identify the location of the entry into the area of the cemetery. A tree in such a location would have been particularly prominent in a relatively flat and probably very open landscape. In an urban context, it is possible that the location of the Lankhills cemetery was also significant in topographical terms. The location of this site at the furthest extremity of the northern cemetery of Winchester has been seen as a result of chronological progression as burial grounds closer to the north gate of the city became fully used, but it is clear from the evidence of sites such as Victoria Road west and Andover Road (Ottaway *et al.* 2012) that there were suitable sites much closer to the town west of the road to Cirencester. It is possible, therefore, that the location of Lankhills owed more to deliberate selection than simple spatial constraint. More speculatively, it can be suggested that factors influencing the selection could have included the position of the site overlooking Winchester (admittedly a characteristic shared by several cemeteries of the town as a consequence of its basic topographical setting) and the potential for display involving extended funeral processions. It follows additionally that such choices might reflect the interest and influence of a specific community within the population of late Roman Winchester, reflected in part by the well-known distinctive character of the site exemplified by the frequency and nature of grave good deposition (see above).

The different circumstances of these few sites make it clear that present data are insufficient to support generalisations about the topographical associations of cemeteries and settlements which might underline the importance of the funeral procession. Was the procession more important in some social contexts than others or, to put it crudely, was it a special feature of high status settlements or communities? There is no particular reason to think so, but all these possibilities may repay further investigation. Again, however, regional variation must be expected.

That the funeral procession was important is hardly a novel conclusion, but its almost universal applicability prompts important questions about the focus of funerary activities – the extent to which they were centred in the home of the deceased rather than in the cemetery, and at intervening locations, for example, which can only be raised rather than considered in detail here. The clearest evidence is for activities at the graveside. Those related to cremation burial rites will not be considered here as the topic has been examined by Jackie McKinley elsewhere in this volume. The first and most obvious point is that, as already mentioned, the lid was probably not yet fastened to the coffin. We cannot know if this was general practice, but the careful placement of items within the coffin, including upright vessels, as at first to third century Pepper Hill and fourth century Lankhills, and widely elsewhere, can only have occurred after the coffin had been lowered into the grave. Items of jewellery, however, are not as easy to interpret. Those which were worn by the deceased were presumably not put in place at the graveside, whereas the piling of bracelets and necklaces beside the deceased may have occurred after the coffin had been lowered into the ground. A whole series of formal and informal acts may have taken place at this stage. These rarely leave archaeologically detectable traces, but at Welford on Avon, Warwickshire, the recovery of joining fragments of a set of fourth century glass beakers both within the lead-lined coffin and in the fill of the grave pit strongly suggest the drinking of a valedictory toast, followed by the smashing of the glasses and then, presumably, the closing of the coffin and the backfilling of the grave (Booth 1993/4). Such evidence is unusual and may reflect an uncommon practice rather than a widely observed one. Either way, it provides a recognisable glimpse of human actions. In the case of Welford, we have no evidence for any subsequent actions at the site of the grave, either in the form of tangible memorials or subsequent transient visits. The former appear to have been extremely rare in Roman Britain, even in the case of burials which may appear to be relatively 'wealthy' in other respects. Nonetheless it is clear that in many cases, for example in the late Roman cemeteries of Winchester, not only were graves marked, thus avoiding disturbance by later burials, at least not until a reasonable amount of time had elapsed, but also that the marking identified the graves as belonging to specific individuals. This is suggested by the relatively small but significant number of cases in which graves were directly recut for the insertion of later burials in such a way as to indicate that it is highly unlikely that the superimposition of the later burial was fortuitous. Unfortunately, in none of the Winchester examples is it possible to establish the length of time elapsed between the original and secondary burials. This could have been quite short, perhaps invalidating the suggestion about named grave markers (above), but there is no reason to assume that was so in every case. Further ceremonies may have ensued, but in the case of Lankhills there is no evidence for this, except in one instance where two complete imbrices were located in a grave fill. It is possible that these merely served as packing between the coffin and the side of the grave pit, but it is also possible that they were originally set upright and formed an opening for communication with the deceased or the pouring of offerings into the grave, in the manner of the well known but uncommon class of pipe burials (e.g. Boon 1972, 108). Even if this interpretation is correct, however, such evidence remains extremely rare and the nature of activities associated with periodic visits to the grave purely speculative. The rest is silence.

Acknowledgements

This paper draws heavily upon the results of projects carried out by a number of colleagues in Oxford Archaeology, to all of whom I am extremely grateful for information and discussion of their work. In addition, thanks are owed to colleagues at Wessex Archaeology and Cotswold Archaeology, particularly for information about Boscombe Down and Denham, respectively. I am also very grateful to John Pearce and Jake Weekes for inviting me to participate in the RAC conference session on which this paper is based and for helpful comments on an earlier draft. Defects in the present version are, of course, the sole responsibility of the author.

Bibliography

Allen, T. G., Darvill, T. C., Green, L. S. and Jones, M. U. (1993) *Excavations at Roughground Farm, Lechlade, Gloucestershire: a prehistoric and Roman landscape.* Oxford Archaeological Unit Thames Valley Landscapes: the Cotswold Water Park, Volume 1, Oxford.

Allen, T., Donnelly, M., Hardy, A., Hayden, C. and Powell, K. (2012) *A road through the past: Archaeological discoveries on the A2 Pepperhill to Cobham road-scheme in Kent.* Oxford Archaeology Monograph No. 16, Oxford.

Andrews, P., Booth. P., Fitzpatrick, A. and Welsh, K. (2015) *Digging at the gateway: Archaeological landscapes of the south of Thanet. Archaeology of the East Kent Access (Phase II).* Oxford: Oxford Wessex Archaeology Monograph No. 8.

Atkinson, R. J. C. (1952/3) Excavations in Barrow Hills Field, Radley, Berks, 1944–5. *Oxoniensia* 17–18, 14–35.

Barber, B. and Bowsher, D. (2000) *The Eastern Cemetery of Roman London, Excavations 1983-1990.* London: Museum of London Archaeology Service.

Biddulph, E. (2009) The Roman cemetery at Pepper Hill, Southfleet, Kent. *CTRL Integrated Site Report Series.* In S. Foreman *Channel Tunnel Rail Link Section 1.* York: Archaeology Data Service (doi:10.5284/1000230).

Bird, D. (2004) Roman religious sites in the landscape. In J. Cotton, G. Crocker and A. Graham (eds) *Aspects of archaeology and history in Surrey,* 77–90. Guildford: Surrey Archaeological Society.

Boon, G. C. (1972) *Isca. The Roman legionary fortress at Caerleon, Mon.* Cardiff: National Museum of Wales.

Booth, P. (1993/4) A Roman burial near Welford-on-Avon, Warwickshire. *Transactions of the Birmingham and Warwickshire Archaeological Society* 98, 37–50.

Booth, P. (2001) Late Roman cemeteries in Oxfordshire: a review. *Oxoniensia* 66, 13–42.

Booth, P. (2011) The late Iron Age and Roman period. In P. Booth, T. Champion, S. Foreman, P. Garwood, H. Glass, J. Munby and A. Reynolds *On Track The archaeology of High Speed 1 Section 1 in Kent,* 243–340. Oxford: Oxford Wessex Archaeology Monograph No. 4.

Booth, P. and Simmonds, A. (2009) *Appleford's earliest farmers: archaeological work at Appleford Sidings, Oxfordshire, 1993-2000.* Oxford: Oxford Archaeology Occasional Paper No. 17.

Booth, P. and Simmonds, A. (forthcoming) *Later prehistoric landscape and a Roman nucleated settlement in the lower Windrush valley at Gill Mill, near Witney, Oxfordshire.* Oxford: Oxford Archaeology Monograph.

Booth, P., Simmonds, A., Boyle, A., Clough, S., Cool, H. and Poore, D. (2010) *Excavations at Lankhills Roman cemetery, Winchester, 2000-2005.* Oxford: Oxford Archaeology Monograph No. 10.

Boylston, A., Knusel, C. J., Roberts, C. M. and Dawson, M. (2000) Investigation of a Romano-British rural ritual in Bedford, England. *Journal of Archaeological Science* 27, 241–254.

Burnham, B. C., Hunter, F., Booth, P., Worrell, S. and Tomlin, R. S. O. (2008) Roman Britain in 2007. *Britannia* 39, 263–390.

Carver, M., Hills, C. and Scheschkewitz, J. (2009) *Wasperton. A Roman, British and Anglo-Saxon community in Central England.* Woodbridge: Boydell.

Chambers, R. A. (1987) The late- and sub-Roman cemetery at Queenford Farm, Dorchester-on-Thames, Oxon. *Oxoniensia* 52, 35–69.

Chambers, R. A. and McAdam, E. (2007) *Excavations at Barrow Hills, Radley, Oxfordshire,1983-5. Vol. 2: The Romano-British cemetery and Anglo-Saxon settlement.* Oxford: Oxford Archaeology Thames Valley Landscapes Monograph 25.

Clarke, G. (1979) *The Roman cemetery at Lankhills.* Winchester Studies 3: Pre-Roman and Roman Winchester Part II, Oxford: Clarendon Press.

Coleman, L., Havard, T., Collard, M., Cox, S. and McSloy, E. (2004) Denham, The Lea (TQ 0490 8600) interim report. *South Midlands Archaeology* 34, 14–17.

Cool, H. E. M. (2004) *The Roman cemetery at Brougham, Cumbria: Excavations 1966-67.* Britannia Monograph Series No. 21, London: Society for the Promotion of Roman Studies.

Crummy, N. and Crossan, C. (1993) Excavations at Butt Road 1976–79, 1986, and 1988. In N. Crummy, P. Crummy and C. Crossan *Excavations of Roman and later cemeteries, churches and monastic sites in Colchester, 1971-88*, 4–163. Colchester Archaeological Report 9.

Crummy, N. and Eckardt, H. (2003) Regional identities and technologies of self: nail-cleaners in Roman Britain. *Archaeological Journal* 160, 44–69.

Crummy, P. (1993) The cemeteries of Roman Colchester. In N. Crummy, P. Crummy and C. Crossan *Excavations of Roman and later cemeteries, churches and monastic sites in Colchester, 1971-88*, 257–275. Colchester Archaeological Report 9.

Crummy, P. (2007) Aspects of the Stanway cemetery. In P. Crummy, S. Benfield, N. Crummy, V. Rigby and D. Shimmin, *Stanway: an elite burial site at Camulodunum*, 423–456. Britannia Monograph Ser No. 24, London: Society for the Promotion of Roman Studies.

Cummings, C. (2009) Meat consumption in Roman Britain. In M. Driessen, S. Heeren, J. Hendriks, F. Kemmers and R. Visser (eds) *TRAC 2008. Proceedings of the Eighteenth Annual Theoretical Roman Archaeology Conference Amsterdam 2008*, 73–83. Oxford: Oxbow.

Davies, M. (2001) Death and social division at Roman Springhead. *Archaeologia Cantiana* 121, 157–169.

Davison, C. (2000) Gender imbalances in Romano-British cemetery populations: a re-evaluation of the evidence. In J. Pearce, M. Millett and M. Struck (eds) *Burial, society and context in the Roman world,* 231–237. Oxford: Oxbow.

Dawson, M. (2004) *Archaeology in the Bedford region.* British Archaeological Reports (British Series) 373, Oxford: Archaeopress.

Dean, M. and Hammerson, M. (1980) Three inhumation burials from Southwark. *London Archaeologist* 4, 17–22.

Egging Dinwiddy, K. and Schuster, J. (2009) Thanet's longest excavation: archaeological investigations along the route of the Weatherlees-Margate-Broadstairs wastewater pipeline. In P. Andrews, K.E. Dinwiddy, C. Ellis, A, Hutcheson, C, Phillpotts, A.B. Powell and J. Schuster *Kentish sites and sites of Kent A miscellany of four archaeological excavations*, 58–174. Wessex Archaeology Reports 24, Salisbury.

Ellis, P. and King, R. (2014) Gloucester: The Wootton cemetery excavations, 2002. *Britannia* 45, 53–120

Esmonde Cleary, S. (1992) Town and country in Roman Britain? In S. Bassett (ed.) *Death in towns: urban responses to the dying and the dead, 100-1600,* 28–42. Leicester University Press.

Esmonde Cleary, S. (2000) Putting the dead in their place: burial location in Roman Britain. In J. Pearce, M. Millett and M. Struck (eds) *Burial, society and context in the Roman world,* 127–142. Oxford: Oxbow.

Esmonde Cleary, S. (2005) Beating the bounds: ritual and articulation of urban space in Roman Britain. In A. Mac Mahon and J. Price (eds) *Roman working lives and urban living,* 1–17. Oxford: Oxbow.

Esmonde Cleary, S. (2009) Life and death in a Roman city; excavation of a Roman cemetery with a mass grave at 120–122 London Road. Gloucester [Review]. *Britannia* 40, 389–390.

Evans, J., Stoodley, N. and Chenery, C. (2006) A strontium and oxygen isotope assessment of a possible fourth century immigrant population in a Hampshire cemetery, southern England. *Journal of Archaeological Science* 33, 265–272.

Farwell, D. E. and Molleson, T. I. (1993) *Poundbury Volume 2: The cemeteries.* Dorset Natural History and Archaeological Society Monograph Series No. 11, Dorchester.

Fiorato, V., Boylston, A. and Knüsel, C. (eds) (2000) *Blood Red Roses. The archaeology of a mass grave from the Battle of Towton AD 1461.* Oxford: Oxbow.

Ford, W. J. (2002) The Romano-British and Anglo-Saxon settlement and cemeteries at Stretton-on-Fosse, Warwickshire. *Transactions of the Birmingham and Warwickshire Archaeological Society* 106, 1–116.

Foundations Archaeology (2003) 124-130 London Road, Gloucester: excavation report and post-excavation assessment, http://www.foundations.co.uk/reports/gloucestershire/lrg.shtml.

Fulford, M. G. and Rippon, S. J. (1994) Lowbury Hill, Oxon.: a re-assessment of the probable Romano-Celtic temple and Anglo-Saxon barrow. *Archaeological Journal* 151, 158–211.

Gowland, R. (2002) Examining age as an aspect of social identity in fourth to sixth century England through the analysis of mortuary evidence. Unpublished Ph.D. thesis, University of Durham.

Greep, S. (2004) Bone and antler veneer. In H. E. M. Cool, *The Roman cemetery at Brougham, Cumbria: Excavations 1966-67,* 273–282. Britannia Monograph Series No. 21, London: Society for the Promotion of Roman Studies.

Hamilton, S. (2007) Cultural choices in the 'British Eastern Channel Area'. In C. Haselgrove and T. Moore (eds) *The later Iron Age in Britain and beyond,* 81–106. Oxford: Oxbow.

Harman, M., Molleson, T. and Price, J. L. (1981) Burials, bodies and beheadings in Romano-British and Anglo-Saxon cemeteries. *Bulletin of the British Museum Natural History (Geology)* 35.3, 145–188.

Hayden, C., Booth, P., Dodd, A., Smith, A., Laws, G. and Welsh, K. (forthcoming) *Horcott Quarry: Prehistoric, Roman and Anglo-Saxon settlement and burial.* Oxford Archaeology Thames Valley Landscapes Monograph.

Heighway, C. M. (1980) Roman cemeteries in Gloucester District. *Transactions of the Bristol and Gloucestershire Archaeological Society* 98, 57–72.

Hey, G., Bayliss, A. and Boyle, A. (1999) Iron Age inhumation burials at Yarnton, Oxfordshire. *Antiquity* 73, 551–562.

Hill, J. D. (2007) The dynamics of social change in later Iron Age eastern and south-eastern England c. 300 BC–AD 43. In C. Haselgrove and T. Moore (eds) *The later Iron Age in Britain and beyond,* 16–40. Oxford: Oxbow.

Hiller, J. and Wilkinson, D. R. P. (2005) *Archaeology of the Jubilee Line Extension: Prehistoric and Roman activity at Stratford Market Depot, West Ham, London 1991-1993.* Oxford: Oxford Archaeology.

Holbrook, N. (2006) The Roman period. In N. Holbrook and J. Jurica (eds) *Twenty-five years of archaeology in Gloucestershire; A review of new discoveries and new thinking in Gloucestershire South Gloucestershire and Bristol 1979-2004,* 97–131. Bristol and Gloucestershire Archaeological Report No. 3, Cotswold Archaeology.

Holbrook, N. and Bateman, C. (2008) The south gate cemetery of Roman Gloucester: excavations at Parliament Street, 2001. *Transactions of the Bristol and Gloucestershire Archaeological Society* 126, 91–106.

Lambrick, G. and Allen, T. (2004) *Gravelly Guy, Stanton Harcourt: the development of a prehistoric and Romano-British community*. Oxford Archaeology Thames Valley Landscapes Monograph No. 21, Oxford.

Lawrence, S. and Mudd, A. (1999) Field's Farm. In A. Mudd, R. J. Williams and A. Lupton, *Excavations alongside Roman Ermin Street, Gloucestershire and Wiltshire: The archaeology of the A419/A417 Swindon to Gloucester Road Scheme*, 99–113. Oxford: Oxford Archaeological Unit.

McKinley, J. (2008) Ryknield Street, Wall (Site 12). In A. B. Powell, P. Booth, A. P. Fitzpatrick and A. D. Crockett, *The archaeology of the M6 Toll, 2000-2003*, 87–190. Oxford-Wessex Archaeology Joint Venture Monograph No. 2, Oxford.

Miles, D., Palmer, S., Smith, A. and Jones, G. P. (2007) *Iron Age and Roman settlement in the Upper Thames Valley: Excavations at Claydon Pike and other sites within the Cotswold Water Park*. Oxford Archaeology Thames Valley Landscapes Monograph 26, Oxford.

Montgomery, J., Evans, J. and Chenery, C. (2009) Oxygen and strontium isotopes. In Carver *et al.* 2009, 48–49.

Moore, T. (2006) *Iron Age societies in the Severn-Cotswolds. Developing narratives of social and landscape change*. British Archaeological Reports (British Series) 421, Oxford: Archaeopress.

Ottaway, P. J., Qualmann, K. E., Rees, H., and Scobie, G. D. (2012) *Roman Cemeteries and Suburbs of Winchester: Excavations 1971-85*. Winchester Museum Service/English Heritage Reports, Winchester.

Parfitt, K. (1995) *Iron Age burials from Mill Hill, Deal*. London, British Museum Press.

Parnell, G. (1985) The Roman and medieval defences and later development of the inmost ward, Tower of London: excavations 1955–77. *Transactions of the London and Middlesex Archaeological Society* 36, 1–79.

Pearce, J. (1999) The dispersed dead: preliminary observations on burial and settlement space in rural Roman Britain. In P. Baker, C. Forcey, S. Jundi and R. Witcher (eds) *TRAC 98. Proceedings of the eighth annual theoretical Roman archaeology conference, Leicester 1998*, 151–162. Oxford: Oxbow.

Pearce, J. (2013) *Contextual archaeology of burial practice: case studies from Roman Britain*. British Archaeological Reports British Series 588. Oxford: Archaeopress.

Philpott, R. (1991) *Burial practices in Roman Britain*. British Archaeological Reports (British Series) 219, Oxford.

Powell, K., Laws, G. and Brown, L. (2009) A Late Neolithic/Early Bronze Age enclosure and Iron Age and Romano-British settlement at Latton Lands, Wiltshire. *Wiltshire Archaeological and Natural History Magazine* 102, 22–113.

Simmonds, A., Marquez-Grant, N. and Loe, L. (2008) *Life and death in a Roman city; excavation of a Roman cemetery with a mass grave at 120–122 London Road, Gloucester*. Oxford Archaeology Monograph No. 6, Oxford.

Simmonds, A., Anderson-Whymark, H. and Norton, A. (2011) Excavations at Tubney Wood Quarry, Oxfordshire, 2001-2009. *Oxoniensia* 76, 105–172

Staelens, Y. J. E. (1982) The Birdlip cemetery. *Transactions of the Bristol and Gloucestershire Archaeological Society* 100, 19–31.

Stanley, C. (1972) Bray Roman cemetery, Berkshire. *Council for British Archaeology Group 9 Newsletter* 2, 12–13.

Struck, M. (1993) *Busta* in Britannien und ihre Verbindungen zum Kontinent. Allgemeine Überlegungen zur Herleitung der Bestattungssitte. In M. Struck (ed.) *Römerzeitliche Gräber als Quellen zur Religion, Bevölkerungsstruktur und Sozialgeschichte*, 81–94. Mainz: Johannes Gutenberg Institut für Vor- und Frühgeschichte.

Struck, M. (1995) Integration and continuity in funerary ideology. In J. Metzler, M. Millett, N. Roymans and J. Slofstra (eds) *Integration in the early Roman west*, 139–150. Dossiers d'Archéologie du Musée National d'Histoire et d'Art IV, Luxembourg.

Taylor, A. (2008) Aspects of deviant burial in Roman Britain. In E. M. Murphy (ed.) *Deviant burial in the archaeological record*, 91–114. Oxford: Oxbow.

Taylor, J. (2007) *An atlas of Roman rural settlement in England*. Council for British Archaeology Research Report 151, York.

Tomlin, R. S. O. and Hassall, M. W. C. (2005) Roman Britain in 2004: III Inscriptions. *Britannia* 36, 473–497.

Toynbee, J. M. C. (1971) *Death and burial in the Roman world*. London: Thames and Hudson.

Warwick, R. (1968) The skeletal remains. In L. P. Wenham, *The Romano-British cemetery at Trentholme Drive, York*, 111–176. London: Her Majesty's Stationery Office.

Webster, C. J. and Brunning, R. A. (2004) A seventh-century AD cemetery at Stoneage Barton Farm, Bishop's Lydiard, Somerset and square-ditched burials in post-Roman Britain. *Archaeological Journal* 161, 54–81.

Weekes, J. (2007) A specific problem? The detection, protection and exploration of Romano-British cremation cemeteries through competitive tendering. In B. Croxford, N. Ray, R. Roth and N. White (eds) *TRAC 2006. Proceedings of the sixteenth annual Theoretical Roman Archaeology Conference, Cambridge 2006*, 183–191. Oxford: Oxbow.

Wells, C. (1982) The human burials. In A. McWhirr, L. Viner and C. Wells, *Roman-British cemeteries at Cirencester*, 135–202. Cirencester Excavations II, Cirencester Excavation Committee.

Wessex Archaeology (2008) Boscombe Down Phase VI excavation, Amesbury, Wiltshire, 2006–7, Interim assessment on the results of the Byway 20 Romano-British cemetery excavations. unpublished Wessex Archaeology Report 56246.04.

Wheeler, H. (1985) The Racecourse cemetery. *Derbyshire Archaeological Journal* 105, 222–280.

Whiting, W., Hawley, W. and May, T. (1931) *Report on the excavation of the Roman cemetery at Ospringe, Kent*. Report of the Research Committee 8, London: Society of Antiquaries.

Williams, R. J. and Zeepvat, R. J. (1994) *Bancroft, A Late Bronze Age Settlement, Roman Villa and Temple Mausoleum 2*. Buckinghamshire Archaeological Society Monograph 7, Aylesbury.

Woodward, A. (1993) Discussion. In Farwell and Molleson 1993, 216–239.

Wright, N. (2008) A lead-lined stone coffin cremation burial from Harnhill, Gloucestershire. *Transactions of the Bristol and Gloucestershire Archaeological Society* 126, 83–90.

Chapter 8

Funerary complexes from Imperial Rome: A new approach to anthropological study using excavation and laboratory data

Paola Catalano, Carla Caldarini, Flavio De Angelis and Walter Pantano

As a consequence of large-scale infrastructural projects and the unstoppable urbanisation of the periphery of Rome, the excavation activity undertaken by the Soprintendenza Speciale per i Beni Archeologici di Roma, especially in the last ten years, has been extraordinarily intense. This work has brought to light c. 6000 burials in total, in the great majority of cases dating to the period from the first to third centuries AD. In this article, we examine inhumations derived from Imperial period cemeteries of equivalent size as expressed by numbers of graves, situated in diverse zones of the *Suburbium* and for which particular characteristics may be hypothesised (Fig. 8.1). What aspect did the *Suburbium* assume in the Imperial period? As Rodolfo Lanciani describes: "Villas, vineyards, and rural properties with houses peopled the Campagna, forming smaller and larger centres so as to create around Rome a great populated park." (Lanciani 1909 (1980, 27, Italian translation)). In recent years, the many extensive excavations undertaken by the Soprintendenza are finally shedding light on extensive areas within this zone, where hitherto unknown archaeological sites are now almost beyond counting (Musco 2001).

A significant contribution to the understanding of how funerary rituals and everyday life proceeded (Buccellato *et al.* 2008) in the *Urbs* and its surroundings is provided precisely by anthropological evidence, gathered first during the course of systematic cemetery excavation (Catalano *et al.* 2006) and later through laboratory investigation (Catalano *et al.* 2003). We are thus endeavouring to satisfy the need to analyse both funerary practices and living conditions from a chronological and historical-social perspective which has long been neglected (Heinzelmann 2001). The anthropologist identifies skeletal elements, their exact position and the

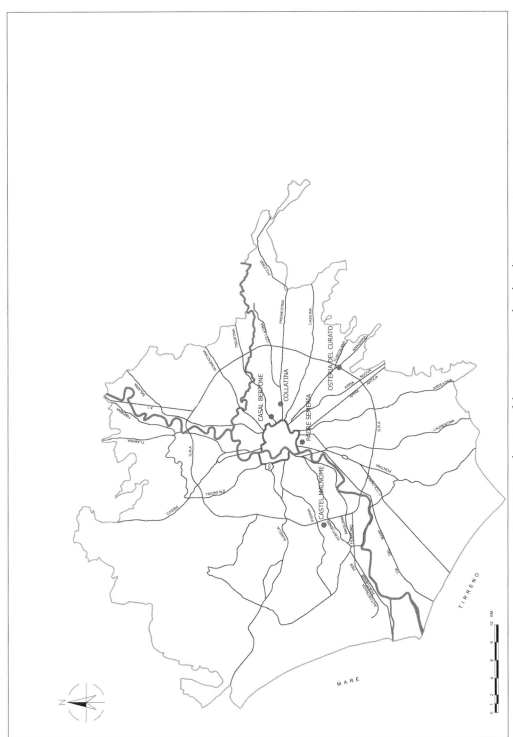

Fig. 8.1. Rome: location of the cemeteries analysed in this paper.

relationships with the other components of the tomb, records the measurements and the observations necessary for an initial determination of sex and age at death and contributes to defining the structure of the tomb by establishing the mode of decomposition and recognizing the effects of transverse compression on the skeleton (Duday 2006). The noting of the separation of skeletal articulations during the decomposition of the corpse demonstrates the possible primary character of a burial and its evolution over time. The data are then digitized: as well as those referred to above, data related to the grave cut, type and covering, the depositional mode of the body and its spatial orientation, the types of grave furnishing and their placing are entered into a database specifically created for this purpose (Minozzi *et al.* 2008).

The demographic profile of the skeletal populations has been established through the determination of sex, possible for individuals whose bone growth is complete (> c. 18 years), and the estimation of age at death: to achieve this, standardised metrical and morphological skeletal indicators have been used. Determination of sex uses the technique proposed by Acsàdi and Nemeskéri (1970), taking account of the further information provided by Ferembach *et al.* (1977): observation is made of the dimorphic traits of the cranio-facial block, the pelvic girdle and, where possible, also other discriminating characteristics, like the robusticity of the pilaster of the femur [a longitudinal bony ridge on the back of the femur] and the diameters of the humeral and femoral heads (Mall *et al.* 2000; Olivier 1955; Steel 1962). The diagnosis of age at death in subadults has been determined from the stage of tooth eruption (Ubelaker 1989), the state of fusion between epiphysis and diaphysis of the principal long bones and the dimensions of the metaphysis (Stloukal and Hanakova 1978; Fazekas and Kosa 1978). For the adults, the following characteristics have been assessed, evaluating the degree of obliteration of the exocranial sutures (Meindl and Lovejoy 1985; Lovejoy *et al.* 1985a; Richards and Miller 1991), tooth wear and the morphology of the surface of the sternal extremities of the ribs (Burns 1999) and vertebral bodies.

The condition of the dentoalveolar complex has also been documented, since in general interesting relationships exist between diseases of the oral cavity and the living conditions and health of a population (Catalano *et al.* 2007). The presence/absence of caries, calculus or tartar and abscesses in adult subjects has therefore been noted.

The formation of caries is a pathological process characterised by a demineralisation of dental hard tissue by acids, as a result of the fermentation of soft foodstuffs with a carbohydrate base, sugars in particular. The caries has been classified under the following headings: location (occlusal, buccal, lingual, mesial, distal etc); severity (superficial, in the dentine, perforation, destruction of the crown); the part of the tooth affected (crown, neck, root). The cariogenic lesions have been documented – following Metress and Convay (1975), Marafon (1979) and Mezl (1985) – using a four degree scale: the first degree is a superficial lesion which affects only the enamel; the second also affects dentine more or less extensively; the third attacks the pulp directly or indirectly, through the smallest of openings which allows the passage of infective agents; the fourth sees necrosis of the pulp, originating directly as a consequence of third degree caries and presents itself as a very wide cavity; if necrosis is complete, only the root of the tooth will be saved. The presence of double or triple caries is also noted.

Abscesses are acute inflammations which establish themselves rapidly and are located in the periapical region of the tooth (Kelley and Larsen 1991). The inflammation of the periodontal tissues happens through the entry of bacteria through the tooth canal because of caries, a breakage or excessive wear. The bacterial attack first causes inflammation and the death of the tooth; then, through the root, it reaches the apex of the root and the tissues beneath, causing inflammation (periapical granuloma, i.e. a mass of granulation tissue around the apex of the tooth) with a consequent accumulation of pus (the abscess) (Hillson 1986; 1996; Clarke and Hirsch 1991). Calculus comprises calcareous deposits formed through the mineralisation of bacterial plaque which is deposited on the surface of the teeth, aided by the presence of saliva. The teeth mainly affected by this deposit are those nearest the salival ducts (lingual surface of the front teeth and buccal surface of the molars). The calculus deposits have been noted by quantity according to the stages set out by Brothwell (1981).

Among the non-specific stress indicators, important for evaluating the general state of health of a population (Caldarini *et al.* 2006), the following have been recorded: presence/absence of porotic hyperostosis and enamel hypoplasia, alterations which can be attributed to infectious diseases and/or to periods of malnutrition during the first years of life. Porotic hyperostosis manifests itself as porosity, the consequence of hypertrophy of the diploe and of the thinning of cortical bone tissue. It is observable on the roof of the orbits (*cribra orbitalia*) and/or on the external surface of the cranial vault (*cribra cranii*). An alteration of this type is the consequence of hyperactivity of the bone marrow that could be attributed to the increased production of red blood cells in anaemia or non-specific infections (Angel 1966; Hengen 1971; Walker 1986; Palkovich 1987; Stuart-Macadam 1987a, b).

Enamel hypoplasia manifests itself as lines or pits on the surface of the teeth; infections, fevers or lack of vitamin D can influence calcium and phosphorus metabolism and slow the rate of growth, producing alterations in the formation of tooth enamel (amelogenesis) (Goodman and Armelagos 1985; Skinner and Goodman 1992). It is possible to measure the distance of the hypoplasic lines from the neck on the buccal surface of all the permanent teeth, in order to trace back to the age at which the defect arose (Buikstra and Ubelaker 1994; Corrucini *et al.* 1985).

Moreover, the condition of the auditory meatus is assessed for the possible presence of exostoses and osteophytes [bony growths]. Even though in some cases these are simple anatomical variants, for some authors the presence of auricular exostosis can be linked to environmental factors, in particular to repeated and prolonged exposure of the auditory canal to cold water (Frayer 1988; Manzi *et al.* 1991).

Inflammatory processes on the periosteum have also been observed, either in acute or chronic form, which lend the bones a porotic and spongy appearance (Steinbock 1976). Such an alteration (periostitis) is attributable to numerous factors, endogenous and exogenous, which are able to cause a localized hypervascularization (Mensforth *et. al.* 1978).

Additionally fractures, and traumas, which are to be understood as modifications related to the joints (arthropathies) and to the muscle/tendon insertions (enthesopathies), have been considered (Canci and Minozzi 2005). The documentation of these morphological variations allows us to understand which are the body parts most affected by functional stress: for example, if there is a prevalent use of the upper limbs compared to the lower limbs; if there is a sexual dimorphism that may or may not be classed within the norm of somatic difference; if a muscle involvement is present which decreases in a cranio-caudal and/or medio-lateral direction and allows the recognition of a prevalent use of some muscle groups compared to others, antagonistic or otherwise (for example flexors compared to extensors or supinators).

The presence of discal hernias is recorded, which are the consequences of degenerative processes attributable to age and/or to single or repeated trauma linked to occupational activity (Capasso *et al.* 1999). The hernia causes a shrinking of the discal space, allowing contact between two vertebrae. The most widespread type of this lesion is related to the disc spreading: the disc spreads out on the vertebral surface, damages the cartilaginous layers, extends into the trabecular bone. The osteolytic lesion which results from this generally has a characteristic half-moon shape and is defined as a Schmörl's node.

The presence of fractures and fusions of vertebral bodies has also been recorded. Fusion can manifest itself as an ossification (a 'flow' of bone) of the longitudinal anterior ligament, either without involving the intervertebral discs, the space of which is spared (diffuse idiopathic skeletal hyperostosis – DISH) or involving the complete vertebrae, with a loss of the intervertebral spaces (anchylosing spondylitis) (Rogers and Waldron 1995).

Considering the problems deriving from the processing of such a complex mass of data, it was decided to test the application of correspondence analysis. This technique allows a matrix of qualitative data to be analysed and is therefore particularly adapted to the evaluation of variables gathered from the study of ancient cemeteries (Pitts 2005; Kelley 2012). The principal characteristic of this technique lies in the capacity to show a multivariate reality, derived from the data matrix, in a space which can be represented graphically, with a minimal loss of information. For a correct interpretation of the results, it is necessary to bear in mind that two points will tend to be closer in this spatial representation the more they are 'associated', i.e. that they tend to vary in the same way; conversely, the further apart they are, the less association there is between them: it is therefore intuitively apparent that a cemetery will be placed near the variable by which it is most influenced and vice versa. For a proper understanding of the method, it should also be specified that if two points are 'associated' (i.e. near in space), this need not imply that they have similar frequencies, but only that their frequencies tend to vary in an analogous way. In other words, the visualisation of a single variable [i.e. characteristic] occurs only once and among the various frequencies of the variable, only that which shows an association, or – if you will – a correspondence with a particular cemetery is of interest to us.

In this article, the first assessment was made using, in their totality, the data gathered from five selected cemeteries, in total 19,320 variables, representing the whole of the features observable in 1,048 individuals. These derive from several cemeteries; 101 are taken from Via di Castel Malnome (XV Municipio, the archaeologist responsible being Laura Cianfriglia), in the far western *Suburbium*, near Via Portuense, c. 5 km outside the Grande Raccordo Anulare (Fig. 8.2). The Castel Malnome necropolis is situated near Ponte Galeria, covering c. 3000 m² (Catalano *et al.* 2010) and its excavation included 290 earth-cut tombs, furnished with tile coverings in 43% of cases (mostly set *a cappuccina*: 87% of the total). From these tombs 292 buried bodies were recorded, most in a good condition of preservation. The grave

Fig. 8.2. Aerial view of the cemetery on Via di Castel Malnome.

Fig. 8.3. Panorama of the cemetery on Via Padre Semeria.

goods are poor and mostly represented by ceramics, lamps, glass unguent bottles and coins; jewellery is very rare. 127 samples were from Via di Padre Semeria (XI Municipio, the archaeologist responsible being Rita Paris), at the junction with Via Cristoforo Colombo, close to the Aurelianic walls (Fig. 8.3) (Benassi *et al.* 2011). The funerary complex consists of open air burial areas, a mausoleum with an enclosure and three rooms used for burials which frequently overlap one other and are placed close to the walls. The covering of the graves, when present, takes various forms including tiles laid flat or set *a cappuccina*, or fragments of amphorae. Most of the tombs lack grave goods, especially because the tombs of the mausoleum had been subjected to systematic robbing.

Seventy-three skeletons derived from Via di Casal Bertone (V Municipio, the archaeologist responsible being Stefano Musco: Musco *et al.* 2008), *c.* 1.5 km from the Aurelianic walls, between Via Tiburtina to the north and Via Prenestina to the south (Fig. 8.4): in this report, we consider the anthropological data from the mausoleum, consisting of a quadrangular hypogeal room roofed by a cross-like vault and with a black and white mosaic floor (Musco *et al.* 2008). Data from area

Fig. 8.4. Via Casal Bertone – Area Q: mausolea and the fullonica.

Q is also presented, in which a short section of the ancient Via Collatina and seven funerary buildings were discovered. In Casal Bertone only 16% of the tombs contained grave goods, mostly small ceramic vessels, glass unguent bottles, coins and nails.

A large sample of 563 inhumations derives from the Collatina necropolis, between Via della Serenissima and Via Basiliano (V and VI Municipio, the archaeologists responsible being Stefano Musco and Anna Buccellato: Buccellato *et al.* 2008), c. 3.5 km from the Aurelianic walls, between the urban section of the A24 to the north and Via Prenestina to the south (Fig. 8.5). The excavation brought to light several mausolea, inhumations and cremation urns (Musco *et al.* 2010). The inhumations were buried in earth cut graves and covered by sloping or flat tiles. They were mostly furnished with a small pottery flagon. In some cases, a lamp or an unguent bottle is present.

The last sample of 184 skeletons derived from the area of Osteria del Curato (X Municipio, the archaeologist responsible being Roberto Egidi: Egidi *et al.* 2003), in the south-eastern *Suburbium*, along Via Tuscolana, immediately within the Grande Raccordo Anulare (Figures 8.6 and 8.7). The necropolis of Osteria del Curato is sited near

Fig. 8.5. Aerial view of the Collatina cemetery.

Fig. 8.6. Aerial view of the cemetery on Via Lucrezia Romana, Osteria del Curato.

Fig. 8.7. Aerial view of the cemetery on via Falconara Albanese, Osteria del Curato.

the Villa dei Sette Bassi, built as a refurbishment of an ancient estate during the reign of Antoninus Pius. This villa was near the Via Latina and supplied by a private aqueduct derived from the Aqua Claudia. It comprised three contiguous buildings. Related to this estate and to the one identified close by at Via Lucrezia Romana were the numerous tombs of Osteria del Curato, mostly dated from the first to third centuries A.D.

The complexity of the situation has made it necessary to use three dimensions in order to be able to display graphically the whole of the variability – represented at an optimum level of significance – and thus being able to interpret 92.5% of the variability as a whole ($p < 0.001$) [i.e. the probability that this is due to chance factors alone being less than 0.1%].

From this overview of the data as a whole it is possible, supported by statistical interpretation, to define affinities and differences between the various cemeteries analysed and avoid subjective interpretation.

In our case, as may be seen in Figure 8.8, the five cemeteries tend to lie at some distance from one another and some (Castel Malnome, Casal Bertone-Area Q and Osteria del Curato) are situated at the extremes of the three-dimensional space, showing particular features which make them very different from the others. But what are these characteristics differentiating the individuals of the cemeteries so significantly? The Castel Malnome cemetery certainly lies at one extreme because its characterisation is strongly influenced by the substantial discrepancy in the ratio of men to women (equivalent to 2:1: 62 males vs 30 females) and by the very remarkable occurrence of trauma due to biomechanical stress (59% on average across males and females). The Casal Bertone-Area Q cemetery is instead especially characterised by a noteworthy frequency of individuals who died at a young age (62%), probably indicative of the difficult environmental conditions suffered by the children. Conditions that were certainly better should be noted for the population of Osteria del Curato (*sex ratio* = 1,05; 24% of infants), which is probably connected to the monumental villa dei Settebassi, with all the advantages that such a location would have brought in terms of access to, and exploitation of, resources (in this sample the incidence of trauma and pathologies related to biomechanical stress was scored on only 11% of skeletons). Indeed, these last two aspects differentiate the Collatina cemetery, for which a probable exploitation by an urban sub-population can be hypothesised and which therefore must have accommodated the remains of very disparate population strata. The graveyard at Via di Padre Semeria, sited in an intermediate position on the graph near the Collatina necropolis, does not seem to be strongly characterized by a single qualitative variable.

With the aim of determining the characteristics proper to each individual graveyard, we continued by separating the characteristics – based on criteria chosen *a priori* – seeking to identify those which reflected in a consistent manner a well-established aspect of the anthropological study. A first assessment was carried out using only the demographic data.

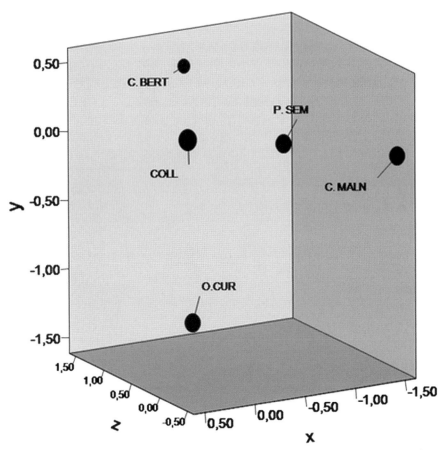

Fig. 8.8. *Correspondence analysis: comparison between cemeteries using all of the archaeological and anthropological variables.*

It is clear in the three-dimensional graph (Fig. 8.9), explaining 94.5% of the variability contained in the data ($p < 0.001$) that the two cemeteries show key features which significantly split from the other sites: the first is Castel Malnome, the position of which can be understood in the light of the unbalanced sex ratio. The second cemetery is Casal Bertone-Area Q, where the high frequency of subadult individuals (62%) distinguishes it from the remaining units analysed.

A second analysis was carried out by considering the data related to oral pathologies, taking into consideration the following variables, presence/absence of caries, calculus and abscesses. For this set of variables it was possible to use only two representational dimensions in order to interpret 97.2% of the variability in a significant way ($p < 0.001$). Figure 8.10 shows the cemeteries tend to be distant from one another in space and each shows particular characteristics. Osteria del Curato is

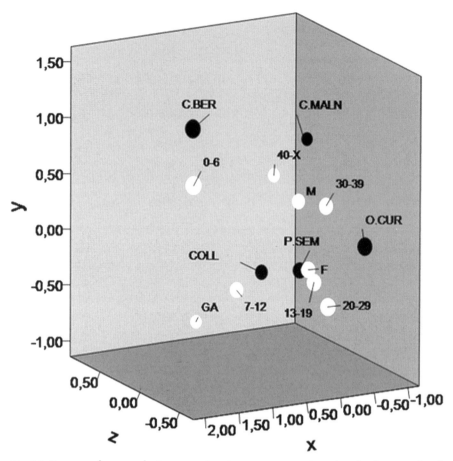

Fig. 8.9. Correspondence analysis: comparison between cemeteries using the demographic data.

located close to the variables for absence of dental pathologies (34% of the individuals do not show dental alterations). The Casal Bertone cemetery is again different, because of the high number of subadult individuals. The positions of the cemeteries of Castel Malnome and Via di Padre Semeria are also peculiar, not being characterised by any variable in particular.

The analysis was also extended to non-specific stress indicators: enamel hypoplasia, porotic hyperostosis (*cribra cranii* and *cribra orbitalia*) and periostitis. Two dimensions suffice to understand the complexity of the data at an appropriate level of significance (92.5% of the total; $p < 0.001$). As we can observe in the graph (Fig. 8.11), the cemetery of Castel Malnome is not characterised by skeletal alteration due to non-specific stress, while the other cemeteries analysed are all uniformly distributed along the first dimension (i.e. the x-axis) (which by itself accounts for 83.5% of the total variance).

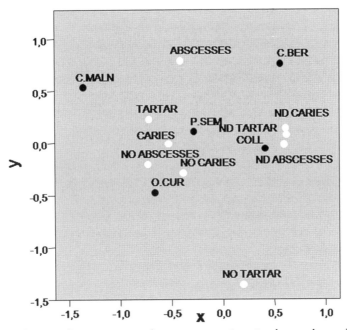

Fig. 8.10. Correspondence analysis: comparison between cemeteries using data on dento-alveolar infections.

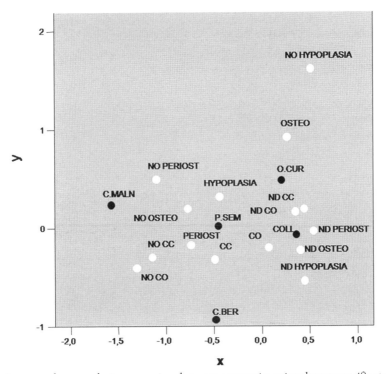

Fig. 8.11. Correspondence analysis: comparison between cemeteries using the nonspecific stress indicators.

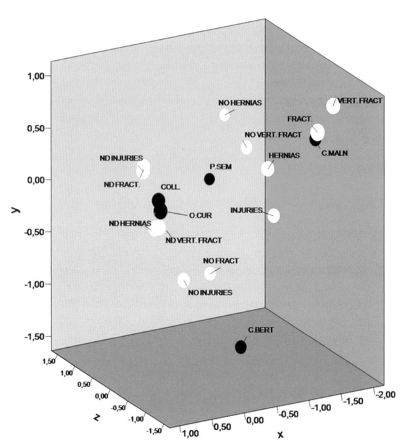

Fig. 8.12. Correspondence analysis: comparison between cemeteries using the data on bone alteration related to biomechanical stress.

Analysis of traumatic events and biomechanical stress also show differences between the five cemeteries. The graph clearly shows in the three-dimensional plot (99.8% of the variability represented; *p* < 0.001) (Fig. 8.12) how the two cemeteries belonging to productive sites, Castel Malnome, near to an area where ancient salt pans have come to light, and Casal Bertone-Area Q, behind an enormous *fullonica/conceria*, are totally separated from the others. It is clear how the individuals buried in these two cemeteries must have experienced biomechanical stress and traumas which characterized activities associated with a high degree of physical stress such as, respectively, the collection and working of salt and the dyeing and/or the tanning of textiles and/or hides.

We hope that the ongoing anthropological investigations, could enable us to depict the most realistic assessment possible of the daily living conditions of the population of the largest city in the ancient world.

Acknowledgements

The authors thank the following: Angelo Bottini (Soprintendente Archeologo di Roma), Anna Buccellato (archaeologist with responsibility for VI Municipio), Laura Cianfriglia (archaeologist with responsibility for XV Municipio), Roberto Egidi (archaeologist with responsability for X Municipio), Stefano Musco (archaeologist with responsibility for V Municipio Sud), and Rita Paris (archaeologist with responsibility for XI Municipio) for having consented to the undertaking of the work and for their precious collaboration; Prof. Franco Vecchi for fundamental guidance in the application of statistical methods; Giordana Amicucci, Chiara Caprara, Loredana Carboni, Simona Minozzi, Romina Mosticone, Alessia Nava, Lisa Pescucci, Gianna Tartaglia and Federica Zavaroni for having contributed to data entry.

Bibliography

Acsàdi, G. and Nemeskéri, J. (1970) *History of Human Life span and Mortality*. Budapest: Akademiai Kiadò.

Angel, J.L. (1966) *Early skeletons from Tranquillity, California*. Smithsonian Contributions to Anthropology 2.1.

Benassi, V., Buccellato, A., Caldarini, C., Catalano, P., De Angelis, F., Egidi, R., Minozzi S., Musco, S., Nava, A., Pantano, W., Paris, R. and Pesucci, L. (2011) La donna come forza lavoro nella Roma imperiale: nuove prospettive da recenti scavi nel Suburbio. *Medicina nei Secoli* 23.3, 291–302.

Brothwell, D.R. (1981) *Digging Up Bones*. Oxford: Oxford University Press.

Buccellato, A., Catalano P., and Musco S. (2008) Alcuni aspetti rituali evidenziati nel corso dello scavo della necropoli Collatina (Roma). In J. Scheid (ed.) *Pour une archéologie du rite - Nouvelles perspectives de l'archéologie funéraire*, 59–88. Collection de l'École Française de Rome 407.

Buccellato, A., Musco, S., Catalano, P., Caldarini, C., Pantano, W., Torri, C. and Zabotti, F. (2008) La nécropole de Collatina. *Les Dossiers d'Archéologie* 330, 22–31.

Buikstra, J. E. and Ubelaker, D. H. (1994) *Standards for Data Collection from Human Skeletal Remains*. Arkansas Archeological Survey Research Series 44, Fayetteville.

Burns, K. R. (1999) *Forensic Anthropology Training Manual*. Englewood Cliffs: Prentice Hall.

Caldarini, C., Caprara, M., Carboni, L., De Angelis F., Di Giannantonio, S., Minozzi, S., Pantano, W., Preziosi, P. and Catalano, P. (2006) Vivere a Roma in età imperiale: evidenze antropologiche da recenti scavi nel Suburbio. *Medicina nei Secoli* 18.3, 799–814.

Canci, A. and Minozzi, S. (2005) *Archeologia dei resti umani*. Rome: Carocci.

Capasso, L., Kennedy, A. R., and Wilczack, C. A. (1999) *Atlas of Occupational Markers on Human Remains*. Teramo: Edigrafital.

Catalano, P., Minozzi, S. and Pantano, W. (2003): Le necropoli romane di età imperiale: un contributo all'interpretazione del popolamento e della qualità della vita nell'antica Roma. *Atlante tematico di topografia antica* 10, 127–137.

Catalano, P., Amicucci, G., Benassi, V., Caldarini, C., Caprara, M., Carboni, L., Colonnelli, G., De Angelis, F., Di Giannantonio, S., Minozzi, S., Pantano, W. and Porreca, F. (2006) Gli insiemi funerari d'epoca imperiale: l'indagine antropologica di campo. In M.A. Tomei (ed.) *Roma. Memorie dal sottosuolo. Ritrovamenti archeologici 1980/2006*, 560–563. Catalogo della Mostra, Rome.

Catalano, P., Caldarini, C., De Angelis, F., Di Giannantonio, S., Minozzi, S. and Pantano, W. (2007) La carie dentaria a Roma: indicazioni da alcuni sepolcreti suburbani d'epoca imperiale. *Medicina nei Secoli* 19.3, 745–761.

Catalano, P., Benassi, V., Caldarini, C., Cianfriglia, L., Mosticone, R., Nava, A., Pantano, W., Porreca, F. (2010) Attività lavorative e condizioni di vita della comunità di Castel Malnome (Roma, I–II sec. d.C.). *Medicina nei Secoli* 22.1–3, 111–128.

Clarke, N. G. and Hirsch, R. S. (1991) Physiological, pulpal, and periodontal factors influencing alveolar bone. In M. A. Kelly and C. S. Larsen (eds) *Advances in Dental Anthropology*, 241–266. New York: Wiley–Liss.

Corrucini, R. S., Handler, J. S. and Jacobi, K. B. (1985) Chronological distribution of enamel hypoplasia and weaning in a Caribbean slave population. *Human Biology* 57, 699–711.

Duday, H. (2006) *Lezioni di Archeotanatologia*. Programma Europeo 'Cultura 2000', Rome.

Egidi, R., Catalano, P. and Spadoni, D. (eds) (2003) *Aspetti di vita quotidiana dalle necropoli della Via Latina. Località Osteria del Curato*. Catalogo della Mostra 2003–2004, Rome: Museo Nazionale Romano.

Fazekas, G. and Kosa, F. (1978) *Forensic Fetal Osteology*. Budapest: Akadémiai Kiadò.

Ferembach, D., Schwidetzky, I. and Stloukal, M. (1977) Raccomandazioni per la determinazione dell'età e del sesso sullo scheletro. *Rivista di Antropologia* 60, 5–51.

Frayer, D. W. (1988) Auditory exostoses and evidence for fishing in Vlasac. *Current Anthropology* 29, 346–349.

Goodman, A. H. and Armelagos, G. J. (1985) Factors affecting the distribution of enamel hypoglasia within the human permanent dentition. *American Journal of Physical Anthropology* 68a, 479–493.

Heinzelmann, M. (2001) Culto dei morti e costumi funerari romani. *Palilia* 8, 21–28.

Hengen, O. P. (1971) Cribra Orbitalia: pathogenesis and probable etiology. *Homo* 22, 57–75.

Hillson, S. W. (1986) *Teeth*. Cambridge: Cambridge University Press.

Hillson S. W. (1996) *Dental Anthropology*. Cambridge: Cambridge University Press.

Kelley O. (2012) Beyond intermarriage: the role of the indigenous Italic population at Pithekoussai. *Oxford Journal of Archaeology* 31, 245–260.

Kelley, M. A. and Larsen C. S. (1991) *Advances in Dental Anthropology*. New York: Wiley–Liss.

Lanciani, R. (1909) *Wanderings in the Roman Campagna*. Boston/New York: Houghton-Mifflin (*Passeggiate nella Campagna Romana*, Rome, 1980).

Lovejoy, C. O. (1985) Dental wear in the Libben population: its functional pattern and role in the determination of adult skeletal age at the death. *American Journal of Physical Anthropology* 68, 47–56.

Lovejoy, C. O., Meindl, R. S., Mensforth, R. P. and Burton, T. J. (1985) Multifactorial determination of skeletal age at death: a method and blind tests of its accuracy. *American Journal of Physical Anthropology* 55, 529–541.

Mall, G., Graw, M., Gehring, K.D. and Hubig, M. (2000) Determination of sex from femora. *Forensic Science International* 113, 315–321.

Manzi, G., Sperduti, A. and Passarello, P. (1991) Behaviour-induced Auditory Exostoses in Imperial Roman Society: Evidence from Coeval Urban and Rural Communities near Rome. *American Journal of Physical Anthropology* 85, 253–260.

Marafon, G. (1979) *Odontoiatria*. Collana Università, Rome: Almes.

Meindl, R. S. and Lovejoy, C. O. (1985) Ectocranial suture closure: a revised method for the determination of age at death based on the lateral-anterior sutures. *American Journal of Physical Anthropology* 68, 57–66.

Mensforth, R. P., Lovejoy, C. O., Lallo, J. W. and Armelagos, G. J. (1978) The role of constitutional factors, diet, and infectious disease in the etiology of porotic hyperostosis and periosteal reactions in prehistoric infants and children. *Medical Anthropology* 2, 1–59.

Metress, J. F. and Convay, T. (1975) Standardized system for recording dental caries in prehistoric skeletons. *Journal of Dental Research* 5, 908.

Mezl, Z. (1985) *Patologia dentaria*. Milan: Masson Italia Editori.

Minozzi, S., Zabotti, F., Torri, C., Pantano, W., Catalano, P., Buccellato A. and Musco, S. (2008) Elaborazione di metodiche comuni. Una banca dati archeologica ed antropologica per la gestione dei dati di scavo. In J. Scheid (ed.) *Pour une Archéologie du Rite. Nouvelles perspectives de l'Archéologie funéraire*, 337–349. Collection de l'École Française de Rome 407, Rome.

Musco, S. and Petrassi, S. (2001) *Luoghi e paesaggi archeologici del suburbio orientale di Roma.* Rome: Pracchia.

Musco, S., Catalano, P., Caspio, A., Pantano, W. and Killgrove, K. (2008) Le complexe archéologique de Casal Bertone. *Les Dossiers d'Archéologie* 330, 32–39.

Musco, S., Catalano P., Benassi, V., Buccellato, A., Caldarini, C., Caspio, A., De Angelis, F., Di Giannantonio, S., Mannino, M., Pantano, W., Zabotti, F. (2010) Tombes d'enfants de l'époque impériale dans la banlieue de Rome: les cas de Quarto Cappello del Prete, de Casal Bertone et de la nécropole Collatina. In A. M. Guimier-Sorbets and Y. Morizot (eds) *L'enfant et la mort dans l'Antiquité* 1, 387–400. Paris: De Boccard.

Olivier, G. (1955) Anthropologie de la clavicule. X. La clavicule des hommes néolithiques. Le problème de la différentiation sexuelle. *Bulletin et Mémoires de la Société d'Anthropologie de Paris* 6, 290.

Ortner, D. J. (2003) *Identification of Pathological Conditions in Human Skeletal Remains.* London: Academic Press.

Palkovich, A. M. (1987) Endemic disease patterns in paleopathology: porotic hyperostosis. *American Journal of Physical Anthropology* 74, 527–537.

Pitts, M. (2005) Pots and Pits: Drinking and deposition in late Iron Age south-east Britain. *Oxford Journal of Archaeology* 24, 143–161.

Richards, L. C. and Miller, S. L. J. (1991) Relationships between age and dental attrition in Australian aboriginals. *American Journal of Physical Anthropology* 84, 159–164.

Rogers, J. and Waldron, T. (1995) *A Field Guide of Joint Disease in Archaeology.* Chichester: Wiley.

Skinner, M. and Goodman, A. H. (1992) Anthropological uses of developmental defects of enamel. In S. R. Saunders and M. A. Katzenberg (eds) *Skeletal Biology of Past Peoples: Research Methods,* 153–174. New York: Wiley–Liss.

Steel, F. L. D. (1962) The sexing of the long bones, with the reference to the St. Bride's series of identified skeletons. *Journal of the Royal Anthropological Institute of Great Britain and Ireland* 92, 212.

Steinbock, T. R. (1976) *Paleopathological Diagnosis and Interpretation.* Springfield: C. C. Thomas.

Stloukal, M. and Hanakova, H. (1978) Die Länge der Längsknochen altslawischer Bevölkerungen – unter besonderer Berücksichtigung von Wachstumsfragen. *Homo* 29, 53–69.

Stuart-Macadam, P. (1987a) A radiographic study of porotic hyperostosis. *American Journal of Physical Anthropology* 74, 511–520.

Stuart-Macadam, P. (1987b) Porotic hyperostosis: New evidence to support the anemia theory. *American Journal of Physical Anthropology* 74, 521–527.

Ubelaker, D. H. (1989) *Human Skeletal Remains: Excavation, Analysis, Interpretation.* Washington D.C.: Taraxacum.

Walker, P. L. (1986) Porotic hyperostosis in a marine-dependent California Indian population. *American Journal of Physical Anthropology* 69, 345–354.

Weaver, D. S. (1986) Forensic aspects of fetal and neonatal specimens. In K.J. Reichs (ed.) *Forensic Osteology: Advances in the Identification of Human Remains,* 90–100. Springfield: C. C. Thomas.

Chapter 9

Animals in funerary practices: Sacrifices, offerings and meals at Rome and in the provinces

Sébastien Lepetz

The analysis of ancient texts reveals the central place held by sacrifice among religious acts. In relation to funerals and the cult of the dead, the rites are known to have unfolded within the framework of a common tradition (of which the meaning is well understood), although the form they took was not perfectly identical and differed according to the families, social groups or regions involved. Since these practices are not detailed in the texts, they leave the researcher wishing to describe them somewhat bereft of evidence. So, if we consider the place of animals, the sacrifice of a sow to Ceres, destined to purify the family (Festus p. 296–298 L and Cicero, *Leg.* 2, 57) and confine the soul of the deceased to the tomb (Virgil, *Aen.*, 3, 66–68) seems a commonly practiced act, as was the shared meal between the deceased, the living and the gods (Fig. 9.1). This funeral meal (*silicernium*, Nonius Marcellus, *De comp. Doctr.* 1, 48), eaten close to the tomb was destined to honour the deceased. Assistants threw on the pyre gifts (*munera*), clothes, provisions, bread and perfumes 'to parade their pain' (Pliny, *Ep.* 4, 2). This act is also described by Virgil for the pyre of Misenus (*Aen.* 6, 224–225): 'gifts (*dona*) were heaped on the flames, of incense, foodstuffs (*dapes*, cf. Festus, p. 59 L)'. An anecdote of Apuleius shows that these offerings were arranged on the pyre with the dead man not yet cremated (Apul. *Flor.* 19). But if we know that food was eaten during the funeral meal, we actually have little detail on the composition of these meals, the range of species consumed, the favoured cuts and the treatment of foodstuffs. We also possess little information on the place and the frequency of food deposits on the pyre or in the graves. The texts say nothing on this subject.

This is not a simple problem. Analysis in fact reveals a multitude of processes (sacrifices, offerings, meals, visits to the deceased) related to different stages and following distinct rhythms (Fig. 9.2), while the archaeological data show remains of different origins, resulting from these various acts. To these ritual remains rubbish is added, more or less erratically, from human activity and by animals living nearby and

Fig. 9.1. Funerary enclosure 7 ES, Porta Nocera, Pompeii. Note the triclinium built against the columbarium (photo, A. Gailliot, MFP/FPN).

it is difficult to distinguish and recognize this. In fact, if the meaning of the acts that produced these traces is precise, particular and defined, these remains should have the same form, as they should usually come from the same species, cut into pieces and eaten (or edible); otherwise they are often mixed and the discovery contexts do not help much to show up their specificity. The approach to these remains therefore is not easy and depends mostly on the conditions of preservation, of excavation, of collection and of study offered by the sites concerned.

If archaeozoological analysis does not allow us to give a meaning to the acts practised, on the other hand it does allow us to describe them objectively: are pieces of animals deposited on the funeral pyre and in the graves and, if so, what are they and what are their characteristics? Why are unburnt bones present around the graves? Can we recognize here the remains of meals? What is the meaning of the dog and equid remains so often found close to funerary structures?

The place of animals in funerary practices in Pompeii

Recent excavations from 2003 to 2007 were focused on a funerary area of the Porta Nocera cemetery, comprising several enclosures belonging to freed slaves (Lepetz and Van Andringa 2006 and 2011; Van Andringa and Lepetz 2008; Van Andringa,

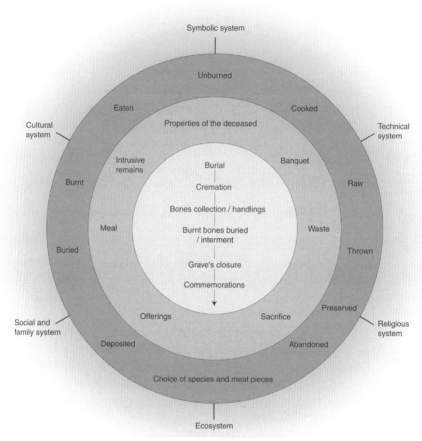

Fig. 9.2. Organization of components, systems and practices involved in the composition of faunal assemblages excavated in cemeteries and connected, or not, to the stages of the funeral.

Duday and Lepetz 2013). These tombs are at the south-eastern exit of the city, aligned along the road running to the Stabian Gate and to the harbour located at the mouth of the River Sarno. In the northern area of the enclosure and facing the road a funerary monument was built, comprising a high podium topped by a pedimented aedicula, within which were three statues (Fig. 9.3). A vaulted niche set in the podium faced the interior of the enclosure (Fig. 9.4). The inscription on a stele within this niche indicates that this space was destined to receive the remains of the enclosure's owner, Publius Vesonius Phileros. The tomb's dedication also specifies that he is the emancipated slave of Gaia Vesonia whose cremated remains were deposited, and discovered, in the same enclosure. By their side another stele indicates the presence of the probable son of the enclosure's owner, Publius Vesonius Proculus: excavation of his remains revealed he was about eight years old when he died. The concession founded by Phileros (enclosure 23 OS) is

Fig. 9.3. Funerary enclosure 23 OS, Porta Nocera, Pompeii. Note the high podium with three statues. Left, Publius Vesonius Phileros, the owner, centre, Vesonia, his patron; right, Marcus Orfelliius Faustus, friend of Phileros. In the background the graves under excavation (photo, author).

located within a funerary area which is well documented by inscriptions. It is thus possible to establish that the joint tomb 21 OS also received the burial of a slave, Stallia Haphe, emancipated by a woman.

There are in total 64 individuals whose graves (dated to the first three quarters of the 1st century AD) were excavated in the different enclosures: according to the anthropological study done by H. Duday (CNRS), 40 are graves of adults (of which 17 were women and 16 men), 13 are cremations of children between three months and 13 years old and there are five inhumations of children less than three years old.

Globally the graves were not very rich in material remains. No ceramic deposit was found in the graves near the cinerary urns; only glass unguentaria from libations are often present. Mixed with the burnt human bones were coins, always present, worked bone fragments, carpological remains and many faunal remains. Most of the latter are burnt bones coming from cuts of meat placed on the funeral pyre and collected with the cooled ashes from the pyres which were situated on cremation areas near the graves.

Fig. 9.4. The tomb of Publius Vesonius Phileros in the niche under the podium. In front of Phileros, the tomb of his son, Proculus; to the right of Proculus, the tomb of Vesonia (photo, A. Gailliot, MFPN/FPN).

Not all the animal remains discovered in the urns are directly linked to funerary practices. The case of grave 21 (enclosure 23) gives an example of this diversity (Fig. 9.5). In the amphora used as funerary receptacle three individuals were successively deposited: first the cremated remains of a new born (subject C), then those of two adolescents (subject A and subject B). In sorting 21A 114 animal bones were collected of which 40% were recognized: the five bones of mammals noted (two teeth and a fragment of pig humerus, a fragment of sheep/goat tooth and a fragment of beef rib) are out-numbered by fish bones (41 instances). The size of the bones (of very small anguilliforms (eels) and bogue) allows us to envision the deposit of fish dishes, perhaps as fried fish or fish sauce. However, it is possible that all or part of these elements was ingested by the deceased before his death and that these bones are less an 'offering' than the remnants of his last meal. It is interesting to note that individual 21B was also accompanied by a large number of fish bones (23 bones), including a perciform (perch-like) fish of very small size associated with some mammal bones including a pig metatarsal. These remains are therefore offerings carried to the funerary pyre or remains of the meal taken around the blaze or ingested remains. But among them other bone categories were collected. Sorting of the sieved residues also allowed 37 elements of the internal skeleton of a slug to be identified, mixed with the remains of subject 21A. These slugs illustrate the presence of invertebrates among the wood used as fuel.

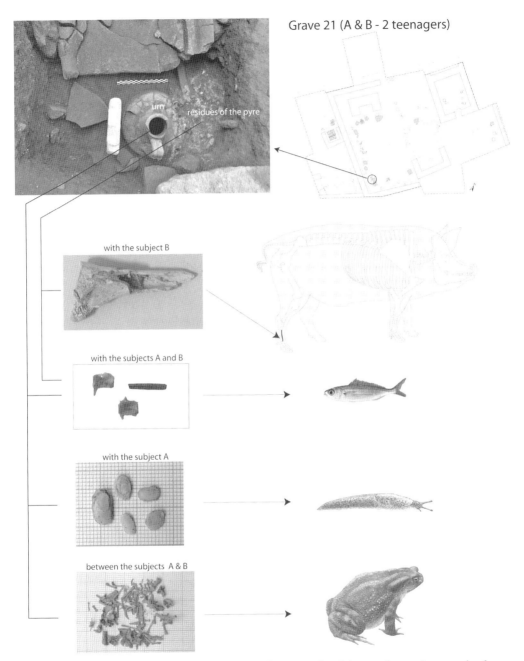

Grave 21 (A & B - 2 teenagers)

urn
residues of the pyre

with the subject B

with the subjects A and B

with the subject A

between the subjects A & B

Fig. 9.5. Pompeii, Porta Nocera. Animal remains found associated with human bones: the example of grave 21 (photos, A. Gailliot, MFPN/FPN; drawing, author).

Furthermore, systematic water sieving with a 500µ mesh allowed recovery, within the contents of the urn of 450 bones of anurans (amphibians including frogs and toads) and some lizard remains. The form of the amphibian bones sometimes allowed recognition of the species. The majority of the recognized remains (NISP (Number of Identified Specimens) = 72) comes from a toad of the genus *Bufo*. Among these 18 could be attributed to the common toad (*Bufo bufo*). The counting of the better represented skeletal parts allowed a minimum of six individuals to be recognized. This species is universally found among European amphibians and likes relatively dry terrain, even if its habitat is quite varied. The collection includes long bones of very young individuals as well as small bones of hands and feet that have not been identified. Alongside the toads, frogs are represented by the green frog (*Rana lessonae/ esculenta*) and the green tree frog (*Hyla arborea*). These bones began to appear at the interface between individuals 21A and 21B and their number increases noticeably in the underlying spits where they are systematically mixed with the remains of subject 21B. It appears very clearly that the lower strata are richer in small-sized bone: the small bones of the big individuals (phalanges, tarsals) and the long bones of the smaller individuals have infiltrated the entire thickness of deposit B (which was not therefore sealed) and the more substantial bones have remained instead in the upper level of this stratum. At least 12 small animals of mixed species fell in the amphora on top of the remains of individual 21B. The animals were not able to escape from this trap and perished there. Their presence shows that the interior of the grave was not totally sealed from intrusive agents. Even if it is difficult to estimate the interval between deposits B and A, the number of animals shows a considerable length of time had elapsed.

If some graves are rather rich in animal remains, others are particularly poor. This is the case for instance for grave 1 (Phileros, the owner of the enclosure), 2 (Vesonia, his mistress) and 5 (Proculus, son of Phileros), where we might have expected more abundant deposits. In fact, these cremations delivered respectively only one, three and seven bones, all very small and among which only one fish element was recognized. The pyres were therefore devoid of meat-based food remains and there is not a direct link between the wealth (or poverty) of the deposit on the pyre and the social position of the deceased.

On the other hand, a little more was found in the grave of a young six-year-old slave (grave 201). Not long before the eruption of Vesuvius (undoubtedly less than 10 years), this child, named Bebryx, was burnt on a pyre built behind enclosure 21. The cremation area had been used for several other ceremonies and particularly for a certain Stallia Haphe, more than thirty years old and a slave freed by a matron of the Stallii, whose grave is located close to that of Bebryx. The anthropological study clearly showed that the two individuals had been cremated on the same platform. The bones of the child were collected in a fabric bag and placed (upside down) in a cinerary urn: this was placed in a pit dug at the site of an older grave (207) in enclosure 21; residues of the pyre were tipped in the pit and a stele was erected. Few material elements were collected among the human bones. In the pyre debris, small fragments of chicken bone were collected (Fig. 9.6). In the cinerary urn, the bones were exclusively of pig, not an entire animal but only some elements (fragments of head, forelimb and vertebra). It is interesting to note that two fragments of vertebra found in the urn and another fragment, found among the remains collected on the cremation area located nearby, come from the same bone (and join). The anthropological analysis showed otherwise that this pyre, situated three metres away, belonged to the child Bebryx (from the connection between pyre and grave of fragments of bone and of matching elements from left and right sides of the body).

A grave located close to this last one (205, an adult woman) produced many animal remains (NISP = 44) and is the richest of the site (Fig. 9.7). Material was found in the cinerary urn (NISP = 14) and in the pyre residues. The two species represented are pig (29 bones) and chicken (eight bones). The pig assemblage is characterized by a significant proportion of bone from the lower part of at least three feet (anterior and posterior) from two individuals, one of which was especially young. In the cinerary urn, only pig was present: chicken was outside the urn where the bones of the younger suid (the smallest) were found. The effects of differential collecting in antiquity explain this phenomenon, with the bigger remains in the cinerary urn collected by hand and the remains outside the urn originating from the cleaning of the cremation area with the smaller bones (fragmented bird bones and small bones of piglets). For the chicken, given that no bones are present within the urn, we can surmise that, as for the pig, pieces were deposited on the pyre or thrown there after consumption but not entire animals.

Grave 39 (a four-year-old child), is rich in several tens of bones, but all of them had particular characteristics. Pig is represented by a charred milk-tooth; we might be surprised to have no trace of any other element from the head and this therefore raises the question of the significance of this evidence: is it the remnant of a deposit of all or part of a piglet head on the pyre or is it a stray bone? The other question concerns a left hemi-mandible of an equid (probably horse) mixed with the human bones (Fig. 9.8). Seventy fragments (deriving from three principle fragments) were collected and the degree of calcination leaves no doubt that they burned in an intense fire: the mixture with the human remains indicates that it was the pyre of the child. Was the deposit of this element alongside the child deliberate? What might its significance be? It is first necessary to note that a horse mandible is strongly linked to the cranium by connecting muscles and ligaments. The separation of one from the other on a fresh head is therefore a task demanding technical competence and determination. It also could not fail to leave knife traces on the mandibular condyles. If the deteriorated state of the bone did not allow us to find these, we might wonder whether the family of the child had wished to disarticulate a fresh left mandible of a horse and to deposit it by its side at the funeral. We must also note that the surface of the equid bone is split, cracked and peeled and this damage is not to be attributed to fire. We also noted deteriorations on the dry bone exposed to the open air for a certain

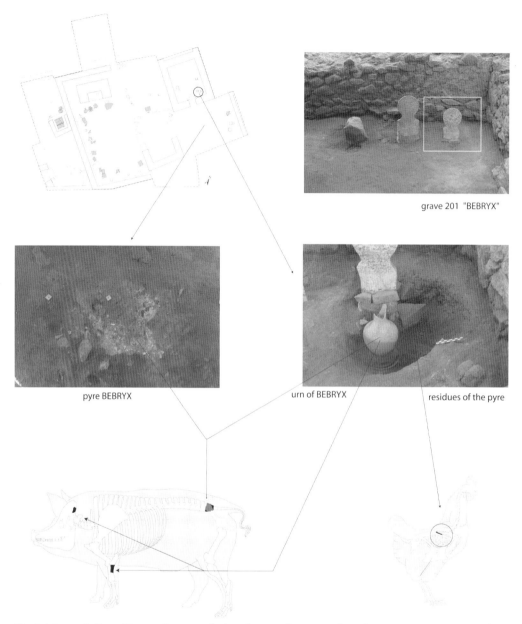

grave 201 "BEBRYX"

pyre BEBRYX   urn of BEBRYX   residues of the pyre

Fig. 9.6. Pompeii, Porta Nocera, Grave 201 (Bebryx). Animal remains from the cinerary urn, pyre residues and the pyre of Bebryx (photos, A. Gailliot, MFPN/FPN; drawing, author).

time (weathering): this bone was thus burned. But was the act deliberate or not? And is it linked with the funerary ceremony? In the latter case, it would be necessary to imagine a special link between the object and the child (a toy for instance) since such an object has no place in funerary ritual. If this is not a link, we can consider

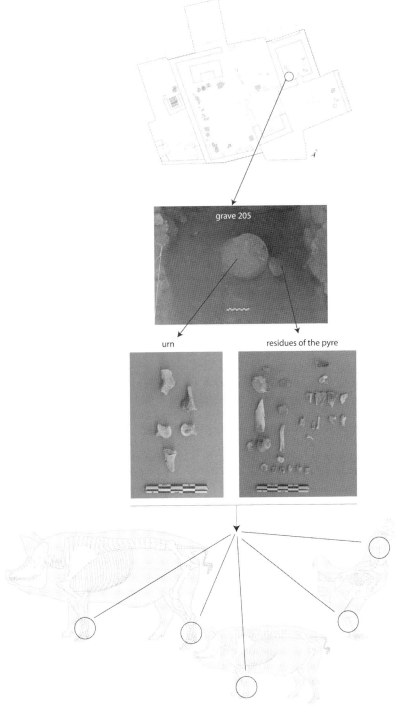

Fig. 9.7. Pompeii, Porta Nocera. Grave 205, the richest in animal remains (photos, A. Gailliot, MFPN/FPN; drawing, author).

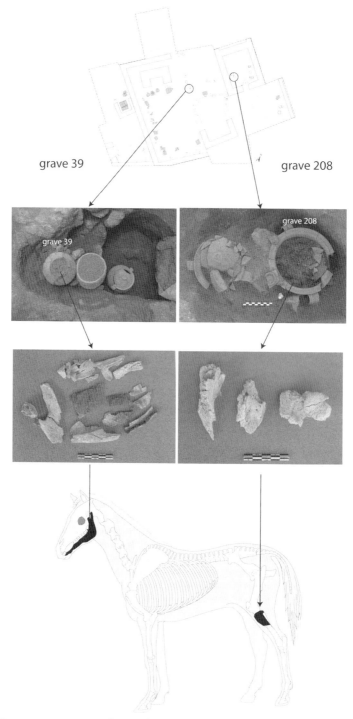

grave 39

grave 208

Fig. 9.8. Pompeii, Porta Nocera. Burnt bones of equids, graves 39 and 208 (photos, A. Gailliot, MFPN/FPN; drawing, author).

that the bone was located near the pyre, since the study of occupation levels showed the presence of waste in the enclosures, and that the person who led the cremation would quietly undertake the cleaning of the area. It is nevertheless surprising that the person who collected the human bones was able to confuse the horse bone with those of the deceased. Could we imagine that it was done in full knowledge of the facts and that the officiant who collected the human bones did not judge it useful to make a difference during the collection?

A parallel can be made with grave 208 (an adult) where a charred rib fragment from a small cow was found. This might constitute the remains of an offering deposited on the pyre or of a meal, where the main assemblage comprises 35 fragments from a single tibia of an equid and three indeterminate others. The fragments were found in large majority in the cinerary urn, in the pit fill. In the cinerary urn they are distributed throughout the fill, indicating that they were collected at the same time as the human bones and were therefore burnt at the same time as them. We can also note here and there traces of damage probably linked to weathering due to the bone's being left on the ground. We can also consider here the meaning of such a deposit. Given the status of horse during the Roman period, which was not eaten and not used in funerary rites, we could be inclined to consider this the chance presence of such a bone in the inferno.

If we therefore exclude horse, the taxa mixed with cremations are pig, sheep, goat, chicken and fish. These bones come from the pyre areas, from heaps of cremated bone in the graves, from pyre residues deposited in the pits or from circulation levels (Fig. 9.9). They relate to the acts linked to the cremation phase of the corpse. We notice that pig, chicken and fish are more frequent as their bones comprise c. 30% each of the sample.

The fish bones are not very numerous (82 bones), considering the great care taken to identify them in excavation and the systematic sorting of a large quantity of sediment. The list of species is relatively diverse and does not appear in any case to be due to special choices; the taxa are those usually met on the region's coasts and can be eaten fresh or preserved in brine or in sauce: sparids

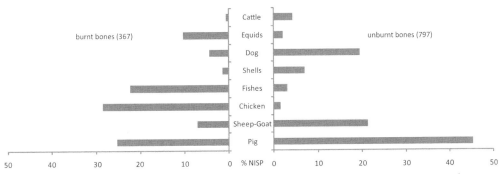

Fig. 9.9. Pompeii, Porta Nocera. Representation of burnt bones and unburnt bones for each animal species.

(a denticinae (*Dentex* sp.), bogue (*Boops boops*)), triglid (*Triglidae*, sea robin), peacock wrasse (*Symphodus tinca*), European conger (*Conger conger*), scombrid (the Atlantic mackerel, *Scomber scombrus*), an anguilliform (European eel, *Anguilla anguilla*) and other indeterminate perciformes.

Half of these bones (NISP = 41) come from grave 21, which is the only one for which it is possible to assert that one or more fish dishes were deposited on the pyre alongside the deceased, the bones' size allowing us to dismiss the hypothesis of ingested remains. Elsewhere, in the seven other graves and pyres where they are present, we have discovered the isolated remains of small fish. The case of fish is a little special. Undoubtedly the bones burned and passed through the pyre, but were they placed with the deceased or were they stomach contents? To ask this question is justified by the fact that the concentrations are very slight and the bony pieces for the most part are extremely small. Some vertebrae are only a few millimetres long and come from animals of a few centimetres in length. It is possible that they were ingested by the deceased just before their death.

As for pig, pieces of meat were undeniably deposited on some pyres. The practice is far from systematic since only 30% of graves contained them. We recognize some pieces of relative importance: grave 205 shows that at least three pig feet were placed and burnt near the deceased and the cremation of the individual from burial 207 also included a piglet skull. But the majority of the cases comprise lone elements relating to small size pieces of meat, perhaps culinary preparations. The analysis of the anatomical distribution of the 93 burnt bones seems to indicate that most parts are represented, heads, vertebrae, limb and foot bones, but ribs are under-represented. If there is not, in practice, strict selection then there may have been preferences which are not perceptible. The pigs are young animals: in the rare cases when this is visible the epiphyses have not fused. Sheep and goat remains are also present in the graves but in rather small quantities: only six graves produced some. On the whole site 26 burnt bones indicate nevertheless that some pieces of sheep could be deposited on the pyres.

Chicken (105 bones) was found in the graves and on the surrounding cremation areas. Undeniably chickens were put on at least 11 pyres but they do not appear to have been complete animals because the entirety of the bones were never found; at most c. ten elements were found and this seems insufficient, even if we consider their vulnerability to fire. It is difficult to define preferences in terms of age or of sex as the material does not allow it. We can however observe that heads are absent and that the vertebrae and bones of the carcass are rare. On the other hand, limb elements (thighs and wings) as well as the extremities are strongly present, indicating a preference for these pieces. The food deposit also sometimes included eggs: grave 207 delivered 25 fragments of the same shell and another burnt fragment was found in pyre area 250.

Two-thirds of the graves contained burnt animal bones, 36 of the 53 considered. Overall the faunal assemblages of the graves are rather different. Some burials are rich in pork, others in poultry and others again in fish, without counting those where there is nothing. No strict rule is perceptible in the representation of taxa and therefore no evolution over time is visible, probably because there is none. Pig and chicken seem to be the favourite two species, fish being present more occasionally. Family or personal choices during the setting up of the pyre as well as preferential collection on the cremation area can explain these differences.

The other category concerns the unburnt bone sample (c. 2,000 bones) found scattered on the ground levels or in a modified position in grave fills, on walkways and roads. Many dog bones (NISP = 123) were documented. Their discovery prompted us to consider the meaning of their presence. Do they have a direct link with behaviour linked to death; are they involved in funerary rituals? In Roman tradition, which is common to all families and respects public regulations, it is clear that the dog is not part of these rites and there is no sense to include it. It could be of course possible that, in certain cases, in response to family or local customs, the animal could have a role in some provinces of the Empire, but that would remain an occasional exception in relation to broader Roman habits, and the reality of these gestures has never been proved. We can at this point consider another explanation. Stray dogs, whether or not they had an owner, were able to enter the enclosures to look for shade, quiet or food to scavenge. Several bones gnawed by them reveal their circulation around the graves, and others that were obviously digested and expelled through faeces (14 bones), show that the place was used as an area where canids could relieve themselves. The presence of these traces of coprolites reveals the picture which must emerge from the study of this funerary area of a place that, far from being sealed and preserved from the world of the living, is open to the street and records its activity. The outskirts of the city are therefore a receptacle for various wastes, the disposal of which was not guaranteed by the city or by private individuals. This is the case for instance for the corpses of some dead animals, notably those of dogs whose remains are found frequently in the free spaces of the *suburbium*. Dispersed by their fellow creatures and by a cohort of necrophagous animals, hidden beneath the vegetation and by the successive backfills, or destroyed by the fire of pyres, as the discovery of darkened bones of dog in one of the cremation areas attests, in the end their bones would have gone mainly unnoticed. But the predominant material of this type comes from pig (60%) and sheep/goat (30%). The fragmentation and the cut marks indicate without doubt that we have food remains, but the damage to the bone surfaces and the anatomical distribution shows the existence of a strong taphonomic filter which favours the more solid bones and the bigger species. We are therefore authorized to consider these traces as probable remains of meals taken around the pyre or graves, thanks to their context and the associated finds (finewares, lamps, unguentaria...) but formal proofs are absent.

The comparison of these results with practices met on other Pompeian sites allows the specificity of the Porta Nocera assemblage to be underlined. No other data concerning the cemeteries are available. As for religious contexts, the discoveries in the house of Amarantus (Powell, in Fulford *et al.* 1995/6, 102–105; Robinson 2002) show the presence of many burnt sheep and chicken bones (all males) in pits dug in the garden. These are probably traces of domestic rituals, perhaps related to the feast of the *Lares*, where it was possible to eat the major part of the meat of the sacrificed animals and reserve, through burning, a part to the gods. If so, this latter part does not present any special characteristics since all anatomical areas are represented. The strong presence of chicken should be noted here, contrasting with the absence of pig.

Suids (pig family) are often very well represented in the faunal assemblages of Pompeii. The excavations in the domus, gardens or forum produce some rather large quantities of pig remains. If we take into account only the three principal domestic species, they dominate the faunal spectra accounting for 50 to 80% of assemblages (MacKinnon 2004, 62–64; Dickmann *et al.* 2002, 38 *sqq*; see also the comments of King in Jashemski and Meyer 2002, 444), as also observed in the funerary enclosures and in the graves. The proportion of cattle is often higher (it can reach 30%, as on the forum excavations), while it accounts for between 1 and 5% of the Porta Nocera sample: the sheep/goat remains are similar to the town average. Chicken distinguishes itself clearly because it comprises 29% of the faunal sample in the cinerary urns but is never so frequent in the settlement: it most often accounts for between 1 and 8% but we note than many studies do not allow us to know the representation of this species, either because it was not distinguished from other birds or because they do not mention it. However, the rates met on these other sites are very close to what we observe in the occupation backfills of the excavated funerary enclosures (1.5%). The effects of conservation and differential collection are probably behind these dissimilarities: the ratio of domestic birds is often underestimated in studies dedicated to diet during the Roman period and it is difficult to allow for that. The case of fish is also special because it is difficult to measure the frequency and the intensity of the collection of the smaller bones on the comparison sites.

On the other hand, this is not the case for the bones of dogs and equids: the Pompeian sites produce only rare fragments of these (<4% for equids, <2% for dogs), which is perfectly logical when we consider the fact that these animals are not eaten and therefore have no reason to be mixed with culinary waste. These animals die or their corpses are brought outside the walls and it is therefore normal to find some of these in the free areas of the *suburbium* or the cemeteries, as is the case for Porta Nocera.

Taking into account the results of this analysis, it might be tempting to associate the presence of pig bones in large numbers on the Porta Nocera site with the rite of sacrifice to Ceres of the *porca praesentanea* which would be followed by a division of the meat between the living, the dead and the gods (Scheid 2008, 7) Of course nothing contradicts the texts but clearly the analogy is too simplistic. Pig is omnipresent in Pompeian and more generally Roman society. It is a meat appreciated by all and of

Fig. 9.10. Lararium of house VII, 2 or 3, Pompeii (photo, W. Van Andringa).

which we find remains in all the food dumps. The animal is especially appropriate to private rituals. It would have been less expensive than a bigger beast, can be easily bred in the dwelling unit or in proximity to the city, can be easily butchered on the spot, and the cuts as prepared can be distributed easily, salted or smoked. This is therefore an obvious victim to designate for the sacrifices executed by the *paterfamilias* or the leader of a small community. We may see a representation of it being sacrificed on the shrine dedicated to the *Lares* of house VII, 2 or 3 (Fig. 9.10). It is also pig of which we see pieces (heads, pieces on skewers, thighs) generally painted near the shrines dedicated to the *Lares* of the kitchens, perhaps to recall the sacrifices celebrated in the house (Fig. 9.11). The pig is therefore intimately linked to the diet and the religious practices of the family. It shares this place with chicken which is rarer on figured representations and in texts, perhaps because it more frequently occurred as the least expensive sacrifice and the one whose butchery is simplest, the minor victim *par excellence*, clearly present on the pyres. The sheep is usually absent from these representations, an important observation.

It would therefore be vain to look on the pyres of Pompeii and around its graves to identify the remains of sacrificed animals, or differentiate the pieces reserved for the gods, for the living and for the dead. The remains offer a blurred picture, one of

Fig. 9.11. Kitchen of house IX, 9c, Casa del Porcellino (also known as Casa di Sulpicius Rufus), Pompeii (photo, author).

sacrifices linked to usual food practices. It is possible that the division is part of the meal eaten in honour of the deceased without always reserving them a special piece. This is what might be inferred from the irregularity of the act of placing pieces of meat on the pyre or in the grave: it is actually far from systematic and sometimes nothing is deposited. When it is carried out, no species and no anatomical element are found more regularly than another as we sometimes find pig, chicken, fish or an egg: practice seems to be free. At the same time, nothing consistently characterizes the remains of meals found around graves: they are archaeologically different from the remains found in the cinerary urns because of the continuous visits of families and of dogs that probably caused the more fragile elements to disappear and to modify the picture that we have of the assemblages (the faunal spectrum and anatomical distribution), but it is probable that these meals were not different from those taken at home.

The picture which emerges from the analysis of animal bones at Porta Nocera and from their link with rituals is therefore of a form of freedom revealed by the relative diversity of the acts. This variety however can only be conceived of in the context of a common tradition to which sacrifice and the sharing and consumption of the usual eaten species contribute. In Pompeii, this is essentially based on pork and chicken.

Provincial practices: some big similarities despite appearances

In the northern provinces of Gaul (Fig. 9.12) the species attested are not different since here too pig and chicken predominate but their importance in funerary practices is much greater. Burials are very rich in cuts of meat (Fig. 9.13): we find them again, of

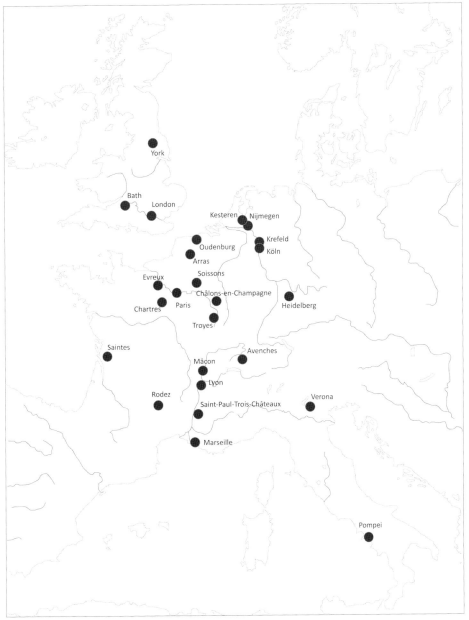

Fig. 9.12. Archaeological sites in Gaul and neighbouring areas mentioned in the text.

Fig. 9.13. Tomb L14001, "Actiparc" site near Arras (France), 1st century AD. Note pig ribs, a half-head of a pig (left side) and many chicken bones (photo, Service Archéologique d'Arras – INRAP).

course, mixed with the bones of the dead, among the ashes of the funerary pyre but also by their side in the burial pit. It is doubtless here that there is a great difference with Pompeii. Certain graves of the first century AD contain half heads of pigs, racks of ribs and of vertebrae (Fig. 9.14), whole limbs and complete unburnt chickens (Fig. 9.15).

The Roman graves are not different to those of the later Iron Age, which are also rich in meat-based deposits (Méniel 2002): there is thus a clear temporal link. On the other hand, it is not necessary to deduce that the rites or beliefs went unmodified between the two periods and that the practices are strictly similar. It would also be very simplistic to see here a major difference in terms of rite between Italy and the northwest provinces. This provincial characteristic is doubtless more revealing of the place held by the meat in Gallic society than significant for possible religious continuity between the two periods. Of course, the precise meaning of these gestures escapes us as well as the detail of the different phases of the funeral, but they are associated with the offerings, meal and sharing of the food which are common to these different religious traditions: their implementation develops in more or less similar ways according to the periods and regions concerned. During the Roman period, we observe in northern France an evolution of the relative proportions of the species involved in burial rituals. While the pig is largely in the majority during the Early Empire, it is relegated to second position in the Late Empire behind chicken, which

Fig. 9.14. Tomb E557, "Actiparc" site near Arras (France), 1st century AD. A pig half-head and cervical vertebrae. The ribs are not linked with the thoracic vertebrae (which are missing) (photo, Service Archéologique d'Arras – INRAP).

is the animal most frequently buried with the dead in the third to fourth centuries (Lepetz and Van Andringa 2004).

It is also necessary to discuss dogs and equids. Their cases are close due to certain similarities. It is clear that neither has a connection to Roman funerary rituals if we follow the texts: these non-edible animals have nothing to do with graves, and the horse, like the dog, is not an edible animal (even if in certain rare cases they were sporadically eaten). Nevertheless excavations, as we saw with Pompeii, reveal remains of these species in funerary contexts. When, as at Porta Nocera, there are isolated traces, we can consider that these are intrusive remains without a connection with funerary practices. But in other cases we are in the presence of complete skeletons or substantial body parts and large numbers of skeletons.

Among some examples of dogs found in cemeteries we can cite the Late Empire cemetery of Oudenburg (2nd-4th century AD, Gautier 1972), where the skull, mandibles and postcranial bones of a dog were found in the fill of grave 76. The animal was situated at the end of the pit but outside the coffin. It is a small dog, 40–50 cm high at the withers. In a Roman cemetery located close to Heidelberg, a dog was buried near a child (Teichert 1987). The size of the animal, 22 cm high at the withers, fits well with a pet dog. When the link between the dog's skeleton and the deceased is proved

Fig. 9.15. Deposit of a chicken and a piece of pig meat near a child's corpse (Tomb 964, Soissons, France, 1st century AD). The chicken was probably complete whereas the pig is only represented by a small piece (photo, B. Gissinger CGA).

(which is not always the case), we can probably see a desire to bring together after death two connected beings, rather than an involvement of the dog in funerary rites.

For the horse, the case of Évreux (*Mediolanum Aulercorum*, Eure, Normandy, France) is a good example to allow us to think about the association of equids and humans on the same site. The discovery in 2007, in the 'Clos au Duc' quartier, of several tens of human skeletons (109 burials) and of very numerous horse remains prompted an excavation which lasted several months (Lepetz *et al.* 2010). Various use phases were determined. The second phase of inhumation covers the second and the earlier third centuries AD. Several tens of dead adults and immature individuals (67) are documented, most often in atypical positions. A third of the adults were buried prone, eleven male adults were buried on their side and 40% of adults were placed on their back. One of the adults had the head flanked by two horse skulls. In at least five cases pairs (adults and/or immature individuals) were buried simultaneously head to toe. Many remains of equids are present in the fills of pits for the human dead. In three cases, the dead are in direct contact with the horse remains (skulls and near complete vertebral columns).

The site presents three types of structures in which horses are involved. Four structures dated to the second use phase of the cemetery produced associations of human skeletons (NISP = 5) and portions of horse carcasses, at least 20 equids, predominantly sections of vertebrae in connection (Fig. 9.16 and 9.17). But all the fills of burials – whatever the chronological phase they are related to – had animal remains in bulk (NISP = 5,100), without anatomical links, looking like reworked rubbish deposits, of which a very high proportion comes from equids. An extensive dump, of which the fill revealed an overwhelming majority of horse bones, also adjoined the necropolis (only 4,400 bones were studied).

The association of men and horses buried in the same place begs the question of the link between them and the nature of the rituals which would be implemented

Fig. 9.16. Évreux (France). General view of a part of the site (photo, INRAP).

Fig. 9.17. Évreux (France). View of 'grave' 7 (photo, INRAP).

in this context. But the human remains are characterized by variety and a lack of regularity or repetition which does not permit us to perceive any recurring gesture or trace of characteristic funerary rituals: many burials were also disturbed and redeposited. The types of horse deposits follow this apparent absence of logic. In certain cases, there are complete sections of vertebrae, sometimes with the head, sometimes without, sometimes with one or more limbs. In other cases, the anatomical elements are disconnected. All this goes in opposition to the notion of rituals which, by contrast, comprise a regular succession of codified gestures carried out in a given and precise order. We should also note how the animals were handled. Some were skinned, others had tendons or flesh removed and the bone worker came to take his raw material from the corpses, as the very numerous sawed bones testify (Fig. 9.18). Dogs also collected their tribute, as the traces of dog teeth spotted on the bone surfaces illustrate, all things not very compatible with a rite.

During the Roman period in Gaul, the horse is not completely absent from certain religious practices (Gaudefroy and Lepetz 2000: Lepetz and Méniel 2008); it is not even absent from the vicinity of cemeteries. We meet it sometimes in the form of complete skeletons near human burials. This is, for instance, the case of the Notre-Dame du Bon Accueil site in Rodez (second to third centuries AD, Lignereux *et al.* 1998), of one of the cemeteries of Saint Rémy-Montlouis in Saintes, (unpublished, excavation J.-P. Baigl) where a horse was discovered near an inhumation, or in Avenches 'En Chaplix' where an individual was found in a pit (St 380) in the southern funerary enclosure (Olive 1999). In the United Kingdom, cases are also mentioned in York, London and Icklingham (Levine *et al.* 2002), always near human corpses. This kind of discovery is rather frequent in the northern continental provinces along the *limes*. Many cases are described, notably in Nimejgen (Netherlands: Haalebos *et al.*, 1995, 27–28), Kesteren (Netherlands: Lauwerier and Hessing 1992) and Oudenburg (Belgium: Hollevoet 1993) where, according to the authors, we could envisage either a military presence, a Batavian origin or a Danubian influence.

However, it is necessary above all to underline that when the contemporaneity of the human burials and horses is proven, the deposits are never directly linked: they are in the same place, of course, but nothing ever indicates that their placing in the earth was carried out simultaneously. From this fact, it therefore cannot be demonstrated that there might be a religious meaning to their presence, even if this hypothesis is often implicitly retained. But an attentive analysis of the context of the discoveries, a fine-grained and rigorous reading of the deposits and skeletal remains and an objective approach to the phenomena often leave the door open to many other possibilities which in several cases are a better fit.

We might consider that a choice was sometimes made to bury the animal to which someone was attached in a privileged place as a form of gratitude or simply as a sign of affection, but we have to recognize that often even the intentionality in the choice of the place is not certain. Thus, at the extensive 'De Prinsehof' cemetery close to Kesteren, where several tens of horse graves and human graves were discovered, the work led by Lauwerier and Hessing (1992: 99) allowed him to conclude that [the] 'explanation of a dumping ground for horses, that later turned into a cemetery, seems to be the most simple and plausible'.

A little further south, in Cologne, several cases of horses buried in cemeteries were similarly signalled for the first century AD (Riedel 1998). Among them is a more equivocal example. In the Friesenstrasse cemetery, the skeleton of a nine-year-old mare was discovered: as the position of the neck shows, its throat was visibly cut and the presence of an iron knife and a characteristic trace on a cervical vertebra

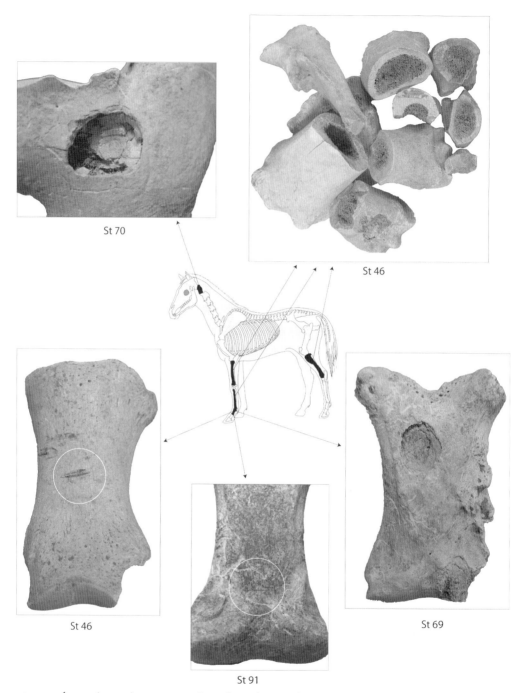

St 70

St 46

St 46

St 91

St 69

Fig. 9.18. *Évreux (France). Various marks on horse bones. The saw marks correspond to what was taken for bone working. The knife marks on phalanges correspond to the removal of tendons (photo, author).*

confirm this. This is dated to the second half of the 1st century AD. It could be a sacrifice, but we can honestly consider that the animal was dispatched following an injury and that its case is similar to those described previously. In any case, the funerary aspect of the gesture is not established. Nevertheless, it is clear that this type of occurrence is rare in the Roman provinces and it is not at all met in Rome (de Grossi Mazzorin 1993).

The case of Gelduba (Krefeld-Gellep, Germany) is a little different. Tens of horses doubtlessly associated with humans were discovered (Pirling and Siepen 2006; also Reichmann 2006); these must be graves of soldiers involved in the battles conducted by the Roman legions against the Batavians during the siege of *Vetera* in AD 69–70. These war burials, without funerary furnishing, therefore have a special status since they resulted from a violent episode necessitating the quick burial of very many human and animal corpses. The situation is unlikely to be very different in Lyon-Vaise for a collective grave of the 3rd century AD in which four horses and four humans presenting extensive injuries were buried together (Billards 1991). The corpses were not disposed of carefully and had even sometimes visibly been dragged by the feet. Similarly, the horses were not arranged in the pit.

The site of Évreux is characterised by the presence of disconnected remains or corpse parts only. It therefore clearly distinguishes itself from the other sites described above. Some anatomical parts of equids, isolated or disconnected, are found elsewhere here or there near graves. The examples are nevertheless difficult to interpret because they are rare, diverse in their aspect and sometimes poorly described in the archaeological literature. Thus, a skull and two mandibles of a horse in grave 134 of the necropolis of Porta Palio (Verona), can be noted but they are poorly dated (Cavalieri Manasse and Bolla 1998). In Bath (Great Britain), a horse skull was discovered in a sarcophagus (Luff 1982, 193). Most often these groups of bones, sometimes partially connected, sometimes disconnected, are situated in the immediate vicinity of burials, in ditches, pits or wells or in backfills. In the northern area of the Valladas cemetery at Saint-Paul-Trois-Châteaux, a ditch section (ditch N, see Olive 1987, 97–101 or Bel 2002, 49–51) produced almost 1400 bones of which a substantial part comes from horses. The state of the bone surfaces did not allow cut marks to be observed but the analysis of the anatomical distribution records an absence of selection. These remains, coming from at least four equids, were mixed up with ceramic fragments and other fragments of objects and of residual fragments of worked stone. In Avenches (Olive 1999), the excavation also produced equid remains mixed with bones of other species and scattered in the boundary ditches.

It seems therefore that the supposed links between burials and horses do not appear as clearly as they are sometimes described. We can only note, for example, the difference with the associations observed for the Merovingian period in northern Gaul or Germania (Dierkens *et al.* 2008). For this period and these regions, the practice of depositing a horse near human remains seems well established, even if recent studies are reconsidering the attribution of a role as psychopomp to these animals:

recent work instead emphasises a social interpretation of the practice, which probably explains its relative rarity.

The case of isolated and scattered horse remains in the surroundings of cemeteries is not clear because the links between the different remains, humans and animals, are not very evident. Indeed, it is possible that they are not connected, their only common point being that they are the product of an activity taking place necessarily outside the city, the deposition in the ground of the dead and the disposal of the corpses of dead animals.

If therefore cases of associations between the dead and horses are rare, on the other hand many peri-urban dumps containing large quantities of horse bones have been discovered in the Three Gauls. Some examples can be observed in Gallia Belgica and in Gallia Lugdunensis: in Mâcon (Barthélémy and Lepetz 1999), Châlons-en-Champagne (Lepetz 2003), rue H. Barbusse, Paris (Poulain 1962), in 17 rue Reverdy, C245, and rue Hubert Latham, C138, Chartres (unpublished), in place Langevin, Troyes (Arbogast *et al.* 2002, p.89) or La Bourse in Marseille (Jourdan 1976, 233–257).

Roman cities produce a significant quantity of waste, texts and archaeology giving many examples (Ballet *et al.* 2003). The management of waste collection and elimination is organized to clean the streets, on the one hand, and to exploit what can be recycled, on the other. Inert waste can be used as construction material and as a road base while *stercus*, the organic waste, (rubbish, excrement, ash, peelings, sweepings etc) forms a secondary raw material useful for agriculture. This was potentially valuable and could be part of the profits that an owner derived from his property (Cordier 2003, 21 and 56). Its removal was a paid service, like road cleaning, which was tendered to private contractors. Areas for depositing waste inside the city were constrained by the spaces available and also undoubtedly by the pressure exercised by the city authorities (Saliou 2003, 46–47). These activities would have involved the suburbs much more and the cemeteries, along the roads and outside the urban boundary, were affected by the uncontrolled dumping of rotting matter. Thus, in Apulia a *lex*, and, in Rome a cippus of L. Sentius erected on the edge of the mass graves of the Esquiline Hill forbid the deposit of *stercus* (Bodel 1986).

The deposits of equids in the suburbs of Mâcon, Châlons-en-Champagne, Chartres, etc. are a response to a similar necessity to move animal corpses outside the walls. Horses were not eaten in antiquity and their remains are not found mixed with butchery waste and food remains. This obligation to move corpses is particularly significant for large beasts which cannot be buried at the bottom of a garden or in a disused well (like dogs, for example) and which, unlike household waste, cannot be used in the fields as manure. They are removed then, transported either on private initiative, or in the context of municipal organization, and dumped in specific reserved areas which had to be accessible by cart and outside the city.

These areas are necessarily located in the vicinity of cemeteries and this is probably the reason why cemeteries are linked to these accumulations of horse bones. This proximity is sometimes very close, as in Mâcon or on the two Chartres sites, where the human burials (children in the two first cases) are situated in the middle of areas of horse bones. In Évreux this proximity is particularly strong, to the point that the rotting carcasses can be in the same pits as the humans. This relative interspersal of human and animal remains and the total absence of rules in treatment of corpses, notably in their disposal, as well as the partial destruction of bodies following repeated excavation, cause us to consider the significance of these assemblages.

Whether street cleaning and waste discharge outside the walls are activities organized by the city or delegated to contractors, we know that the city is in charge of its human dead and corpses present on the public way (see for example Ausonius (*Epigr.* LXXII) for the cross-roads where 'the bare skull of an unburied man' is abandoned or Petronius (*Petr.* 134) when Proselenos asks of Encolpius 'what foul thing or corpse have you trodden on at a cross-road in the dark?'). Furthermore, the *lex Libitina* from Pozzuoli (II, 1–2, Hinard and Dumont 2003) establishes the amount of the fine collected by the contractor, i.e. the undertaker, in the case of abandonment of a corpse on the public highway. The presence of corpses mixed with rubbish is therefore well attested in Roman Italy (Panciera 2000; Saliou 2003, 43–44), including the bodies of destitute adults or abandoned children (Hinard and Dumont 2003, 102) which, for religious, social reasons or hygiene, must be removed. They are then disposed of in places adapted for the task, at the same time as torture victims, by the appropriate services.

With reference to the site of Évreux, the analysis of disposal of the human dead, the absence of associated deposits, the lack of care with which the pits were disturbed, the nature of horse remains, their partial disintegration, the collection of raw materials and the decomposition of rotten carcasses allow us to consider that this part of the Le Clos aux Duc site was one of those places like the *puticuli* of the *atrae Esquiliae* of Rome, i.e. those pits which were cemeteries of the poor and criminals (Belayche 2003, 55) and where it was also possible to get rid of dead animals.

Animals are therefore associated in various ways with funerary rituals in Rome or in the provinces. Their nature and their role differ according to locality or period but their presence demonstrates the importance of practices linked to the sacrifice, the sharing of food and the meal. Deposited on the pyre or thrown in the flames, placed in the earth as whole animals near the dead or eaten by the living around the graves, they participate in the homage to the deceased, in the wish to establish his soul in the grave and in the purification of the family. What is certain is the difficulty of linking food deposits in graves to particular beliefs in the hereafter or to the immortality of the soul (Lepetz, Van Andringa 2004): the food offering was not made to feed the deceased, but showed the separation from the living. Nevertheless, the animals involved could be eaten, with a clear preference for pork and chicken, victims *par excellence* of the sacrifice practiced by the *paterfamilias*. Dogs and horses do not have usually any role in these rites: their remains or their corpses can be discovered intermittently near a grave or in its fill, but the cases where a link is proved are rare and where there is a relationship it is rarely a ritual one.

Bibliography

Arbogast, R. M., Clavel, B., Méniel, P., Yvinec, J. H., Lepetz. S. (2002) *Archéologie du cheval.* Paris: Éditions Errance.

Ballet, P., Cordier, P., and Dieudonné-Glad, N. (eds) (2003) *La ville et ses déchets dans le monde romain. Rebuts et recyclages – actes du colloque de Poitiers, 19–21 Septembre 2002,* 133–163. Poitiers: Editions Monique Mergoil.

Barthélemy, D. and Lepetz, S. (1999) Le site du parking Rambuteau à Mâcon (Saône-et-Loire). Carrières de terres et dépôts de restes de chevaux à l'époque gallo-romaine (Ier–IIIe siècles). *Travaux de l'institut de recherche du Val de Saône-mâconnais* 4, 101–120.

Bel, V. (2002) *Pratiques funéraires du haut Empire dans le midi de la Gaule – La nécropole gallo-romaine du Valladas à Saint Paul-Trois-Châteaux, (Drôme).* Monographies d'Archéologie Méditerranéenne 11, Lattes: CNRS UMR 154.

Belayche, N. (2003) Pouzzoles, Eléments d'histoire et de topographie. In F. Hinard and C. Dumont (eds) *Libitina, Pompes funèbres et supplices en Campanie à l'époque d'Auguste,* 45–55. Paris: De Boccard.

Billard, M. (1991) Violent traumatic injuries on human skeletal remains buried with horses in a Gallo-Roman collective grave (Lyon-Vaise, France, AD 200–300). *International Journal of Osteoarchaeology* 1, 259–264.

Bodel, J. (1986) Graveyards and Groves. A Study of the Lex Lucerina. *American Journal of Ancient History* 11, 1–133.

Cavalieri Manasse, G. and Bolla, M. (1998) La necropoli di Porta Palio. In P. Fasold et al (eds) *Bestattungssitte und kulturelle Identität. Grabanlagen und Grabbeigaben der frühen römischen Kaiserzeit in Italien und den Nordwest-Provinzen,* 116–141. Xantener Berichte 7, Cologne: Rheinland Verlag.

Cordier, P. (2003) Le destin urbain du stercus et de l'urine. In P. Ballet, P. Cordier, and N. Dieudonné-Glad (eds) *La ville et ses déchets dans le monde romain. Rebuts et recyclages – actes du colloque de Poitiers, 19–21 Septembre 2002,* 51–60. Poitiers: Éditions Monique Mergoil.

De Grossi Mazzorin, J. (1993) Analisi zooarcheologica dei resti equini della tomba n.3. *Relazione su scavi, trovamenti, restauri in Roma suburbio, 1990–1991,* 141–242. Bull. della Comm. Arch. di Roma 94, Rome: L'Erma di Bretschneider.

Dickmann, J.-A. *et al.* (2002) Die Casa dei Postumii in Pompeji und ihre Insula: fünfter Vorbericht. *Mitteilungen des Deutschen Archäologischen Instituts Römische Abteilung* 109, 243–316.

Dierkens, A., Le Bec, C. and Périn, A. (2008) Sacrifice animal et offrandes alimentaires en Gaule mérovingienne. In S. Lepetz and W. Van Andringa (eds) *Archéologie du sacrifice animal en Gaule romaine. Rituels et pratiques alimentaires,* 279–299. APA 2, Montagnac: Editions Monique Mergoil.

Fulford, M. and Wallace-Hadrill, A. (1995/6) The house of Amarantus at Pompeii (I, 9, 11–12): an interim report on survey and excavation in 1995–96. *Rivista di studi pompeiani – Associazione internazionale amici di Pompei* 7, 77–113.

Gaudefroy, S. and Lepetz, S. (2000) Le dépôt sacrificiel de Longueil-sainte-Marie "L'Orméon": un culte de tradition locale sous l'Empire? In W. Van Andringa (ed.) *Archéologie des sanctuaires en Gaule romaine,* 157–192. Mémoire du centre Jean-Palerne 22, Saint-Étienne: Publications de l'Universitaire de Saint-Étienne .

Gautier, A. (1972) Dierenresten van het laatromeins grafveld te Oudenburg (prov. West-Vlaanderen, België). *Helinium* 12, 162–175.

Haalebos, J. K. *et al.* (1995) *Castra und canabae Ausgrabungen auf dem Hunerberg in Nijmegen 1987–1994,* Libelli Noviomagensis 3, Nijmegen: Kathholieke Universiteit.

Hinard, F. and Dumont, C. (eds) (2003) *Libitina, Pompes funèbres et supplices en Campanie à l'époque d'Auguste*. Paris: De Boccard.

Hollevoet, Y. (1993) Ver(r)assingen in een verkaveling. Romeins grafveld te Oudenburg (prov. West-Vlaanderen), *Archeologie in Vlaanderen* III, 207–216.

Jashemski, W. F. and Meyer, F. G. (2002) *The Natural History of Pompeii*. Cambridge: Cambridge University Press.

Jourdan, L. (1976) *La faune du site gallo-romain et paléo-chrétien de la Bourse (Marseille)*. Aix-en-Provence, Marseille: CNRS 338.

Lauwerier, R. C. G. M. and Hessing, W. A. M. (1992) Men, horses and the Miss Blanche effect; Roman horse burials in a cemetery at Kesteren, The Netherlands. *Helinium, 32*, 78–109.

Lepetz, S. (2003) Gérer les rejets de boucherie et les cadavres animaux dans les villes de Gaule romaine. In P. Ballet, P. Cordier and N. Dieudonné-Glad (eds) *La ville et ses déchets dans le monde romain. Rebuts et recyclages – actes du colloque de Poitiers, 19-21 Septembre 2002*, 209–217. Poitiers: Éditions Monique Mergoil.

Lepetz, S. and Méniel, P. (2008) Des sacrifices sans consommation: les dépôts d'animaux non consommés en Gaule romaine. In S. Lepetz and W. Van Andringa (eds) *Archéologie du sacrifice animal en Gaule romaine – Rituels et pratiques alimentaires*, 147–156. Archéologie des plantes et des animaux 2, Montagnac: Editions Monique Mergoil.

Lepetz, S. and Van Andringa, W. (2004) Caractériser les rituels alimentaires dans les nécropoles gallo-romaines: l'apport conjoint des os et des textes. In L. Barray (ed.) *Archéologie des pratiques funéraires – approches critiques. Actes de la table ronde de Glux-en-Glenne des 7 et 9 juin 2001)*, 161–170. Bibracte 9, CAE européen Mont-Beuvray.

Lepetz, S. and Van Andringa, W. (2006) Pour une archéologie de la mort à l'époque romaine: Fouille de la nécropole de Porta Nocera à Pompéi. *Compte rendu de l'Académie des Inscriptions et Belles Lettres, Séances de l'année 2006 - Avril-juin* 150, 1131–1161.

Lepetz, S., Bémilli, C., and Pluton-Kliesch, S. (2010) Le site antique du "Clos au Duc" à Évreux (Eure). Sépultures de privilégiés ou trous à ordure? In A, Gardeisen, E. Furet, and N. Boulbes (eds) *Histoire d'équidés: des textes, des images et des os*, 28–56. Monographies d'Archéologie Méditerranéenne 4, Lattes.

Lepetz, S. and Van Andringa, W. (2011) *Publius Vesonius Phileros vivos monumentum fecit*: Investigations in a sector of the Porta Nocera cemetery in Roman Pompeii. In M. Carroll and J. Rempel (eds) *Living through the Dead: Burial and Commemoration in the Classical World*. 110–133. Oxford: Oxbow Books.

Levine, M. A. *et al.* (2002) A Romano-British horse burial from Icklingham, Suffolk. *Archaeofauna* 11, 63–102.

Lignereux, Y. *et al.* (1998) Un cheval gallo-romain inhumé dans le cimetière du site de Notre-Dame du Bon Accueil (Ile-Ille siècle après J.-C., Rodez, Aveyron). *Revue de médecine vétérinaire* 149.5, 379–386.

Luff, R. M. (1982) *A Zooarchaeological study of the Roman north-western provinces*. BAR International Series 338, Oxford.

MacKinnon, M. (2004) *Production and Consumption of Animals in Roman Italy - Integrating the zooarchaeological and textual evidence*. JRA Supplementary Series 54, Portsmouth, RI: Journal of Roman Archaeology.

Méniel, P. (2002) Les animaux dans les rites funéraires au deuxième Age du fer en Gaule septentrionale. *Anthropozoologica* 35, 3–16.

Olive, C. (1987) Premières observations sur les offrandes animales des nécropoles de Saint-Paul-Trois Châteaux (Drôme) et d'Avenches (Suisse). *Nécropoles à incinération du Haut Empire, Actes de la table-ronde de Lyon, 30-31 mai 1986*, 97–101. Lyons: Direction des Antiquités Historiques – Région Rhône-Alpes.

Olive, C. (1999) Etude des ossements d'animaux. In D. Castella (ed.) *La nécropole gallo-romaine d'Avenches "En Chaplix" : Fouilles 1987-1992, Vol. Etude des sépultures*, 137–147. Cahier d'Archéologie romande 77, Lausanne.

Panciera, S. (2000) Nettezza urbana a Roma. Organizzazione e responsabili. In X. Dupré Raventos and J. A. Remolà (eds) *Sordes urbis. La eliminacion de residuos en la ciudad romana. Actes de la reunion de Roma (15-16 de noviembre de 1996)*, 95–105. Rome: Bretschneider.

Pirling, R. and Siepen, M. (eds) (2006) *Die Funde aus den römischen Gräbern von Krefeld-Gellep: Katalog der Gräber 6348-6361*. Stuttgart: Steiner.

Poulain, T. (1962) Les chevaux gallo-romains découverts à Paris rue Henri-Barbusse (IIIè–IVè siècles). *Zeitschrift für Tierzüchtung und Züchtungsbiologie/Journal of Animal Breeding and Genetics* 76, 238–242.

Reichmann, C. (2006) Kriegsgräber. In R. Pirling and M. Siepen (eds) *Die Funde aus den römischen Gräbern von Krefeld-Gellep: Katalog der Gräber 6348-6361*, 497–512. Stuttgart: Steiner.

Riedel, M. (1998) Frühe römische Gräber in Köln. In P. Fasold *et al.* (eds) *Bestattungssitte und kulturelle Identität. Grabanlagen und Grabbeigaben der frühen römischen Kaiserzeit in Italien und den Nordwest-Provinzen*, 307–318. Xantener Berichte 7, Cologne: Rheinland Verlag.

Robinson, M. (2002) Domestic burnt offerings and sacrifices at Roman and Pre-Roman Pompeii, Italy. *Vegetation History and Archaeobotany* 11, 93–99.

Saliou, C. (2003) Le nettoyage des rues dans l'Antiquité: fragment de discours normatif. In P. Ballet, P. Cordier and N. Dieudonné-Glad (eds) *La ville et ses déchets dans le monde romain. Rebuts et recyclages - actes du colloque de Poitiers, 19-21 Septembre 2002*, 37–49. Poitiers: Éditions Monique Mergoil.

Scheid, J. (2008) De l'utilisation correcte des sources écrites dans l'étude des rites funéraires. In J. Scheid (eds) *Pour une archéologie du rite. Nouvelles perspectives de l'archéologie funéraire*, 5–8. Collection de l'École Française de Rome 407, Rome: École Française de Rome.

Teichert, M. (1987) Brachymel dogs. *Archéozoologia* 1, 69–75.

Van Andringa, W. and Lepetz, S. (2008) I riti e la morte a Pompei: nuove ricerche archeologiche nella necropoli di Porta Nocera. In P. G. Guzzo and M. P. Guidobaldi (eds) *Nuove ricerche archeologiche nell'area vesuviana (scavi 2003-2006), Atti del Convegno Internazionale organizzato dalla Soprintendenza archeologica di Pompei con l'Istituto Nazionale di Archeologia e Storia dell'Arte e con la Soprintendenza Speciale per il Polo Museale Romano, 1-3 febbraio 2007*, 377–388. Rome: L'Erma di Bretschneider.

Van Andringa, W., Duday, H., and Lepetz, S. (eds) (2013) *Mourir à Pompei - Fouille d'un quartier funéraire de la nécropole romaine de Porta Nocera (2003-2007)*. Collection de l'École Française de Rome, 468. Rome: École Française de Rome.

Chapter 10

"How did it go?" Putting the process back into cremation

Jacqueline I. McKinley

Introduction

The Romano-British mortuary rite of cremation, as in former and subsequent periods, comprised a series of related acts which could leave various forms of physical evidence. These include archaeological features and deposits directly related either to the primary rite of cremation or the secondary rite of burial, and potentially to other forms of post-cremation deposition. Other evidence pertaining to various parts of the rite may be ascertained or deduced through analysis of the various archaeological components (e.g. cremated bone, fuel ash, pyre and grave goods) recovered from such cremation-related contexts.

Although specific areas of research may focus on one particular part of the rite, those engaged in mortuary studies will generally aim to ascertain details of the sequence of events involved. To this end, it is imperative that those recovering the primary data in excavation avoid the assumption that every deposit containing cremated bone is 'a cremation'; an expression which invariable translates as 'a burial' though the two are not synonymous. A variety of deposit types can be associated with the rite and more than one such deposit may be represented within a single feature. While this may appear a statement of the obvious, experience illustrates that many of those undertaking the recovery of this material in the field are not cognizant of the 'process'. Despite advances in our understanding of the subject, and improvements in excavation methods and recording over the last few decades (including publication of guidance on field recovery; e.g. McKinley 1998; 2000a) many excavators remain unaware of what they may encounter or how best to collect the surviving data. If archaeologists are to understand the formation process of a deposit it has to be excavated and recorded in such a way that pertinent information, often not evident/visually accessible during excavation, can be recovered at a later stage in analysis.

Whilst the study of cremation may draw on a wide and varied field of information, the subject of this paper is the archaeological manifestation of the rite recoverable in excavation, with a particular focus on the cremated remains themselves. The first section presents a review of the various types of deposit regularly encountered on Romano-British sites together with a discussion of their form and nature. This is followed by advice on excavation and recording procedures to help maximise data recovery.

Deposit types and frequency of recovery

The current Romano-British archaeological record includes around 6,000–7,000 cremation burials excavated country-wide from a combination of large urban cemeteries, small rural grave groups and as singletons. Many of these burials (up to 78% from any one cemetery) also had associated deposits of pyre debris (Fig. 10.1) made somewhere within the grave. The grave may represent the most commonly recorded location for pyre debris in the Romano-British period, where its presence appears deliberate rather than incidental, forming part of the mortuary rite, but it may also be found elsewhere – either redeposited or, more rarely, *in situ.*

Pyre sites

In situ deposits of pyre debris require the survival of the pyre site. Such finds are rare at any temporal stage in the British archaeological record, but a relatively high proportion are of Romano-British date, probably largely due to the presence of pyre-related features. Reviews by Philpott (1991, 48–49) and Struck (1993) innumerated most of the c. 22 sites at which such features had been found. They lay mostly in central-eastern England and the northern frontier area. The pyre sites comprised '*busta*-style' features; rectangular pits over which the pyres were constructed and into which the debris fell as the cremation progressed (Struck's *Grubenbusta*) or, less frequently, flat pyre sites where the remains were left *in situ* and over which a mound was raised (*Flächenbusta*; Struck 1993), though barrows were also raised over *Grubenbusta* (Charlton and Mitcheson 1984). Such features appear to conform to the definition of a *bustum* given by Festus (2nd century AD) as '...where the dead person had been cremated and buried' (see Toynbee 1996, 49 and 219; McKinley 2000b, 39). It should be noted, however, that elsewhere Servius (4th century AD) specifies that the *bustum* is '...where the deceased is cremated and his bones are buried next to it' (*Serv. A.* 11,201, in Hope 2007, 113), which carries slightly different connotations indicating a separate if adjacent feature for burial of the bones. The two need not have been mutually exclusive and could represent temporal or geographic variations. The dual function of some of the British examples of these '*busta*-style' pyre sites as the place of burial has been questioned elsewhere by the writer, largely on the basis of the extreme paucity of cremated bone recovered (McKinley 2000b). Possible pyre sites of a different form were recorded from four of the cemeteries listed in the

---

Fuel ash from pyre structure;
analysis may demonstrate deliberate selection of specific species for use in pyre construction in contrast with known common species from an area or in relation to an individual's age/sex/status (e.g. Tacitus *Ger.* 17.1; Campbell 2004, 270; Cool 2004, 441)

Human bone not collected for burial;
despite contemporaneous literary allusions to the necessity of burying all the bone remaining after cremation (Noy 2005, 368) this was clearly not the case in practice, at least Britain. The quantity of the bone included in the burial could vary widely (e.g. 24–1324.6g from undisturbed adult burials at Brougham (McKinley 2004a) i.e. *c.* 1.5–83% of the expected average weight of bone from an adult cremation (McKinley 1993)) with sometimes substantial quantities of bone being deposited with the cleared pyre debris. Analysis may show links between specific debris deposits and burials or demonstrate selected exclusion of certain skeletal elements from the burial

---

Pyre goods (animal remains, artefactual material, other organics);
materials incidentally or deliberately excluded from burial (see e.g. Polfer 2000; Cool 2005, 456)

Incidental 'natural' inclusions indicative of ground on which the pyre was constructed, such as:

burnt flint – pyre constructed over a naturally flinty soil
fuel ash slag – pyre constructed over a highly siliceous soil
burnt clay – pyre constructed over a clay/silty clay soil

Fig. 10.1. Common components of pyre debris.

reviews by Philpott (1991, 48–49) and Struck (1993), including at least one of those with possible *busta* (see below). In all these cases the number of pyre sites within any one cemetery was low, with generally only one example from each site, the highest number (eight) being recorded at Petty Knowes, High Rochester in Northumbria (Charlton and Mitcheson 1984).

In the last decade or so the number of pyre sites has more than doubled, but the overall figure remains low at around 100. Most of the *c.* 10 locations from which these recent examples have been recovered lie, again, in eastern England and the numbers of pyre sites from each, as with the previously recorded examples, is generally between one and five (see below). Larger numbers are emerging, however, with a possible 10 recorded at St. Stephens, St. Albans in Hertfordshire (McKinley 1992; Niblett *pers.*

comm.), 16 from Denham in Buckinghamshire (Coleman *et al.* 2001; 2004; 2006), and the highest number currently recorded from any one cemetery, 23, found at Springhead in Kent (Biddulph 2006).

The general paucity of pyre sites is probably largely due to their inherently ephemeral nature, and the extensive, often deep ploughing undertaken in the later 19th and much of the 20th centuries across large parts of the British Isles, particularly in England. Other factors could include possible regional variations in the form of pyre sites (see below) and the focus of modern archaeological investigations within certain parts of the country (see e.g. McKinley 2008a).

Most pyres were probably constructed on a flat ground surface; the distinctive rectilinear, layered form of the structure providing a support for the corpse within a high temperature (800–1,000°C), oxygenating environment for several hours (Fig. 10.2; McKinley 1994a, 72–81; Vitruvius 2.19.5 in Hope 2007, 111). In these cases, the necessary under-pyre up-draft would be encouraged by the form of the structure itself. Such a pyre is likely have been built on a prepared area of ground, which may, for example, have been de-turfed to avoid accidental spread of fire across dry vegetation. The effects of the heat (most of which rises) on the underlying surface is relatively small. In various experimental cremations conducted by the writer, where temperatures of between 700–1,000°C were maintained for up to five hours (McKinley 2008a, fig. 10.1) 'firing' of the ground surface (commonly pink/orange/red reflective of oxygenating conditions and black/grey indicative of reducing conditions) was observed to penetrate c. 50–100 mm into a silty clay soil (Fig. 10.3). Visual changes to other soil types may be less noticeable, however, and in some instances, as demonstrated in one experimental pyre undertaken on calcareous beach sand and another over a silty sand humic garden soil, there may be no observable change at all.

If the debris on a pyre site were left *in situ* following collection of the bone for burial, there could be c. 100–150 mm of fuel ash (*pers. obs.*). This already fairly shallow depth of material could be further diminished by a number of mechanisms. In a windy environment, if the pyre site was not rapidly covered, much of the fine-particle wood ash would be scattered, the quantity and distance moved dependent on the strength and duration of the wind. This could be exacerbated by manipulation of the pyre debris such as tending during cremation and the mode of collection of bone for burial. If, for example, bone fragments were raked off the surface rather than recovered individually by hand, (McKinley 2004a; 2008b, 136), not only would some of the larger fragments of charred wood be removed with the bone but the fine particle-sized fuel ash would be disturbed and potentially blown away.

The frequent recovery of redeposited pyre debris from various locations (see below) demonstrates that pyre sites were often cleared of debris after cremation. Consequently, it is probable that many would not have any associated *in situ* debris which would serve to render their lack of recognition, particularly if disturbed, even more likely.

under-pyre draught enters here

Fig. 10.2. Pyre construction: an experimental cremation showing the open rectangular framework of the pyre providing a stable platform for the corpse in the hottest (at/towards the top) part of an oxygenating fire. In this case a combined fuel source of peat and wood was used hence the comparative scarcity of the latter.

What these observations serve to illustrate is that even were a flat pyre site, whether cleared or uncleared, covered relatively rapidly after cremation, a fairly minimal level of damage could remove much or all of the evidence of its existence. Even normal ploughing of c. 0.20 m, let alone the c. 0.30–0.60 m deep steam-ploughing of the mid 19th-20th centuries, could do a lot of damage and potentially eradicate the remains.

Fig. 10.3. An experimental bustum-style *pyre site, cleared of debris and showing the characteristic salmon-pink (oxygenated) [the lighter area on the margins and sides of the grave pit] and black (reduced) in situ burning to the silty clay soil.*

In situ debris has, however, been recovered from a variety of under-pyre features. These appear to take two basic forms, under-pyre draft scoops/pits and larger *busta* or '*busta*-style' pits (*Grubenbusta*), though there is likely to have been some overlap and blurring between them in terms of size/form and function. Interpretation of such features is often complicated and rendered inconclusive by uncertain levels of truncation, and, particularly with respect to those excavated more than 15–20 years ago, a lack of recorded detail of their form, contents and the distribution of the various archaeological components within them. The cremated bone itself, for example, often receives little other than a passing comment as to its existence. The use of the term 'ashes' is often non-specific as to the type of ash – wood ash or bone ash (NB 'ashes' are the inorganic remains surviving following the removal by oxidation of the organic components of the material being burnt, the term is not indicative of particle size or any specific material type).

Flat pyre sites

Evidence for flat pyre sites is extremely rare and interpretation is often inconclusive due to incomplete, imprecise or insufficiently detailed records. For example; a mid-1800s description of masses of burnt material on the old ground surface surrounded by stones at Harley Hill in Derbyshire (Marsden 1986, 46) may represent evidence for a flat pyre site, but lack of detail of the context and the archaeological components renders the interpretation inconclusive.

The excavator of the deposit sealed by a stone cairn at High Torres, Wigtownshire stated that he believed '...the cremated remains had been gathered together and buried on the spot' (Breeze and Ritchie 1980). The finds were fairly spectacular (including weaponry) and the presence of the cairn suggests a 'burial mound', but none of the 140 g of burnt bone recovered was identified as human (cattle and sheep), and although fuel ash (charcoal) was recovered there is no record of *in situ* burning marking the pyre site. Could this have been a pyre site? Can it really even be described as a 'burial' given the absence of cremated human remains below the mound? (NB cremation of an average-sized adult male produces c. 2,500 g of bone; McKinley 1993). Both seem unlikely in the face of the reported evidence, but it may be that some vital details pertaining to the deposit are missing. Alternatively, the feature may represent a cenotaph such as those found at Lankhills in Hampshire (MacDonald 1979) and elsewhere (McKinley 2000b; 2004a; Toynbee 1996, 54).

Jessup's 1959 (19) review of Romano-British walled cemeteries lists three sites where the antiquarian excavators made what he termed '...uncritical records of burnt soil areas ...'. At Litlington in Cambridgeshire the '... sites of pyres [were] in angles of enclosures'; at Sutton Valence, Kent '...the site of the funeral pyres was found near the wall on the north side'; and at Langley (or Lockham *Archaeologia Cantiana* 15 (1883) 76–88) in Kent a 50mm deep layer '...containing charcoal, ashes, burnt bone and pottery fragments ... was thought to mark the site of the funeral pyres'. Each of these examples seems to describe surface spreads of pyre debris but there is no record of discrete areas of *in situ* burning. This suggests that the observed deposits may have represented the *ustrina,* the areas of the cemetery in which cremations were undertaken, rather than the pyre sites *per se* (see below).

There is limited detail pertaining to the six 'pyre sites' recorded in the 19th century at Bayford, Sittingbourne, Kent, other than a mention of 'burnt earth' and, occasionally, charred wood (Payne 1886). There are descriptions of the mass of material – including animal bone, pottery and fragments of metal objects (with no mention of whether these materials were burnt or not) – recovered from amongst the 'burnt earth', but there is no mention of human bone; how much there was (if any?) or how it was distributed. On the strength of this evidence it is not even conclusive that the material recovered was pyre debris and it could, for example, have been associated with debris from funerary feasts or banqueting undertaken within the confines of the cemetery (Alcock 1980; Hope 2007, 115–116, 234–44; Toynbee 1996, 62 and 95).

Evidence for cleared pyre sites is extremely rare but there are convincing indications for at least one such site at Corbridge in County Durham (Casey and Hoffmann 1995). Here a 1.5 m² cobbled platform capped by burnt clay lay in the southern part of a c. 75 m² area of *in situ* burning (ibid, context 13, fig. 2). A charcoal spread of about 1 m² (with unspecified inclusions) lay to the south. A series of charcoal-rich negative features set in the north-eastern part of the area may also have been related to pyre structures but it is less clear in what capacity. Further deposits of pyre debris are likely to have been lost as a result of ploughing and machine stripping of the site which substantially reduced the surviving depths of features (ibid).

At least some of the remains from Herd Hill, Beckfoot in Cumbria (Bellhouse 1954) may have originated from similar cobble-construction cremation platforms, but wind erosion and a lack of detail concerning the surviving bone renders the interpretation inconclusive (but probable). Bellhouse (ibid, fig. III.2) recorded a spread (no dimensions) of cremated bone and burnt cobbles lying on 'clean sand' together with fragments of various artefacts. More burnt cobbles and oak charcoal were recovered from the overlying (Roman) ground surface from which he believed the material to have derived by the mechanism of wind erosion. The form of other deposits recovered from the same site by both Bellhouse and other workers is debatable due to the lack of context data (ibid; Hogg 1950; Bellhouse and Moffat 1959). Some may have represented *in situ* debris left after the removal of most of the bone for burial and others redeposited pyre debris, but a number of pyre sites clearly did (and probably still do) survive at this site (see below).

The remains of a possible pyre site of similar form – with a cobble base or surround – were found at Sandhead, Glenluce in Galloway (Curle 1932, 375) but again, the information is imprecise (unquantified calcined human bone, fuel ash and burnt stone below undressed boulders) and the cairn could be covering the remains of a burial with redeposited pyre debris.

Further potential evidence for the use of a stone platform on which to construct the pyre comes from Bourn in Cambridgeshire (Liversidge 1977). Here burnt wood and 'traces of human bone' lay on a layer of stones, which themselves were central to a 3.6 × 4.6 m area of 'black earth' sealed below barrow 1. There is no direct mention of *in situ* burning but such is perhaps implied by the 'black earth'. Given that the site was located on boulder clay this is intriguing since, if correct, it indicates a reducing atmosphere across the pyre site with none of the expected reddened areas indicative of an oxygenating atmosphere (e.g. Fig. 10.3). The mention of mere 'traces' of human bone (though the term is highly subjective) suggests most of it was not there; i.e. this may have been the pyre site but was not, as the excavator suggested, the place of burial. A similar deposit, but without the stone 'platform' was recovered under barrow 2 at the same site, and was again interpreted as a pyre site doubling as grave by the excavator. However, the recovery of only 'a few fragments of human bone' in circumstances where 1,000–3,000 g may be anticipated (on an uncleared, unmanipulated pyre site) argues against this being the place of burial.

An area of cobbling (F24) found within the cemetery at Brougham in Cumbria, although apparently overlain by 'dark sooty earth', showed no evidence of *in situ* burning. There was, however, an adjacent deposit of pyre debris (Cool 2004, 440; McKinley 2004a, 306). In this case the platform may have served a different purpose other than to support the pyre but again, incomplete records (from excavations in the 1960s) render interpretation inconclusive.

Cut features: under-pyre draft pits

The shallower of the two pyre-related features for which we have evidence in Britain, the under-pyre draft pit or scoop, has clear Late Iron Age precursors. The best-known examples are from Westhampnett in East Sussex, where a variety of shallow linear, T-, L- and cross-shaped features containing pyre debris were found to lie on the margins of the small Late Iron Age cremation cemetery (Fitzpatrick 1997, fig. 7). These features – similar to some for which there is anthropological evidence from Asia and Australia (Dubois and Beauchamp 1943, 485; Hiatt 1969) – appear to have been there to assist the efficient functioning of the pyre, providing a 'flue' for oxygen drawn in at the base. The scoop/pit will doubtless have also served as a repository for fuel ash in the early stages of cremation, ensuring it did not block the essential oxygen supply.

There is relatively little evidence for such features from Roman Britain. Their interpretation is at times open to debate and their function, in at least some cases, may have been two-fold; serving both as pyre site and subsequently as the place of burial. As with the flat sites, their survival could be assisted or secured by the raising of a mound over the *in situ* debris; i.e. a variation on the *bustum* with some of the characteristics of both *Grubenbusta* and *Flächenbusta*.

A potential example was found at Beckfoot (Bellhouse 1954). Here a shallow (0.12 m) pit, c. 1.82 × 0.76 m, contained a charcoal-rich (pine ash) fill overlain by a layer of cremated bone with associated charred oak wood and other materials indicative of the deceased having been cremated on a bed. The deposit survived due to being covered by a mound of sand (and presumably some other stabilising material), possibly before having fully cooled (ibid), which could have taken two-three days (*pers. obs.*). The description is consistent with that of an un-manipulated pyre site, but there are, unfortunately no details related to the quantity of bone or its distribution, so it is not possible to deduce further details of the formation process. Other pyre sites from Beckfoot appear to have been constructed on the flat, possibly on stone platforms (see above), and it may be that further variations in the form of pyre sites remain to be found there. The under-pyre feature at Risehome in Kent is similarly described as a 'small draft-trench' (Jessup 1959, 6) as opposed to the larger grave-like features commonly visualised as *Grubenbusta*. Here too, a mound was subsequently raised over the feature.

A more recent example comes from The Lea, Denham in Buckinghamshire (Coleman *et al.* 2001; McKinley 2000c), where the remains of a cremated juvenile were recovered from a 1.0 × 0.70 m, 0.07 m deep feature with *in situ* burning, sealed below a colluvial deposit. The bone was confined to the upper levels of the charcoal-rich fill and the identifiable skeletal elements lay in anatomical order (ibid). The site is believed not to have comprised the grave (only 98.6 g of bone was recovered, the remains of a 5–10-year-old juvenile) but to represent the *in situ* remains of the bone fragments not collected for inclusion in the burial, predominately smaller bone fragments including

Fig. 10.4. Experimental cremation showing the effects of a veering strong wind; a windbreak was required on the windward side to counter the breeze and avoid uneven burning and collapse of the pyre.

shattered tooth enamel. The layout suggests that the fragments recovered for burial were collected individually by hand rather than raking material off the upper levels of the debris. A potentially associated feature here was a series of possible stake-holes on the western windward side of the pyre site, which may indicate the presence of a windbreak (Coleman *et al.* 2001), though the excavator is now uncertain of this interpretation and the 'stake-holes' could be the result of bioturbation (Coleman *pers. comm.*). It is, however, possible that such structures were sometimes employed. The writer was obliged to construct a windbreak during one of the experimental cremations undertaken on a windy day in Shetland (Fig. 10.4); the wind was too strong, causing uneven burning and consequent uneven collapse of the pyre, and had to be subdued. The construction of a temporary windbreak, such as those commonly used by sunbathers on beaches, was sufficient to regain control of how the pyre burnt.

Cut features: busta

Busta, or *Grubenbusta* (Struck 1993), relatively deep (*generally* 0.50 m or more) rectangular pits (around 1.50 m × 1.0 m plus) over which the pyre was built and into which the debris fell, appear to represent a largely Roman phenomenon in Britain (Figs. 10.5 and 10.6). There is, however, evidence from two features at Puddlehill in Bedfordshire suggestive of late Iron Age examples (Whimster 1981, 154 and 354). Bone

Fig. 10.5. An experimental bustum-style pyre site showing in situ *burning to the sides of the pit/grave and the build-up of debris in the base with the cremated remains laying in the upper levels. The charred logs from the base of the pyre structure remain along the sides and at either end of the cut.*

and charcoal from the base of one grave showed restriction of skull fragments to one end lending support to the interpretation, but there is no mention in either case for evidence of *in situ* burning. Recently, evidence for Bronze Age examples of this form of feature have been emerging from Cambridgeshire (Dodwell 1998; 2009; in prep.).

Ten of the sites listed by Philpott (1991, 48–49) and Struck (1993, table 1) had one or more features designated as *Grubenbusta;* the writer has reassigned those from Wroxeter (see below) and Beckfoot (see above) as variations on the *busta* theme for the purposes of the current discussion. Features of this form have been found at a minimum of seven other locations in addition to those previously listed including two in London (Barber and Bowsher 2000, 309; Mackinder 2000, 10–13), and one each in Berkshire (Crockett 1996, 132; McKinley 1996, 156), Bedfordshire (Albion Archaeology 2008), Hampshire (Clarke 1979, 77 and 129), Hertfordshire (Niblett *pers. comm.*) and Kent (Biddulph 2006, fig. 6). Although the features appear of similar form and – as indicated by the *in situ* burning in each case – all seem to have functioned as pyre sites, examination and consideration of the materials recovered from the debris, particularly the cremated bone, indicates that not all also functioned as the grave. Even where the pit did form the place of burial, post-cremation manipulation of the remains appears to have occurred in numerous cases. In some instances, some of the

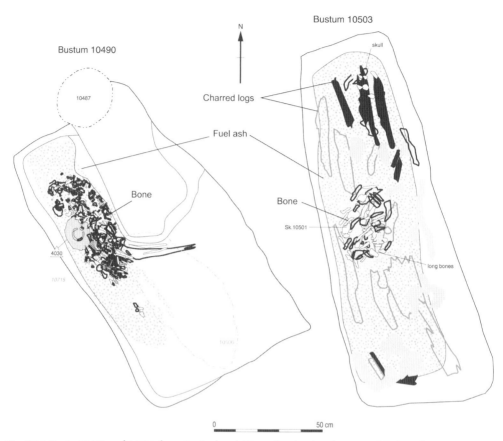

Fig. 10.6. Busta 10490 and 10503 from Springhead, Kent showing in situ *pyre debris and post-cremation manipulation of the cremated remains.*

bone was collected and placed in a vessel which was returned to the pit/grave and in other cases the bone was raked-together into a heap (Fig. 10.6). In essence, the formation process did not end with cremation, a variety of ritual activities followed and although externally the features may appear similar, their function and the form of the final 'burial' may have varied.

For example, the bone (no weight given) and pyre goods (hobnails) in grave 359 at Lankhills, Winchester in Hampshire had been formed into two piles on top of the charcoal-rich fill of the feature, the sides and base of which showed evidence of *in situ* burning (Clarke 1979, 77 and 129). Four of the features designated *busta* at Petty Knowes contained no bone at all and most of the others appear to have held very little (subjective statements only by way of quantification, no weights given; Charlton and Mitcheson 1984). The recorded data indicates the removal of most of the bone for burial elsewhere (McKinley 2000b). Few, if any, of the 10 pits (1.10 m × 0.60 m to 1.75 m × 0.70 m, and 0.07–0.50 m in depth; some possibly under-pyre draft pits) with *in situ* burning and charcoal-rich fills from St. Stephen's in St. Albans probably functioned as the grave. Most of the bone in each case had been removed, presumably for deposition elsewhere, leaving 98–410 g bone *in situ*; i.e. c. 6%–26% by weight of the total expected from the average adult cremation (McKinley 1993). The archaeological components within the remaining fills were all thoroughly mixed, further demonstrating human manipulation of the remains, presumably during recovery of the bone for burial.

At least two of the three undisturbed *Grubenbusta* from Springhead had formed the place of burial (the lack of disturbance is important as it will have ensured the preservation of the original formation process which assists in a more accurate interpretation). Each contained >1,000 g of bone, which lay on a bed of fuel ash towards the base of the cut with no overlying pyre debris (Biddulph 200; Witkin and Boston 2006; McKinley 2006). In one case the bones had been raked together to form a heap, in the other it appears to have remained roughly in anatomical order as it would have fallen.

The *bustum* excavated in plot 3 of the East London cemeteries was in many ways a classic form; 1.80 × 1.0 m, and 0.68 m deep, with a primary charcoal-rich fill inclusive of lumps of charred wood and clear (red) *in situ* burning to the upper margins of the feature. The exact location of the human bone was not recorded and the feature had, unfortunately been subject to later disturbance probably resulting in the loss of some of the bone, but there was evidence to suggest that the 838.1 g of bone recovered had been subject to some post-cremation human manipulation (McKinley 2000d, 61–63).

The 1.0 × 0.80 m (no depth given) feature interpreted as a *bustum* by the excavator at Great Dover Street, London, contained a charcoal-rich fill incorporating 'food' offerings (vegetables, nuts and a chicken) and grave goods including an impressive collection of lamps (Mackinder 2000, 10–13). A substantial quantity of cremated human bone was recovered (c. 1,000 g) from a discrete location within the grave leading the excavator to suggest it may have derived from elsewhere. The grave was cut through brickearth, which in experimental pyres has been demonstrated to show striking colour changes (Fig. 10.3). In the absence of any stated scorching of the sides of the cut, together with the excavator's suggestion that the bone was brought in, the interpretation of this feature as a *bustum* is rendered questionable.

Cut 337, a slightly irregular pit (1.8 × 0.90 m, 0.20 m deep) excavated at Wickhams Field, nr. Reading in Berkshire, had an incomplete 'lining' of 'fired clay' and a thick (0.15 m) charcoal-rich fill (Crockett 1996, figure 74), the upper levels of which contained 140 g of cremated bone sealed by the backfill (McKinley 1996, table 22). Due to the excavation methodology, the horizontal distribution of the bone could not be discerned but the form of the deposit and quantity of bone indicate that the feature represents the pyre site but not the place of burial.

Pyre sites incorporated within a larger cut features
The pyre site at Wroxeter appears to represent a rare variation on the *busta* theme with the pyre site constructed on the base of a large square pit (Philpott 1991, 48).

There are few further details but the setting, at least, has similarities with an early 19th century report from India, of the king of Tanjore having been cremated on a pyre erected within a shallow pit c. 4.30–5.40 m^2 (Dubois and Beauchamp 1943, 364).

Cremation-related structures

The use of cobble platforms as a base on which to construct pyres has been discussed above (*Flat pyre sites*). Such features would presumably provide a stable, relatively dry and vegetation-free surface on which to build. They could be reused several times and would presumably be fairly easy to clear after use.

A more prominent cremation-related structure is represented by the unusual and slightly enigmatic – since they are poorly defined in the published image – 'cremators' reported from St. Stephen's (Davey 1935). These three brick-lined chambers, 1.83 m × 0.76 m to 2.44 m × 1.65 m in plan and 1.20–1.67 m deep, are recorded as having contained large quantities of wood, bone ash and pyre goods (though discussion with the excavator revealed that, in fact, no bone was recovered from the fills; Davey *pers. comm.*). There is limited detail on the relative distribution of the archaeological components and the formation process of these 'contained' deposits is unclear. Currently the only other British examples are from Colchester, where three rectangular structures of tile and stone construction described similarly sized areas to the features at St. Stephen's (2.1 m × 0.61–0.76 m; Black 1986, 210–211).

Several stone or tile structures, mostly square but occasionally circular, of probably similar form, are known from contemporaneous sites elsewhere in Europe (Polfer 2000, fig. 3.1). It is not clear how Polfer's examples of 'permanent *ustrina*' compare in size with those from St. Stephen's and, as in the latter case, it is also uncertain quite how they would have functioned. If the funeral bier was set upon a ledge of some sort and the fuel for cremation lay in the chamber below, it is unclear how the draft was provided and how the ash was prevented from building-up around the fuel. Alternatively, the pyre structure may have been built over the chamber as normal, possibly even set on a grille of some sort. The chambers presumably had to be emptied from above as there appears to have been no other access point. The absence of any human bone from amongst the mass of presumed pyre debris in the St. Stephen's examples is somewhat strange given the often substantial quantities of bone recovered from dumps of redeposited pyre debris (see below). This suggests that after each cremation the upper layers of debris, in or on which the bone would lie, were completely removed from the structure, the bone for burial separated out and the rest of the debris (inclusive of human bone) disposed of elsewhere as normal. If this were so, however, one would also expect that most of the pyre goods, which would rest at the same level as the bone, to have also been removed. Yet Davey (1935) describes at least some artefactual material as having been recovered amongst the material from the chambers, though this was predominantly iron nails which could have been from old timbers used as fuel or potentially associated with the pyre structure.

Although clearly of a different nature, a pyre site from High Rochester, Northumbria, reported by Bosanquet (1935; though excavated in the mid-1800s), also has a slightly unclear association with a permanent structure, in this case a circular stone tomb. The latter, c. 5 m in diameter, contained the remains of 'a quantity of ashes [fuel ash] containing some fragments of calcined bone' over which a 'tent-like' structure of stone slabs had been set. A further description states that the 'natural soil' had been 'acted upon by fire' (ibid). There is no indication of the size of the heat-altered area or its relationship to the burial remains. There is certainly no clear evidence in the original statements to suggest this is a *bustum* as defined by Festus (i.e. the same feature serving as pyre site and grave), though it could qualify by Servius' definition (i.e. pyre site and grave adjacent; see above, *Pyre Sites*). Bosanquet (ibid) was of the impression that the foundations for the tomb 'can hardly have been laid before the cremation' but the structure acted to contain the remains of both the primary and secondary parts of the rite, adding yet another dimension to the formation process.

Redeposited pyre debris

Most Romano-British burial grounds appear to have served a dual function; the *ustrina* – or area in which the cremations were undertaken – being within the 'confines' of the cemetery. The quantities of pyre debris produced within urban cemeteries, in which hundreds of cremations may have been undertaken, would have been prodigious. Experimental cremations have shown that a pyre constructed using, for example, 900 kg of wood will produce 3.8 kg of wood ash (i.e. about 10 litres or 4% of the fuel by weight) if left to burn for around 18 hours (i.e. overnight; *pers. obs.*). Although there are likely to have been variations dependent on the size of the pyre, diameter of the logs/poles, wood species and weather (see above), these figures give some guide to the large amounts of material which could potentially accumulate.

The frequent presence of redeposited pyre debris within cremation graves has been discussed above (*Deposit types and frequency of recovery*). Although the quantities are sometimes substantial, they are rarely sufficient to suggest they represented the totality of what remained from an individual cremation (compared, for example, with the quantities recovered from *Grubenbusta*). Consequently, large amounts of debris were presumably deposited elsewhere and some has been recovered both as surface spreads and within cut features other than graves (McKinley 2000b).

The location of this material gives some insights as to how cemeteries were organised and functioned. The large urban cemeteries in particular would have wanted to control where cremations were undertaken and it is from such sites that we have evidence for extensive *ustrina*, designated areas, generally within one part of the cemetery in which most cremations would have been undertaken. There was, however, always room for variation. Individual surface spreads of pyre debris in more dispersed locations within cemeteries, even excluding the recovery of a pyre site, conjure up the image suggested by Servius' definition of a *bustum* (*Serv. A.* 11,201 in Hope 2007, 113; see above), with individual pyre sites situated adjacent to the grave.

Grubenbusta are also found within the larger cemeteries, either as singletons or in small groups (e.g. Biddulph 2006), possibly reflecting zonation of a specific mortuary practice followed by particular groups within the population and/or as a result of temporal variation.

Surface spreads of pyre debris

As with flat pyre sites, material discarded on the surface was probably gradually denuded and dispersed, and there is limited evidence for deposits of this type. Examples include those from the walled cemeteries in Kent and Cambridgeshire mentioned previously (Jessup 1959, 19; see discussion of flat pyre sites above). The area within the walled cemetery enclosure at Derby Racecourse, Derbyshire, was covered with large quantities of charcoal and scattered cremated bone, though there is no quantification of the material or details of the formation process (Wheeler 1985). It is debatable how many of the 39 'discrete deposits' of cremated bone recovered from the cemetery were the remains of burials and how many comprised redeposited pyre debris, some are certainly likely to have been the latter on the basis of the site descriptions and archaeological components (McKinley 2008b, 137–138). Even if only half were burials, however, if the cremations were undertaken within the confines of the enclosure this would still produce a substantial quantity of debris (potentially 200 litres).

Part (approximately 4 × 9 m, maximum 0.39 m deep) of what was believed to be a large surface spread of pyre debris (representing about one-eighth of the whole) was excavated on the north side of the cemetery at Trentholme Drive in York (Wenham 1968, 21–26). The debris comprised fuel ash, cremated human and animal bone, and other pyre goods. Only the latter was quantified though some analysis of the fuel ash was also undertaken. Consequently, our understanding of the formation process is again limited, though it was suggested that the material was cleared from a masonry 'cremator' (ibid), presumably similar to those mentioned above. Discrete deposits of pyre debris were also recovered across much of the site but were assumed to have been redeposited from disturbed graves.

The 27 or so surface spreads of pyre debris located around the northern margins of the Baldock Area 15 cemetery, Hertfordshire, each contained the cremated remains from a single individual and were probably located close to the individual pyre sites (Burleigh and Stevenson *pers. comm.*; McKinley 1991).

It should, however, be noted that not all evidence for grave-side burning and deposition of artefactual material need be related to the pyre. There is plenty of contemporaneous written evidence to indicate post-burial offerings and festivities being undertaken in or around tombs which could have produced both incidental and deliberate deposits of materials similar to those used as pyre and grave goods (e.g. Hope 2007, 66, 115–116, 154–155; Toynbee 1996, 62 and 95). Recent excavations in Roman period Italian cemeteries have uncovered archaeological evidence for such deposits in the environs of graves (Ortalli, this volume).

Pyre debris deposits within existing cuts

Redeposited pyre debris dumped on surfaces would undoubtedly have been spread, eroded and lost unless sealed by later deposits or protected by some other means. It is probably more than just fortuitous that most known examples are from walled cemeteries or urban sites where later occupational build-up sealed at least parts of the deposits. The dumping of material into pre-existent features – e.g. pits, ditches, wells – some or all of which may eventually have over-flowed into spreads, is more likely to survive but there is similarly limited evidence for such deposits. In the cemeteries of East London several features, including redundant quarry pits were used (Barber and Bowsher 2000, figure 56). The material from one of the latter comprised approximately 26 kg of bone including the remains of a minimum of 19 individuals (McKinley 2000d). The feature had originally been at least three times its surviving size but was truncated by medieval activity; i.e. one deposit had potentially contained the remains from 60 cremations. The excavation methodology, which involved recovery of the material as a single context/sample, placed limitations on the analysis of the formation process of the deposit.

A recent, and as yet not fully analysed, example comes from Amesbury Down in Wiltshire (Wessex Archaeology 2008; McKinley in prep.). Here most of the upper 0.20–0.30 m depth of a squared mortuary enclosure ditch around a central grave containing a sarcophagus burial was full of pyre debris with the remains of one urned and a minimum of 14 unurned burials made within it (Fig. 10.7). The 'whole-earth' recovery of this largely undifferentiated fill in spits and blocks has enabled concentrations of bone not visible in excavation to be ascertained, and it should prove possible to deduce further details of the formation processes.

Pyre debris deposits within specifically excavated cuts

In addition to the common redeposition of pyre debris within cremation graves, and the less frequent recovery of material as spreads and from pre-existent features, debris also appears to have been subject to deposition within what where probably deliberately excavated features. One impressive example comes from Holborough in Kent, where pyre debris had been deposited in a series of pits adjacent to the grave and all the features then sealed below a barrow mound (Jessup 1959, 6–7).

Deposits of this type, albeit more mundane in terms of contents and location, are likely to be fairly common, but may be difficult to distinguish from unurned burials made in graves which also incorporated pyre debris in their fills. To date, a minimum of 13 sites from various parts of England are known to include such deposits. The numbers are relatively low (c. 150 in total) but it is likely that many of the deposits described as 'token' burials, represented by a mass of charcoal with a very small quantity (<10 g) of bone mixed in it, are actually deposits of pyre debris. The quantity of bone and distribution within the fill are the prime distinguishing characteristics (though the former alone should be treated with caution). An informed decision on the interpretation of these contexts can only be made where the fill has been

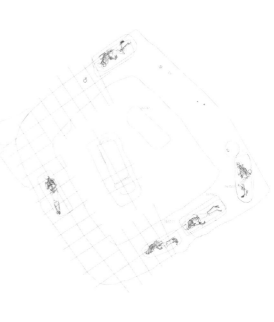

Fig. 10.7. Block excavation of redeposited pyre debris from the upper levels of a Romano-British mortuary enclosure at Amesbury Down, Wiltshire.

excavated in such a way as to facilitate the formation process to be studied, since the distribution of the archaeological components within such a fill can be very difficult, if not impossible, to detect visually in excavation (McKinley 1998; 2000a).

At Wall in Staffordshire, analysis of the ceramic pyre goods suggests a distinction was made between the vessel forms included in the burial and those deposited with the pyre debris (Leary 2008; see also Booth, this volume). A similar selection process applied to pyre goods and their deposition was noted by Polfer (2000) at contemporary sites in Luxembourg. Distinctive forms of artefactual material also seem to have been associated with deposits of pyre debris at Brougham (Cool 2004, 455–457).

Burial remains

With this, the most familiar and frequently recovered type of cremation-related deposit, it is all too easy in excavation to view the remains as an 'object' rather than the remains of an act or sequence of acts. A cremation grave may contain a variety of archaeological components, some related to the act of burial itself and others to subsequent activities which may be part of the mortuary ritual or simply related to the pragmatic action of backfilling the grave. The focus of this section is on the deposition of the cremated bone within that act of burial.

The majority of Romano-British burials were made in a ceramic container, the generic term 'urn' being used to cover a variety of vessel forms used to perform this function (Philpott 1991, 22–25, 35–43). On rare occasions glass vessels also acted as containers for the bones (ibid 26–27). Lids of various forms have been recovered, ceramic vessels being the most common, particularly inverted dishes. Large flat stones were also used for this purpose. In a recent example from Poundbury Farm, Dorchester, in Dorset, the stone had preserved the primary seal to the vessel which comprised a 20–40mm 'plug' of clay, which without the stone to protect it would undoubtedly have washed away and been lost during its time in the ground (Fig. 10.8; Egging Dinwiddy and Bradley 2011, plate 5.10, figure 5.7). Many vessels are likely to have been sealed in this way or with some form of organic 'lid', possibly including wax, all traces of which will have been lost. Even those with more robust lid-forms may have lost them due to disturbance.

The surviving presence of a lid preserves the contents of the vessel from the damage wreaked by soil infiltration and contamination by other archaeological and coarse components originally in the grave fill. Without the protection of a lid, over time soil etc. will filter into the vessel, between the bones and into the dehydration fissures formed in cremation. Acidic burial environments will cause preferential degradation and potentially disintegration of the trabecular bone, and inevitably leads to reduced bone fragment size (McKinley 1994b; 1997, 244–252; 2004a). Understanding the effects of these post-depositional formation processes is imperative to the interpretation of some aspects of the mortuary rite.

Whilst undisturbed vessels, free of any post-depositional contamination in many respects offer the best examples of how the bone would have appeared at the time of deposition (Fig. 10.9), infiltration by soil early-on may at times preserve evidence which would be subsequently lost. Rare examples of charred soft tissues have been recovered from Romano-British cremation graves (McKinley 2008a, fig. 10.7) and it is likely that bone did not comprise the only cremated material to be included in the burial. This organic material would generally be lost with time but the slow filtering of soil into the gaps left as it disintegrated may give some indication of its presence (McKinley 1997 figs. 141 and 142, 252).

The position and angle of the bone within the vessel should give some indication as to how it was inserted. A central cone demonstrates that the vessel was upright and bone was fed-in from above, a concentration to one side indicates that it was laid down and the bone inserted from the side and a level spread suggests that the vessel was shaken to even out the distribution. The orientation of individual bone fragments demonstrates similar influences, i.e. if vertical, angled or laid horizontal. The distribution of skeletal elements within the vessel fill may also indicate details of the burial formation process and possibly point to how material was collected from the pyre site (McKinley 2000e; 2008b, 138). Where more than one individual is identified amongst the remains of a burial, the relative distribution of the elements from each individual may assist in deducing if they were cremated together on one pyre or separately, and how their remains were recovered and deposited. Similar questions surround the distribution of pyre goods and, potentially, grave goods placed with the bone during burial.

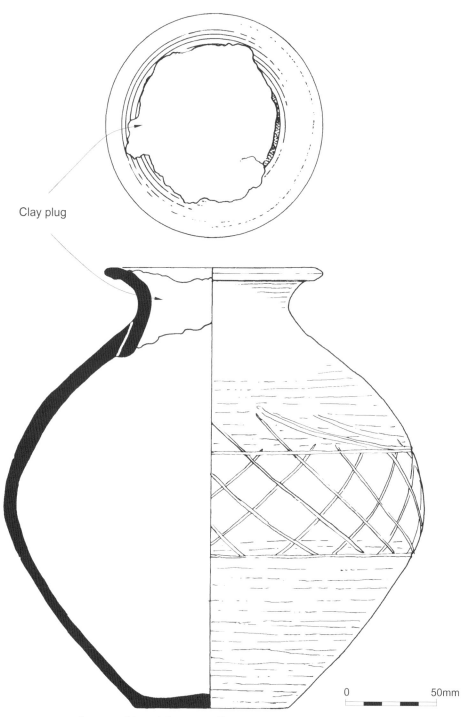

Clay plug

Fig. 10.8. Remains of an urned burial from Poundbury Farm, Dorchester showing the in situ *clay plug sealing the vessel.*

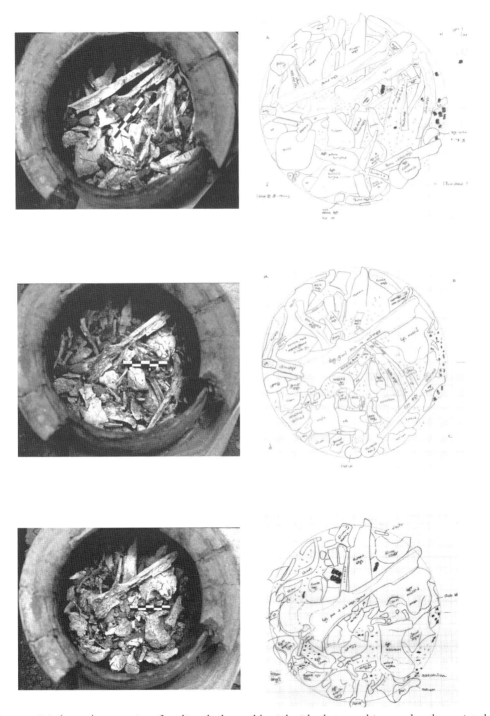

Fig. 10.9. Spit (20mm) excavation of undisturbed urned burial with photographic record and associated annotated working drawing (1:1 scale).

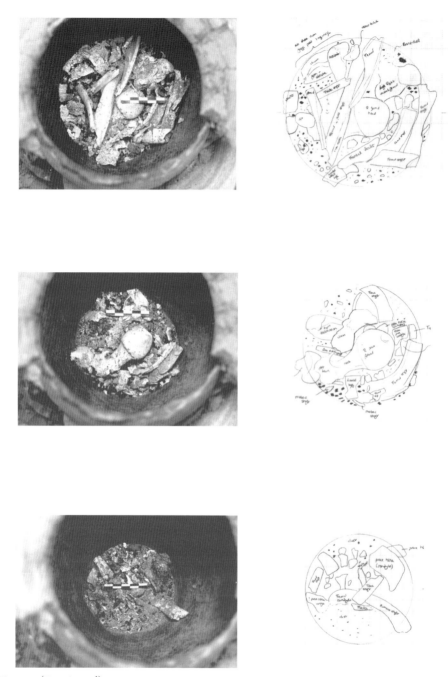

Fig. 10.9. (Continued).

There is rarely any surviving evidence for the containers used in making unurned burials but a variety of organic materials – textiles, skin, possibly basketry – are likely to have been employed. Sometimes wooden caskets were used as containers, their presence being apparent from soil stains or fittings (Philpott 1991, 12–21). In this form of burial infiltration by the grave fill is inevitable. Many graves holding the remains of unurned burials also had deposits of pyre debris made within their fills. It can be difficult to distinguish the two deposits in excavation due to infiltration of the former by the latter giving a homogenous appearance to the whole. The position of the greatest concentration of bone within the grave and its place within the formation processes, can often only be deduced during analysis of the cremated remains.

The presence and location of any bone is also of great significance in interpretation of another form of deposit substituting for the burial, the cenotaph. Such features can have all the outward appearance of a grave but include little or no bone (Cool 2004, 457–460; McKinley 2000b; 2004a, 306–307; Toynbee 1996, 54; Wheeler 1985).

Excavation

What will hopefully be apparent from the preceding text is the variety in the type and form of cremation-related features and deposits hitherto observed on Romano-British sites. What will also be apparent is that in many cases details of form and function are unclear and our understanding of the formation processes, and at times the nature of some of these features and deposits, is questionable. Our comprehension can only be enhanced by more detailed and accurate data recovery during excavation; this affects our understanding and interpretation of all the archaeological components recovered from a context.

All cremation-related deposits should be subject to 'whole-earth' recovery (i.e. 100% 'sampling'). Unlike an inhumation burial, the archaeological components of a cremation-related deposit are not easily visible to the excavator. It is essential to the analysis of all these components – be they human bone, animal bone, artefactual or organic material – as well as the overall interpretation of the deposit, that all that has survived is available for analysis.

Analysis of the formation process of any one of the deposit types described often requires a more detailed break-down than may be achieved by the undifferentiated recovery of what appears to represent (and may indeed do so) a single context. The whole-earth recovery is, therefore, often undertaken in a series of related 'samples' (see below). These 'samples' will thereafter be subject to environmental processing comprising flotation (500 micron for recovery of charred plant remains and charcoal) and wet sieving to 1mm fraction-size (McKinley 2000e, 414; 2004b). Extraneous coarse components (e.g. stone/pea-grits) should be fully extracted from the larger sieve fractions (5 mm and above) and the residues from the smaller fractions (2 mm and below) retained for scanning by the various specialists (primarily the osteologist).

Pyre sites

In situ material on a pyre site may lie as it fell as the pyre collapsed or it may have been subject to one or more form of post-cremation manipulation. The latter may be apparent where the bone has been pulled together into a clear concentrated heap, or possibly collected and placed in an organic or ceramic container. Elsewhere, such details are likely to be detectable only during analysis of the archaeological components. In this context, the human remains provide the framework around which the other material will have been laid, and by which distribution and formation process can be judged.

An undifferentiated, homogenous *in situ* deposit (i.e. a single context) from either a flat site or a cut feature should be excavated as a series of blocks and, should the deposit be sufficiently deep, spits. Since the size of such features/deposits will vary one cannot be too prescriptive but as a guide, an area c. 1.40 m long by 0.60 m wide should be divided down the long axis and collected in a series of 0.20 m² blocks to either side, each block being allocated a separate sample number within a consecutive series (Fig. 10.10). If the blocks are made too large it will not be possible to recover sufficient spatial detail, whereas there is likely to be limited advantage in making them

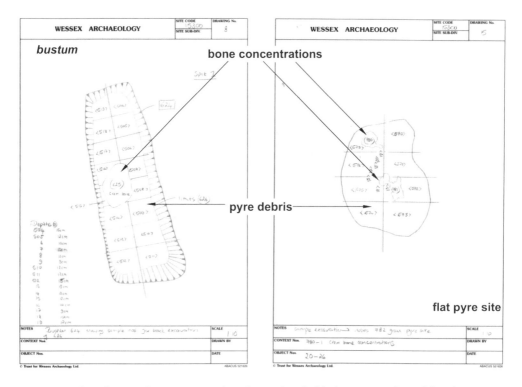

Fig. 10.10. Working drawing demonstrating advised procedure for block-recovery of pyre debris from various forms of pyre site comprising whole-earth recovery as a series of samples. KEY: [cut number], (layer/fill number), <sample number>.

much smaller. The block excavation allows monitoring of the distribution of human skeletal elements and the location of artefactual remains not evident and subject to 3D recording in excavation. Depths of >0.10 m should be excavated in 0.10 m spits.

Plans and, where relevant, sections showing the excavated segments with clear labelling of spits/blocks and visible concentrations of bone (which would be allocated a separate context number, representing, as they do, separate stratigraphic entities) or other material types, object/special finds locations etc are essential. Levels, relative to Ordnance Datum (m aOD) should also be marked on the plans. Written records describing dimensions, form, components and their distribution as far as is evident should be clearly stated on the context record sheets, together with a note of the sample numbers pertaining to that context. It is particularly helpful to record what was evident within the deposit at surface level and the maximum bone fragment size observed prior to excavation. A full photographic record should also be made before, at pertinent stages within and post-excavation.

More detailed recording of the distribution of individual skeletal elements – each separately numbered and recorded on plan or subject to 3D digital recording – has been undertaken on some pyre sites from other periods. This is very time consuming in excavation and analysis. Since many of the bone fragments, particularly the smaller elements, cannot be distinguished in excavation, it is generally only the larger fragments that are recovered in this way and it is still necessary to undertake whole-earth recovery described to ensure the spatial distribution of the smaller fragments is adequately recorded. Whilst there may be some cases where this level of detail may add to our knowledge, in most cases it would be of limited benefit given the additional time and cost.

Redeposited pyre debris

Spreads of redeposited pyre debris, debris deposited in large features (ditches, large pits etc.) and that from within structures, should be treated in a similar fashion to the pyre sites. Because the detail of formation process being looked for here is generally slightly different, the blocks could be slightly larger, but should not be in excess of 0.50 m² and 0.10 m represents a useful spit depth (Fig. 10.7). Essentially what we would be trying to distinguish in these cases are concentrations of bone and layering/overlapping of deposits not evident visually. In their analysis, the osteologist would be monitoring the distribution of duplicate skeletal elements and checking for joins between bone fragments from the deposit as a whole. This should assist in ascertaining if the debris was deposited straight from the pyre or subject to several deposition episodes, as well as deducing such information as the minimum number of individuals; i.e. the minimum number of cremations represented giving a potentially interesting check against the number of burials in the cemetery.

Formal deposits of redeposited pyre debris and graves containing the remains of unurned burials with additional deposits of pyre debris may look exactly the same, appearing as a charcoal-rich black fill at surface level in which some bone may or may not be evident. Disturbance of the deposit may compound the intrinsic difficulties with interpretation. Details of the formation process are often key to interpretation of deposit type. The cuts containing such deposits are generally relatively small (averaging around 0.30–0.60 m diameter) and whole-earth recovery of the deposit in quadrants is generally sufficient (same context number but as separate samples using a series of consecutive sample numbers). Truncation of such features is common and they rarely survive to more than 0.30 m in depth (average c. 0.20 m), and excavation in spits is often unnecessary though it can be advantageous in some cases

Burial remains

Graves containing the remains of urned or unurned burials may also include pyre debris in large quantities (see above) or as easily distinguished discrete deposits. Where the different stratigraphic entities are easily distinguished, they should be attributed separate context numbers and collected as such, thereby enabling the formation processes to be analysed.

Urned burials

After recording (pre-excavation plans and photographs; similar records at half excavation stage and following full exposure of the burial; written records clearly describing the deposits) complete or near complete vessels should be wrapped with crêpe bandage (flexibility and support) and lifted. Excavation in these cases will continue under laboratory conditions. With badly damaged vessels in which details of the burial formation process are likely to have been lost, the bone from 'outside' the vessel (i.e. disturbed/redeposited from inside) should be kept separate from that 'inside', each being allocated separate but preferably consecutive context numbers.

Laboratory excavation of the urned burials should preferably be undertaken by an osteoarchaeologist. The fill should be removed in quadrants and spits of 20 mm, a photographic record being made at each spit level together with an annotated scale drawing of the contents where the level of preservation renders this possible (Fig. 10.9). This will enable the horizontal and vertical distribution of the skeletal elements to be recorded and micro-details of the burial formation process to be ascertained. Where, for some reason (often financial) it is not possible for the osteologist to undertake this work, the fill should still be removed in 20 mm spits, with a digital photographic record at each level and the context record sheet should be completed as normal. The vessel may be attributed an object number, but since it, together with its contents, represents the remains of a discrete act, it too should be given a context number as with any stratigraphic entity.

Unurned burials

Such deposits will generally follow the same excavation procedure as the discrete deposits of redeposited pyre debris. The deposit should be quadranted and, if it is a deep deposit (>0.15 m), excavated and collected as 0.10 m spits. Each quadrant and, where appropriate, spit should be collected under the same context number but with an individual sample number, preferably forming part of a consecutive series.

Concluding remarks

The archaeology of cremation is a fascinating and complex study from which much remains to be gleaned. Despite the recovery of a substantial body of tantalising evidence from a variety of features and deposits, there remain voids in our comprehension which can only be filled by better quality excavation and recording of the archaeological evidence. Improved understanding of the cremation process and the taphonomic effects of the burial environment on the archaeological components related to the mortuary rite have enabled a reassessment of important data recovered decades ago, and questioned the assumptions behind their interpretation. Recently recovered data have shed light on the process of cremation, but some forms of evidence remain relatively rare, and a wider spectrum of ritual activity undoubtedly remains to be discovered and illuminated.

Acknowledgements

The writer is grateful to colleagues who were kind enough to give information on projects as yet awaiting publication; Howard Brooks of Colchester Archaeological Unit, Laurie Coleman of Cotswold Archaeological Unit, Natasha Dodwell of Oxford Archaeology East and Natasha Powers of the Museum of London. Thanks are also due to Paul Booth and Oxford Archaeology for permission to reproduce Figure 10.6. Figures 10.7 and 10.8 are reproduced by permission of Wessex Archaeology. Linda Coleman of Wessex Archaeology provided invaluable assistance in production of the figures for publication. The paper was submitted in May 2009.

Bibliography

Albion Archaeology (2008) *Land West of Bedford*. Interim Report 5.
Alcock, J. P. (1980) Classical religious beliefs and burial practice in Roman Britain. *Archaeological Journal* 137, 50–85.
Barber, B. and Bowsher, D. (2000) *The Eastern Cemetery of Roman London*. MoLAS Monograph 4. London: Museum of London Archaeology Service.
Bellhouse, R. L. (1954) Further finds in the Beckfoot cemetery area [and] Roman cremation-sites at Herd Hill. *Transactions of the Cumberland and Westmoreland Antiquarian and Archaeological Society* 54, 51–55.
Bellhouse, R. L. and Moffat, I. (1959) Further Roman finds in the Beckfoot cemetery area. *Transactions of the Cumberland and Westmoreland Antiquarian and Archaeological Society* 58, 57–62.
Biddulph, E. (2006) The Romano-British cemetery at Pepper Hill, Kent. In S. Foreman, *Channel Tunnel Rail Link Section 1. CTRL specialist report series*. York: Archaeology Data Service (http://dx.doi.org/10.5284/1000230).
Black, E. W. (1986) Romano-British burial customs and religious beliefs in South-East England. *Archaeological Journal* 143, 201–239.

Bosanquet, R. C. (1935) The Roman tombs near High Rochester. *Proceedings of the Society of Antiquaries of Newcastle-upon-Tyne* 4 (6), 246–251.

Breeze, D. J. and Ritchie, G. J. N. (1980) A Roman burial at High Torrs, Luce Sands, Wigtownshire. *Dumfries and Galloway Natural History and Archaeological Society* 55, 79–85.

Campbell, G. (2004) Charcoal and other charred plant remains. In H. Cool, *The Roman Cemetery at Brougham, Cumbria*, 267–271. Britannia Monograph Series 21. London: Society for the Promotion of Roman Studies.

Casey, P. J. and Hoffmann, B. (1995) Excavations on the Corbridge bypass, 1974. *Archaeologia Aeliana* 23, 17–45.

Charlton, B. and Mitcheson, M. (1984) The Roman cemetery at Petty Knowes, High Rochester, Northumberland. *Archaeologia Aeliana* 12, 1–31.

Clarke, G. (1979) *Pre-Roman and Roman Winchester. Part II. The Roman cemetery at Lankhills*. Winchester Studies 3. 404–423.

Coleman, L., Havard, T., Collard, M., Cox, S. and McSloy. E. (2001) The Lea, Denham, Buckinghamshire. *South Midlands Archaeology* 31, 17.

Coleman, L., Havard, T., Collard, M., Cox, S. and McSloy. E. (2004) The Lea, Denham, Buckinghamshire. *South Midlands Archaeology* 34, 14–17.

Coleman, L., Havard, T., Collard, M., Cox, S. and McSloy. E. (2006) The Lea, Denham, Buckinghamshire. *South Midlands Archaeology* 36, 27–29.

Cool, H. (2004) *The Roman Cemetery at Brougham, Cumbria*. Britannia Monograph Series 21. London: Society for the Promotion of Roman Studies.

Crockett, A. (1996) Iron Age to Saxon settlement at Wickhams Field, near Reading, Berkshire: excavations on the site of the M43 motorway service area. In P. Andrews and A. Crockett, *Three excavations along the Thames and its tributaries, 1994*, 113–170. Wessex Archaeology Report No. 10.

Curle, J. (1932) An inventory of objects of Roman and Provincial Roman origin found on sites in Scotland not definitely associated with Roman constructions. *Proceedings of the Society of Antiquaries of Scotland* 66, 277–376.

Davey, N. (1935) The Romano-British cemetery at St. Stephens, near Verulamium. *Transactions of the St. Albans and Hertfordshire Architectural and Archaeological Society* 4, 243–275.

Dodwell, N. (1998 unpublished) Report on the cremated human bone. In C. Evans and M. Knight, *The Butcher's Rise Ring Ditches. Excavations at Barleycroft Farm, Cambridgeshire*. Cambridge Archaeological Unit Report 283.

Dodwell, N. (2009 unpublished) Assessment of the human bone. In A. Pickstone and R. Mortimer, *The archaeology of Brigg's Farm, Prior's Fen, Thorney, Peterborough*. Oxford Archaeology East Client Report 1082.

Dodwell, N. (in prep.) The cremated bone from the Bronze Age barrows. In C. Evans and M. Knight, *A book of sites: Excavation at Barleycroft Farm: The archaeology of the Lower Ouse Environs Vol. II*. McDonald Institute Monograph, Cambridge University.

Dubois, J. A. and Beauchamp, H. K. (1943) *Hindu manners, customs and ceremonies*. Oxford: Clarendon Press.

Egging Dinwiddy, K. and Bradley, P. (2011) *Prehistoric Activity and a Romano-British Settlement at Poundbury Farm, Dorchester, Dorset*. Wessex Archaeology Report 28, Salisbury: Wessex Archaeology.

Fitzpatrick. A. P. (1997) *Archaeological Excavations on the Route of the A27 Westhampnett Bypass, West Sussex, 1992 Volume 2*. Wessex Archaeology Report No 12. Salisbury: Wessex Archaeology.

Hiatt, B. (1969) Cremation in Aboriginal Australia. *Mankind* 7(2), 104–120.

Hogg, R. (1950) A Roman cemetery site at Beckfoot, Cumberland. *Transactions of the Cumberland and Westmorland Antiquarian and Archaeological Society* 49, 32–37.

Hope, V. M. (2007) *Death in ancient Rome*. London: Routledge.

Jessup, R. F. (1959) Barrows and walled cemeteries in Roman Britain. *Journal of the British Archaeological Association* 22, 1–32.

Leary, R. (2008) Romano-British pottery. In A. B. Powell, P. Booth, A. P. Fitzpatrick and A. D. Crockett, *The archaeology of the M6 Toll 2000-2003*, 147–168. Oxford Wessex Archaeology Monograph No. 2, Salisbury: Wessex Archaeology.

Liversidge, J. (1977) Roman burials in the Cambridge area. *Proceedings of the Cambridge Antiquarian Society* 67, 11–38.

MacDonald, J. L. (1979) Religion. In G. Clarke, *Pre-Roman and Roman Winchester. Part II. The Roman cemetery at Lankhills*, 404–423. Winchester Studies 3. Oxford: Clarendon Press.

Mackinder. A. (2000) *A Romano-British cemetery on Watling Street. Excavations at 165 Great Dover Street, Southwark, London.* MoLAS Archaeological Study Series 4, London: Museum of London.

Marsden, B. M. (1986) *The burial mounds of Derbyshire* (revised edition). Bradford/Ilkley.

McKinley, J. I. (1991 unpublished) Cremated Bone from the Area 15 cemetery, Baldock, Hertfordshire. Report for G. Burleigh, Letchworth Museum.

McKinley, J. I. (1992 unpublished) Cremated Bone from St. Stephens cemetery, St. Albans, Hertfordshire. Report for Verulamium Museum.

McKinley, J. I. (1993) Bone fragment size and weights of bone from modern British cremations and its implications for the interpretation of archaeological cremations. *International Journal of Osteoarchaeology* 3, 283–287.

McKinley, J. I. (1994a) *Spong Hill Part VIII: The Cremations.* East Anglian Archaeology 69, Dereham: Field Archaeology Division, Norfolk Museums Service.

McKinley, J. I. (1994b) Bone fragment size in British cremation burials and its implications for pyre technology and ritual. *Journal of Archaeological Sciences* 21, 339–342.

McKinley, J. I. (1996) Cremated human bone. In P. Andrews and A. Crockett *Three excavations along the Thames and its tributaries, 1994*, 156–157. Wessex Archaeology Report No. 10.

McKinley, J. I. (1997) The cremated human bone from burial and cremation-related contexts. In A. P. Fitzpatrick, *Archaeological Excavations on the Route of the A27 Westhampnett Bypass, West Sussex, 1992 Volume 2*, 55–73. Wessex Archaeology Report No. 12, Salisbury: Wessex Archaeology.

McKinley, J. I. (1998) Archaeological manifestations of cremation. *The Archaeologist* 33, 18–20.

McKinley, J. I. (2000a) Putting cremated human remains in context. In S. Roskams (ed.) *Interpreting Stratigraphy. Site evaluation, recording procedures and stratigraphic analysis*, 135–139. BAR International Series 910. Oxford: Archaeopress.

McKinley, J. I. (2000b) Phoenix rising; aspects of cremation in Roman Britain. In J. Pearce, M. Millett and M. Struck (eds) *Burial, Society and Context in the Roman World*, 38–44. Oxford: Oxbow.

McKinley, J. I. (2000c unpublished) Cremated bone from Denham, Berkshire. Report for Cotswold Archaeological Trust.

McKinley, J. I. (2000d) Cremation burials. In B. Barber and D. Bowsher, *The Eastern Cemetery of Roman London*, 264–277. MoLAS Monograph 4. London: Museum of London Archaeology Service.

McKinley, J. I. (2000e) The analysis of cremated bone. In M. Cox and S. Mays (eds.) *Human Osteology*. Greenwich Medical Media (London), 403–421.

McKinley, J. I. (2004a) The human remains and aspects of pyre technology and cremation rituals. In H. Cool, *The Roman Cemetery at Brougham, Cumbria*, 283–309. Britannia Monograph Series 21 London: Society for the Promotion of Roman Studies.

McKinley, J. I. (2004b) Compiling a skeletal inventory: cremated human bone. In M. Brickley and J. I. McKinley (eds) *Guidelines to the standards for recording human remains*, 9–13. Institute of Field Archaeologists Paper no. 7. Southampton/Reading: BABAO/Institute of Field Archaeologists.

McKinley, J. I. (2006) Human remains from Section 1 of the Channel Tunnel Rail Link, Kent. In S. Foreman, *Channel Tunnel Rail Link Section 1. CTRL specialist report series*. York: Archaeology Data Service (http://dx.doi.org/10.5284/1000230).

McKinley, J. I. (2008a) In the heat of the pyre: efficiency of oxidation in Romano-British cremations – did it really matter? In C. W. Schmidt and S. A. Symes (eds) *The analysis of burnt human remains*, 163–184. London: Academic Press.

McKinley, J. I. (2008b) Ryknield Street, Wall (Site 12). In A. B. Powell, P. Booth, A. P. Fitzpatrick and A. D. Crockett, *The archaeology of the M6 Toll 2000-2003*, 87–190. Oxford Wessex Archaeology Monograph No. 2. Salisbury: Wessex Archaeology.

Noy, D. (2005) The Romans and cremation. In D. Davies (ed.) *Encyclopaedia of Cremation.* Aldershot: Ashgate, 366–369.

Payne, G. (1886) Romano-British interments discovered at Bayford next Sittingbourne, Kent. *Archaeologia Cantiana* 16, 1–8.

Philpott, R. (1991) *Burial practices in Roman Britain: a survey of grave treatment and furnishings A.D. 43-410.* BAR British Series 219. Oxford: Tempus Reparatum.

Polfer, M. (2000) Reconstructing funerary rituals: the evidence of *ustrina* and related archaeological structures. In J. Pearce, M. Millett and M. Struck (eds) *Burial, Society and Context in the Roman World*, 30–37. Oxford: Oxbow.

Struck, M. (1993) *Busta* in Britannien und ihre Verbindungen zum Kontinent. Allgemeine Überlegungen zur Herleitung der Bestattungssittte. In M. Struck (ed.) *Römerzeitliche Gräber als Quellen zu Religion, Bevölkerungsstruktur und Sozialgeschichte*, 81–93. Mainz: Johannes Gutenberg-Universität.

Toynbee, J. M. C. (1996) *Death and Burial in the Roman World.* London: Johns Hopkins University Press.

Wenham, L. P. (1968) *The Romano-British cemetery at Trentholme Drive, York.* London: HMSO.

Wessex Archaeology (2008) *Boscombe Down Phase VI Excavation, Amesbury, Wiltshire: Interim assessment of the results of the Byway 20 Romano-British cemetery excavations.* Client Report 56246.04.

Wheeler, H. (1985) The Racecourse cemetery. In J. Dool, H. Wheeler *et al.* Roman Derby: Excavations 1968–1983. *Derbyshire Archaeological Journal* 105, 222–280.

Whimster, R. (1981) *Burial practices in Iron Age Britain.* BAR British Series 90 Oxford: BAR.

Witkin, A. and Boston, C. (2006) Iron Age and Roman human remains from Pepper Hill, Southfleet, Kent. In S. Foreman, *Channel Tunnel Rail Link Section 1. CTRL specialist report series*, York: Archaeology Data Service (http://dx.doi.org/10.5284/1000230).

Chapter 11

Afterword – Process and polysemy: An appreciation of a cremation burial

Jake Weekes

Introduction

All funerals are a blend of tradition and innovation. In attempting to interpret this pattern in the archaeological evidence of the Romano-British cremation tradition in South-East England, I have outlined a microcosmic process akin to structuration (Giddens 1984; Weekes forthcoming) whereby the agents of each funerary sequence *translate* traditional expectations into novel versions of the rite; the innovations of these actors (Parkin 1992) may in turn be variously translated into the rituals of subsequent funerals by others. This can be seen as a sort of 'mechanism' whereby traditions can form from apparent one-offs, or could be changed and replaced, while other unique or very seldom repeated ritual actions and objects simply never catch on. In this exploratory essay, I wish to consider the symbolic meanings of such diverse actions as evidenced by a particular burial. This is introduced first, but I then put further consideration of it aside in order to develop an interpretive framework, before returning to the example as a case study in process and polysemy.

Crundale

This small first- to third-century rural cemetery was discovered 30 years ago, at Crundale Limeworks (Bennett *in prep.*), situated to the south of the Crundale–Godmersham road and south-west of Canterbury in south-east England (TR 07404890, centred). The site was excavated under rescue conditions after the owner of the works informed the Canterbury Archaeological Trust that ancient remains had been uncovered as a result of quarrying (the depth of archaeological deposits removed in this operation cannot be reconstructed). Eight cremation burials were found in a typical location for rural burials to the north of parallel

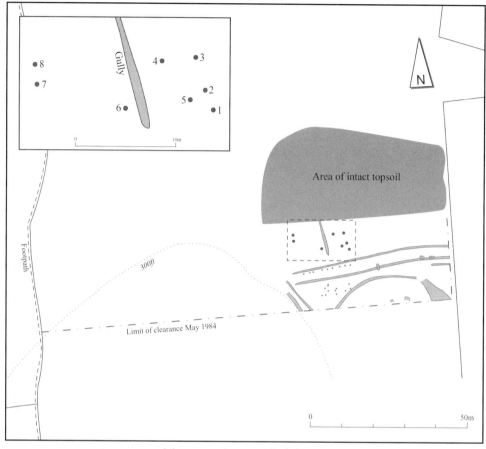

Fig. 11.1. Crundale Limeworks: site paln (after Bennett 1985, fig. 2).

linear features that may have marked a trackway and/or settlement boundary (Fig. 11.1).

The three adjacent late second-century amphora burials on the site (Burials 1, 2 and 3) can be used effectively to illustrate the structuration/translation theory. In all three cases the Dressel 20 amphorae contained loose (or perhaps originally bagged) deposits of cremated bone. Burials 2 and 3 also produced accessory vessels, in Burial 2 contained within the amphora, in Burial 3 placed in the rectangular grave pit near it, along with a pair of hobnailed shoes. Burial 1 produced no accessory vessels; instead the amphora contained, along with some cremated bone of an adult, the objects presented in Table 11.1.

Thus, while it was difficult to establish the chronological sequence in which some of these burials were made, there is unmistakable evidence here of some shared traditions as well as real diversity, to the degree that the Burial 1 assemblage

Table 11.1. Finds in Burial 1 at Crundale Limeworks

Type	Small find number	Material	Description
Brooch	16	Cu alloy	Enamelled, circular, ornate, probably continental.
Brooch	49	Cu alloy	Rectangular, enamelled, probably Romano-British.
Inkwell	1	Cu alloy	Small, cylindrical, inscribed *[E]x of(ficina) Socr[atis(?)]* 'from the workshop of Socra', ornate.
Mesh rings	4–13	Cu alloy	In close proximity, some fragments, 4 round section with 3 with overlapping ends secured with small fe rivet.
Fragments	14–15	Cu alloy	Sheet, 2 fragments.
Beads	16–22	G	Mainly globular.
Hobnails	18–20	Fe	As well as mineralized leather boot remains in situ against edge of amphora base.
Ring-shaped object and fittings	Various	Ag	Various silver pieces of unclear position.
Nails and fittings	Various	Fe	Of unknown function.

is surely unique in its *particular combination* of grave goods (cf. Philpott 1991 for example).

Meaning?

I find the archaeological reconstruction and exploration of traditions and diversity in ritual actions and objects interesting enough (Weekes 2005; 2008), but I must admit that beyond the comparative study of actions and objects of funerals lie far more problematic questions: what do the symbols mean, and how do symbolic rituals relate to the structuration of society and 'normal' everyday life?

Some have been very sceptical that such questions could ever be convincingly researched (e.g. Härke 1997, 194). Yet I would think that a systematic qualitative study might be just as approachable as for, say, that of folk tales; funerals can be considered a form of folk art. And plenty of theoretical frameworks exist within studies of other media, psychology, sociology and anthropology that are yet to be applied to the Roman funerary sphere. Here I would like to focus on polysemy and liminality, two analytical concepts developed by twentieth-century anthropologist Victor Turner which should contribute much to debates around the 'meaning' of funerals, past and present (cf. Gilchrist and Sloane 2005). In doing so, I also propose the division of funerary meanings into:

a. Plural meanings embodied in single acts or objects at any particular time in the funerary process: or synchronic meanings;
b. Meanings that link directly to an overall perceived purpose of the funeral as a whole: or diachronic meanings.

Liminality, a 'betwixt and between' state entered into, lived through and then left behind, contributes to diachronic meanings, whereas polysemy is more clearly synchronic.

Synchronic meanings

Something that should now be realised and embraced within the study of ritual in general and funerals in particular is that symbols are multivocal; single meanings (or schemes of meaning) offered as interpretations of funerary ritual or any particular aspect of it can only tell part, and probably a very small part, of the story. Ritual actions and objects can embody/generate/be invested with multiple meanings by participants and local observers, a fact long recognised by anthropologists (but see particularly Turner 1977 [1967], 50, 52; 1969). In the funerary sphere, it is often especially clear that no single 'logic' persists among those carrying out the ritual as to the symbolic meaning of all actions and objects. This is actually most obvious when the overall perceived 'logic' of the funerary process is questioned by outsiders. The informants of ethnographic and historical accounts often tell us, for example, that the spirit of the deceased is considered to have left the body at the moment of death, and has already travelled to a form of afterlife; yet these same informants nevertheless embark on a protracted sequence of actions apparently expressly designed to bring about or aid the same journey (see for example van Gennep 1960 [1909], 151–2; Metcalf and Huntington 1992, 86–88 and Parker Pearson 1999, 2). And our own experience is equally admissible here. At a modern day funeral anywhere in the world, how many of the participants, if questioned, would give a consistent set of meanings for the mortuary ritual? We would probably expect a variety of meanings to be taken from and invested in the actions and objects involved, and perhaps even logically conflicting exegesis from a single informant.

A participant in funerary activity might equally derive different meanings from the same actions or objects, ranging from understanding how they fit into the overall 'project', perhaps organized into the sorts of diachronic meanings discussed below, to the very personal, private and circumstantial; and to this conscious and unconscious multivalence can be added the nuances that derive from the funerary experience itself, so that the actions and objects become media for a more unrestricted flow of meanings in a stream of consciousness.

Clearly there remains much about which we can only guess from the archaeological record (again cf. Härke 1997), depending on our relative confidence in the universality of things like grief, or even psychodynamics. But it is my contention that, if we at least recognise polysemy and adopt a systematic if qualitative approach to its disentanglement, we might begin to see more than just the limits of such a study, and perhaps begin to benefit from its possibilities and potential. The description of polysemy must of course remain qualitative, and its caveats understood, but this should not preclude its serious investigation.

Critically, polysemy must be understood as an inclusive and organic quality, not a complete system which could overly restrict and determine the meanings of ritual actions in order to fit them to a perceived etic framework. Structuralism for example, a systemized model of the dynamics of symbolic ritual meanings, can be reductive in this way. Structuralist approaches present a very particular disambiguation of the cultural content of ritual, and can be applied both synchronically and diachronically, as exemplified in the French studies of mythology and ritual since Levi-Strauss (see Csapo 2005), and particularly by J.P. Vernant. Scheid's (1984) exposition of Roman funerary ritual as expressly aimed at bringing about separation of the living and the dead is clearly influenced by this school of thought, but happily founded in documentary evidence rather than external reading of perceived symbols in traditional societies. Scheid noticed a dichotomous symbolism in the textual accounts of Roman funerals with the dead being associated with darkness, silence, and so on, and the living represented by the antithetical qualities of light, sound etc. Within a Structuralist framework, such polar opposites are typically adduced as the building blocks of a 'structural' language, sociologically functioning as means of mediating between extremes. And surely the living and the dead are a singularly potent expression of a central sociological problem, i.e. that of culture vs. nature. From a synchronic, polysemic perspective (see below for a diachronic application) the Scheidian approach therefore can offer a Levi-Straussian disambiguation of aspects of Roman funerals at least, and it is surprising that it has only recently become more influential in the archaeological study of the subject (see Lepetz *et al.* 2011; van Andringa *et al.* 2013; Pearce this volume). It should be reiterated, however, that polysemy does not in fact preclude and is not supplanted by such structural patterns of meaning, but can simply include them among many indices.

Beyond and largely since Structuralism lies a broader study of semeiosis. The history of ideas in Semiotics, the 'science of signs', may look like a jungle to those who don't enjoy playing with theory, but the potential insights it offers to the archaeology of funerary practice are too important to ignore (see Cobley and Jansz 2010 for example). Semiotics offers alternative symbolic mechanisms to 'binary opposites', which can become a symbolic straight jacket if used as a cure-all. Consider for example the ideas of the nineteenth- and early twentieth-century American philosopher Charles Peirce, who generated extraordinary pioneer maps of the 'Semeiotic' and built up a complex symbol typology, or the work of Roland Barthes on the connotative element of overtly denotative signs, or Jacques Lacan on the boundary between conscious signifier and unconscious signified. Application of Roman Jacobson's general model of the features and functions of the communication event could supply a further index to understanding of the typology of symbols in funerary practice. Such approaches might certainly be considered before venturing into the polysemic 'forest of symbols' (Turner's nicely polysemic title for a book about people whose symbolism and culture was largely based on trees; 1967), but have so far not been deployed in our subject.

Strictly semeiotic or not, it would in any case seem that a meaningful qualitative appreciation of polysemy in Roman funerals is indeed plausible, but also that it is vital that such a study remain systematic and rooted in the data and their cultural and historical context. While not a study of Polysemy or semiotics per se, John Pearce's recent elucidation of the metaphors and metonymy of idealised *humanitas* in Gallo-Roman and Romano-British funerals is a good example, and a convincing response to a funerary material culture and its contemporary message, rooted in the data and qualitative interpretation. Crucially, Pearce (2015) identifies the complex of potential meanings and how they fit into a general idealised context, and thus provides a credible explanation of the milieu to which complexes of feasting and/or grooming symbols in funerals, for example, pertain.

Actually, a key concept in the specialised funerary context is indeed the fact that the signs are 'idealised', which, since Ucko (1969) many archaeologists should realise more (Pearce 2000). Funerals are not merely to be read as straightforward indices of social sign and cultural value, but must be tentatively deciphered with their specialised qualities firmly in mind. This neatly brings us to the specialised subject of liminality, which is best considered among the diachronic meanings of funerary ritual.

Diachronic meanings

Van Gennep's concept of actions and associated objects of rites of passage, presumably for both the deceased and ritual actors in the case of funerals, is a diachronic sequence (Fig. 11.2) of separation from an initial state (e.g. the world of the living), transition

Van Gennep	Rites emphasize separation		Rites emphasize transition	Rites emphasize incorporation	
Turner	Structure	Anti-Structure			Structure
	Liminal rites (decreasingly) respect *communitas* with the dead until new structure is recognized				
Scheid	Purity	Contamination			Purity
	Rites of reversal (increasingly) promote separation from the dead until new purity is recognized				
Turner and Scheid	The changing relationship between living and the dead is managed through ritual evocations of both the liminal state of anti-structure and desired separation within renewed, stable structure				

Fig. 11. 2. Funerals as social catharsis in response to the nature of death? Comparing and combining schemes of meaning in funerals adapted from the ideas of A. van Gennep (1909), V. Turner (1969) and J. Scheid (1984).

(liminality), and final incorporation in the new state (e.g. the world of the dead); in many cases, the latter would be where the corpse and those sharing its final journey would part company, signified by completion of the burial or other disposal of the human remains.

The scheme has been seen to describe the universal abstract form of all rituals that bring about social recognition of change in life stages. For some it is too abstract, and for Metcalf and Huntington (1992, 112), it is merely a 'vague truism unless it is positively related to the values of the particular culture', a view which archaeologist Michael Parker Pearson (1999, 22) has endorsed. Even so, the meaning of van Gennep's scheme still seems to work at the more universal, super-cultural, super-historical level, reflecting the physical and symbolic changes that ritual participants bring to bear via specialized actions and objects. I also think we should allow participants in funerary rituals to have some version of this format in their own minds, which can be expressed in various ways, and which could include 'selection, change, completion', for example, or just 'beginning, middle, end' (cf. other artistic or theatrical processes). The fact that culture- or situation-specific details hang from such a handy framework may be less significant for participant anthropologists than for archaeologists (or criminologists, for that matter), who must start somewhere when analysing evidence for events requiring reconstruction.

Perhaps this is why van Gennep's model seems to have been especially influential among archaeologists, but it also inspired anthropologist Turner to investigate liminality more deeply. Turner showed that promotion and production of liminality, a state where normal social and cultural structures are suspended, is in fact the most defining aim and characteristic of schemes of ritual action. The drama of ritual creates places and times where only transition and statelessness are emphasized, where subjects are suspended between states, belonging to neither. Turner went much further to produce a wide-ranging argument for this special condition, asserting that it constituted an 'anti-structure' to counter everyday 'structure', used in the European anthropological sense, epitomised by hierarchies and social rules. In fact, Turner argued that the main product of this strange liminal context was an egalitarian and constraint-free state he called '*communitas*', where 'normal' hierarchies are inverted, and a 'levelling' of status is paramount. Turner's examples range from the annual festivals (e.g. Saturnalia), and healing and initiation rituals of traditional and proto-historic societies, for example, to the historic examples. In the latter case, he saw a natural gravitational pull away from spontaneous or existential *communitas*, towards normative and ideological varieties of *communitas* in more rigid historical settings (for example, the medieval European Franciscan movement, which began spontaneously, but necessarily became entrenched in a code).

It is not hard to see how a destruction of social asymmetry could be brought about in the funerary context, not only among living participants in the event but also between the living and the dead (or at least between the family of the deceased and other members of a community; cf. Scheid 1984), only to be concluded when the dead reach their final 'resting' place and normal polarities are resumed. Consider the communal act of wearing black at a funeral, for example, or of silence or hushed voices as opposed to lively chatter. In a liminal levelling of the living and the dead there might be qualities and objects that the living and the dead share, along with and through the liminal funerary context (such as lamps, which allow the living to see in the dark if funerals take place at night, but may be equally symbolic lights for the dead to use; cf. again the Scheidian view). These collective roles and oppositions inform the drama, and it is actually the liminal state of *communitas* that is painted by Turner as the very context that allows the world of symbolic structure in the Levi-Straussian sense to come to the fore. This can easily encompass the funereal purpose as achievement of 'separation' between the living and the dead (note, another generalized element of van Gennep's tripartite scheme) emphasized by Scheid, for example (Fig. 11.2), along with other meanings, because it is by nature polysemic, even fundamentally embracing paradox.

I would suggest that an inclusive ritual perspective could accommodate the Turneresque and Scheidian viewpoints side by side, which otherwise might seem logically mutually exclusive. This is actually easy if a changing dynamic and dialectic is proposed, whereby the 'dangerous' state of *communitas* between the dead and the living, perhaps most extreme at the earliest stages of funerary rites, is gradually reduced as the cathartic rituals of separation take effect.

The semiological discourse can also contribute to thinking about diachronic meanings, like Jacques Lacan's concept of *pointes de capiton* ('upholstery buttons'), the master signs developed through a diachronic sign series in tandem with synchronic sign combination (ibid). This calls to mind Howard Williams' (2004) foray into the various uses and meanings of ceramics associated with cooking and eating Romano-British cremation and burial, a useful and novel essay into what was surely a significant complex of meanings, based around food and drink, in this widespread tradition. As Williams properly recognizes, the concept of eating and drinking seems indivisible from late Iron Age and early Romano-British cremation funerals, and we might add that this could be considered typical of one of Lacan's *pointes de capiton*. Alternatively, Jacques Derrida's work furthers recognition of mediation in the production of signs and thus introduces agency in the development of a post-structuralist emphasis on *parole* rather than the systems dominated concept of *langue* (cf. the work of Giddens and also phenomenology); Derrida's concept of *différance*, the constant deferral and modification of meaning as a sign series progresses, is perhaps especially applicable to the funerary sequence and its repeated translation. Another way of looking at this would be to consider once again the dialectic between *communitas* and separation suggested above. Could one product of this be a form of *différance*, both sociological and personal in effect?

In any case, given polysemy and liminality, the same funeral must at least theoretically accommodate conflicting views as to its overall purpose for participants, which is perhaps especially pertinent in cross-cultural or creole situations, where no particular consensus as to purpose, or 'hieros logos,' necessarily predominates. If this is the case, the interest for archaeologists and anthropologists might lie in the degree to which new ideas and material culture are allowed to redefine the anti-structure, liminality and *communitas* in the funerary context, and how that context in turn may serve as a melting pot for the flux and formation of creole signs, ideas and identities in more everyday life. A factor such as *humanitas*, therefore, as John Pearce describes it through the funerary material culture of late Iron Age and Roman Gaul and Britain, could be seen as a local cultural-historical variant of *communitas* in the funerary sphere, all the more interesting in that it presents 'partisan' version of ritual *communitas,* that of a provincial sub-group representing what it saw as a 'Roman' way of death.

So much for the sociological and cultural implications for meaning; these should not be allowed to crowd out the more personal meanings, which, after all, are likely to be the more emic, weighing heavily in the experiences of those taking part (see Graham 2015, 7–8 on personal responses to and interactions with the corpse; cf. Williams 2004). On top of this, bereavement would seem to be a human condition, however defined and played out in cultural/historical context, and it is plainly wrong to leave it out of a study of the meaning of Roman funerals (as noted by Pearce some time ago (2000, 5–6)). And while the very facts of society and culture can be drawn on by mourners in personal responses to a death, which is thereby given a shared cosmological context, it would be reductive to over-emphasise concepts like 'anti-structure' and especially 'separation', when what is being achieved by a funeral most often seems to a redefining of relationships that are meant to continue: the fact of death ultimately 'deferred'.

Application to the funerary process

In order to contextualise such meanings, we first need to recognise the entire funerary process and not just the burial (Pearce 1998; 2000; 2002; 2013; Cool 2004; Weekes 2008; 2014). In Burial 1 at Crundale we most obviously have evidence of the *deposition* aspect of a funeral, but there is also both certain and potential evidence here of the foregoing actions relating to human remains and other objects (*selection, preparation, modification, location*) as well as subsequent rituals of commemoration. Using this analytical framework, we might begin to realise the data within a more realistic diachronic scheme, and at the same time form an appropriate reconstruction of the synchronic milieu. The latter might include clues to a polysemy made up of broadly cultural references, such as creole 'Romanness' in the funerary sphere, a mixing of late Iron Age, Gallo-Roman and perceived 'Roman' funerary styles. We have here perhaps a version of *humanitas*, alongside what might be more localised binary opposites in the Structuralist sense, as well as regional styles, local, social and even personal variants. We can also consider the patterning of diachronic meanings throughout. So, while van Gennep's scheme of the *rites de passage* should not be jettisoned, further sequential meanings and interpretations can be layered onto it, like *pointes de capiton* (in the

Williams mode pertaining to food and drink, or perhaps Pearce's *humanitas*), Turner's liminality, anti-structure and *communitas*, and Scheidian reversal. We can also afford to admit the emic perspective of personal connections, participation and nuance.

A reading of Burial 1

Selection

While obviously 'selected' by death itself, the person represented by Burial 1 was also selected by the living for this type of funeral, which we know could not have been the same for everyone in Roman Britain (Pearce 2013, 23ff). Nothing about this person's life or death meant that they were given another type of funeral. Alternatively, was a 'bad death' socially and personally alleviated by a good funeral? Whatever the viability, it would also have informed the selection of other ritual materials, such as the wood fuel, and perhaps expertise, for the pyre, and items required for any laying out period of the body, for containing and transportation of the body and for burial, along with eating, drinking and so on. We can propose the beginnings of a polysemy here that included cultural connotations of funerary *humanitas* (such as the probable inkwell and unidentified items?). From a diachronic viewpoint, at this stage we could consider the state of shared liminality between the living and the dead as perhaps being at its most acute, or was the deceased's sustenance, for example, already kept separate?

Preparation

Was there a period of laying out and display of the body? If so, a number of the items later placed in the burial could have formed part of clothing that adorned the corpse (beads and brooches, footwear, unidentified items?). This is certainly a time when *communitas* between the dead and the living might be most obvious in the form of shared space and form, although the body of the deceased would already be changing due to post-mortem factors, with implications for personal experience and interactions. Probably what we could describe as creole events could also be 'shared' between the dead and the living, but a pyre would need to be built or reserved, people, food and drink and transport organised, and a burial pit prepared (if it was to be immediately required), so the process of reversal and separation will equally have started in earnest.

Location

If there is a clear aspect of Romano-British funerary tradition that is most like the Roman, it is the conducting of funerary activity like cremation and inhumation beyond the limit of the 'lived space' of a settlement; this is most obvious in the urban setting. Such boundaries are not so easily defined in a rural setting, but the location by the trackway recurring in this and other examples suggests that a similar distancing applies as in the towns. Possible removal to a different place for cremation is also removal of contamination and towards more normal structural relations; it is also a final shared journey, which in the case of Burial 1 apparently included separate places for cremation and burial (although admittedly truncation of the site through quarrying may have removed evidence of any nearby surface cremation).

Modification

Cremation, even pre-conquest, was a rite shared, along with other cultural factors, with connected (and perhaps related) elites in northern Gaul, already by then part of the Roman Empire. In early Roman South-East England, this treatment of the dead became part of the expression of a developing Romano-British culture *per se*. Yet rural elites, perhaps more likely descendants of those of the pre-Roman Iron Age, may have felt they were adhering to what was by then a long family tradition rather than marking out a divergent cultural identity. This would not preclude expressions of *humanitas* in the cremation rite itself, and it is perhaps interesting in this case that pyre side ritual produced not only the burnt animal bone, but also burnt oyster shells. In the general scheme, corporeal annihilation of the body of the deceased ought to mark a drastic material hiatus in *communitas* between the living and the dead, and, while still highly liminal, would surely be an important milestone on the road to renewed structure and purity. Any materials burnt on the pyre with the corpse would also seem to be clearly consigned to and associated with the province of the deceased rather than those in attendance, emphasising increasing separation? If family and friends of the deceased were present at any stage of the visceral exercise of the pyre a culturally framed response to the body and its destruction would have been unavoidable and was more than likely integral to the burning process. They would be witnesses to (and possibly participants in) an ultimate physical violence inflicted on the corpse as an expression of kindness and correct procedure, no doubt alien to most modern westerners (cf. Graham 2015b; Weekes 2008, 151; this volume). All the senses would be affected.

Deposition

The transformed (and indeed stabilised) material of the deceased was at some point separated from the pyre material, and transferred (perhaps in a bag) to a modified amphora placed in a pit in the ground: a token amount of the cremated bone was placed within the amphora along with accessory items. Many of the elements of this burial are a typically individual expression of a widespread and long lived tradition in Roman South-East England. The symbolism of food and drink in response to death, perhaps idealised *humanitas*, interweaves with local, familial and even personal practice; the item which exemplifies this is the amphora, used in adjacent Burials 2 and 3, but also more regionally. We might also consider the footwear in this light. Beads and brooches, the inkwell and unidentified items seem more likely to have had a personal connotation in life, though still obviously traditionally and culturally affected personal symbols of *humanitas*.

Perhaps it was now, however, that the diachronic meaning of the rite came to the fore, with all objects associated with the deceased throughout the ritual now being deposited along with the remains as a deliberate act of association. As we perhaps know for ourselves, the act of 'proper' deposition of the deceased and any associated objects marks a most important movement beyond the oddly suspended social and personal reality of the funeral (which I now personally identify as liminality) and towards a more everyday separation of the living and the dead. Deposition can be construed as a final act of incorporation, separation and liminality, a full stop marking the end of anti-structure and contamination and the completed structuration of a new purity: of everything, especially the dead and the living, in the right place.

Commemoration

Yet afterwards, perhaps on typical 'days of the dead', or culture-specific events like the *cena novendialis*, for example, or perhaps just at certain times, the living might re-enter the place of the dead, probably acknowledging them through ritual preparation and/or manner which might again invoke the anti-structural ideals of shared liminal polysemy and *communitas*. The amphora in Burial 1 protected its contents, but also most likely would have allowed continued access to them. Indeed, amphora necks may once have protruded from the surface, as at Ostia, for example. Food and drink may therefore have continued to be essential *pointes de capiton* in a new form of dramatic relationship.

Conclusion

These interpretations and ideas obviously make for a particular reading of Burial 1 at Crundale Limeworks. Some aspects of the potential synchronic polysemy of the various stages of this funeral, such as the wider references to cultural, regional and social symbols, are of necessity reconstructed using the comparative evidence of broader study of the traditions within which this particular set of events took its place.

Other factors discussed here, perhaps those with a more diachronic character like ritual liminality, anti-structure, *communitas* and reversal, seem to me at least to tap into something more universal about funerals and their place in human societies. Are some of the deepest connotations of funerals not diffusional, but parallel? In this sense, we might begin to explore ritual generally as a profoundly deep structure *per se*, deeper than any of the meanings with which we may consciously inform it. At such a fundamental level, funerals would be characterized as first and foremost an *active process*, constructing *Death*. In fact, in light of all of the above, I would prefer to give *actions* of the process, rather than whatever *thoughts* we attach to them, the primacy in the meaning of ritual.

Bibliography

Bennett, P. (1985) Crundale Limeworks (N.G.R. TR 074489). *Archaeologia Cantiana* 101, 285–288.
Bennett, P. (*in prep*) Crundale Limeworks, Kent.

Cobley, P. and Jansz, L. (2010) *Introducing Semiotics. A Graphic Guide*. London: Icon Books.

Cool, H. E. M. (2004) *The Roman Cemetery at Brougham, Cumbria. Excavations 1966-67*. Britannia Monograph Series 21. London: Society for the Promotion of Roman Studies.

Csapo, E. (2005) *Theories of Mythology*. Oxford: Blackwell.

Duday, H. (2009) *The Archaeology of the Dead. Lectures in Archaeothanatology* (translated by Anna Maria Cipriani and John Pearce). Oxford: Oxbow.

Giddens, A. (1984) *The Construction of Society. Outline of the Theory of Structuration*. Cambridge: Polity Press.

Gilchrist, R. and Sloane, B. (2005) *Requiem. The Medieval Monastic Cemetery in Britain*. London: Museum of London Archaeology Service.

Graham, E-J. (2015a) Introduction: embodying death in archaeology. In Z. Devlin and E-J. Graham (eds) *Death embodied: archaeological approaches to the treatment of the corpse*, 1–17. Oxford: Oxbow.

Graham, E-J. (2015b) Corporeal concerns: the role of the body in the transformation of Roman mortuary practices. In Z. L. Devlin and E. J. Graham (eds) *Death Embodied. Archaeological Approaches to the Treatment of the Corpse*, 42–62. Oxford: Oxbow.

Härke, H. (1997) Ritual, symbolism and social inference. In C. K. Jensen and K. H. Nielsen (eds) *Burial and Society: The Chronological and Social Analysis of Archaeological Burial Data*, 191–195. Aarhus: Aarhus University Press.

Lepetz, S., Van Andringa, W, Duday, H., Joly, D., Malagoli, C., Matterne, V. and Tuffreau-Libre, M. (2011) *Publius Vesonius Phileros vivos monumentum fecit*: investigations in a sector of the Porta Nocera cemetery in Roman Pompeii. In M. Carroll and J. Rempel (eds) *Living Through the Dead. Burial and Commemoration in the Classical World*, 110–133. Oxford: Oxbow.

Metcalf, P. and Huntington, R. (1992) *Celebrations of Death: the Anthropology of Mortuary Ritual*. Cambridge: Cambridge University Press.

Parker Pearson, M. (1999) *The Archaeology of Death and Burial*. Stroud: Sutton Publishing.

Parkin, D. (1992) Ritual as spatial direction and bodily division. In D.de Coppet (ed.) *Understanding Rituals*, 11–25. London: Routledge.

Pearce, J. (1998) From death to deposition: The sequence of ritual in cremation burials of the Roman period. In C. Forcey, J. Hawthorne and R. Witcher (eds) *TRAC 98: Proceedings of the Seventh Annual Theoretical Roman Archaeology Conference Nottingham 1997*, 99–111. Oxford: Oxbow Books.

Pearce, J. (2000) Burial, society and context in the provincial Roman World, in J. Pearce, M. Millett and M. Struck (eds) *Burial, Society and Context in the Roman World*, 1–12. Oxford: Oxbow Books.

Pearce, J. (2002) Ritual and interpretation in provincial Roman cemeteries. *Britannia* 33, 373–377.

Pearce, J. (2013) *Contextual Archaeology of Burial Practice Case studies from Roman Britain*. British Archaeological Reports S588, Oxford: Archaeopress.

Pearce, J. (2015) A civilised death? The interpretation of provincial Roman grave good assemblages. In H. Brandt, M. Prusac and H. Roland (eds) *Death and Changing Rituals: Function and Meaning in Ancient Funerary Practices*, 223–248. Oxford: Oxbow Books.

Philpott, R. (1991) *Burial Practices in Roman Britain. A Survey of Grave Treatment and Furnishing A.D. 43-410*. BAR British Series 219. Oxford: Tempus Reparatum.

Scheid, J. (1984) *Contraria facere*: renversements et déplacements dans les rites funéraires. *Annali dell' Istituto Orientale di Napoli 6*, 117–139.

Turner, V. (1967) *The Forest of Symbols. Aspects of Ndembu Ritual*. Ithaca, N.Y.: Cornell University Press.

Turner, V. (1969) *The Ritual Process. Structure and Anti-Structure*. Chicago: Aldine.

Ucko, P. J. (1969) Ethnography and the archaeological interpretation of funerary remains. *World Archaeology* 1, 262–280.

Van Gennep, A. (1960) [1909]. *The Rites of Passage* (translated by M. B. Vizedom, and G. L. Caffee). Chicago: University of Chicago Press.

Weekes, J. (2008) Classification and analysis of archaeological contexts for the reconstruction of early Romano-British cremation funerals. *Britannia* 34, 145–160.

Weekes, J. (forthcoming) Evidence for personalisation of cult in early Romano-British cremation burials in south-east England. In R. Haeussler, A. King, G. Schörner and F. Simón (eds), *Religion in the Roman Empire: The Dynamics of Individualisation.* Oxford: Oxbow Books.

Weekes, J. (2014) 'Cemeteries and funerary practice'. In M. Millett, L. Revell, and A. Moore (eds), *The Oxford Handbook of Roman Britain.* Oxford: Oxford University Press. I:10.1093/oxfordhb/9780199697713.013.025.

Williams, H. M. R. (2004) Potted histories – cremation, ceramics and social memory in early Roman Britain. *Oxford Journal of Archaeology,* 23(4), 417–427.